Trends in Mathematics is a series devoted to the publication of volumes arising from conferences and lecture series focusing on a particular topic from any area of mathematics. Its aim is to make current developments available to the community as rapidly as possible without compromise to quality and to archive these for reference.

Proposals for volumes can be sent to the Mathematics Editor at either

Birkhäuser Verlag
P.O. Box 133
CH-4010 Basel
Switzerland

or

Birkhäuser Boston Inc.
675 Massachusetts Avenue
Cambridge, MA 02139
USA

Material submitted for publication must be screened and prepared as follows:

All contributions should undergo a reviewing process similar to that carried out by journals and be checked for correct use of language which, as a rule, is English. Articles without proofs, or which do not contain any significantly new results, should be rejected. High quality survey papers, however, are welcome.

We expect the organizers to deliver manuscripts in a form that is essentially ready for direct reproduction. Any version of TeX is acceptable, but the entire collection of files must be in one particular dialect of TeX and unified according to simple instructions available from Birkhäuser.

Furthermore, in order to guarantee the timely appearance of the proceedings it is essential that the final version of the entire material be submitted no later than one year after the conference. The total number of pages should not exceed 350. The first-mentioned author of each article will receive 25 free offprints. To the participants of the congress the book will be offered at a special rate.

Mathematics and Computer Science

Algorithms, Trees, Combinatorics and Probabilities

Danièle Gardy
Abdelkader Mokkadem
Editors

Springer Basel AG

Editors' addresses:

Danièle Gardy
Université de Versailles-St-Quentin
PRISM
Bâtiment Descartes
45 avenue des Etats-Unis
78035 Versailles Cedex
France

e-mail: gardy@prism.uvsq.fr

Abdelkader Mokkadem
Université de Versailles-St-Quentin
Département de Mathématiques
Bâtiment Fermat
45 avenue des Etats-Unis
78035 Versailles Cedex
France

e-mail: mokkadem@math.uvsq.fr

2000 Mathematical Subject Classification 68Q25; 68Rxx

A CIP catalogue record for this book is available from
the Library of Congress, Washington D.C., USA

Deutsche Bibliothek Cataloging-in-Publication Data

Mathematics and computer science : algorithms, trees, combinatorics and probabilities /
Danièle Gardy ; Abdelkader Mokkadem ed.. - Basel ; Boston ; Berlin : Birkhäuser, 2000
(Trends in mathematics)

ISBN 978-3-7643-6430-4 ISBN 978-3-0348-8405-1 (eBook)
DOI 10.1007/978-3-0348-8405-1

Originally published by Birkhäuser Verlag, Basel in 2000

Printed on acid-free paper produced from chlorine-free pulp. TCF ∞

9 8 7 6 5 4 3 2 1

Contents

V Other Topics 243

Foreword

This book is based on the proceedings of the Colloquium on Mathematics and Computer Science held in the University of Versailles-St-Quentin on September 18-20, 2000. It contains articles on topics that are relevant both to Mathematics and to Computer Science. These articles have been selected throughout a rigourous process; each of them has been submitted to two referees at least. This selection process, which took place under the supervision of the scientific committee, ensures the high level and the originality of the contributions.

We thank the invited speakers, the authors, the participants to the colloquium and, of course, the Scientific Committee whose members were D. Aldous, F. Baccelli, P. Cartier, B. Chauvin, P. Flajolet, J.M. Fourneau, D. Gardy, D. Gouyou-Beauchamps, R. Kenyon, J.F. Le Gall, C. Lemaréchal, R. Lyons, A. Mokkadem, A. Rouault and C. Roucairol.

We also take this opportunity to thank the University of Versailles-St-Quentin, the University of Paris-Sud (Orsay) and the Centre National de la Recherche Scientifique (CNRS), whose financial and material support contributed to the success of the Colloquium.

The Organisation Committee

Preface

Since the emergence of computers, Mathematics and Computer Science have been close partners. The boundary between these two sciences is not well defined, which leads to a very productive emulation. Wide areas are common to Mathematics and Computer Science: analysis of algorithms, trees, combinatorics, optimization, performance evaluation, discrete probabilities, computational statistics, ... These common domains are what we call here Mathematics and Computer Science (MCS). Perhaps it is not the most appropriate name; we shall let the two communities eventually decide for a more convenient appellation.

This book produces a collection of original articles, which reflect the state of the art of MCS. It is divided in five parts.

Trees and Analysis of Algorithms.
Antos and Devroye consider a binary search tree for an ordinary random walk; they prove that the asymptotic probability distribution of the height normalized by $\sqrt{2n}$ is the Erdös-Kac Rényi distribution. The paper of Chassaing, Marckert and Yor gives a simple but subtle proof for the limit law of the couple height-width in the case of binary trees. The next two papers concern other aspects of binary search trees; Dekking, De Graaf and Mester study the positioning of arm nodes and foot nodes on the tree, and Drmota obtains the asymptotic expectation of the saturation level. Fill and Janson prove that the asymptotic distribution of the standardized random number of comparisons used by Quicksort has an infinitely differentiable density f whose derivatives $f^{(k)}$ have superpolynomial decay. Gittenberger generalizes results obtained by Panholzer and Prodinger for binary trees to a larger class of rooted trees, the simply generated trees. The paper of Jacquet, Szpankowski and Apostol is related to information theory; they consider a universal prediction algorithm for mixing sources and prove that this predictor is asymptotically optimal and that the prediction error is $O\left(n^{-\epsilon}\right)$ for any $\epsilon > 0$.

Combinatorics and Random Generation.
The article of Bousquet, Chauve, Labelle and Leroux gives a new bijective proof of the multivariate Lagrange inversion formula; the proposed bijection is easier than former bijections to put into application, and leads to a combinatorial interpretation of various enumeration formulas for rooted trees. Bousquet-Melou and Schaeffer present a method to enumerate paths on the square lattice that avoid a horizontal half-line; they prove that the corresponding generating functions are algebraic of degree 8 over the field of rational functions and give new details on the asymptotic of the number of paths of lenght n. This nice result allows them to solve Kenyon's questions. The paper of Denise, Rocques and Termier presents two interesting alternatives of the recursive method for random generation of words of context-free languages; thanks to this approach, they can generate words according to exact frequencies or expected frequencies. Merlini, Sprugnoli and Verri find a non abelian group structure for a subclass of generating trees; this is achieved via a correspondence with monic, integer proper Riordan Array. In their article, Pergola, Pinzani and Rinaldi introduce well-defined operations on the set of suc-

cession rules, which allow the use of succession rules instead of generating functions in some combinatorial enumeration problems.

Algorithms and Optimization.
Two articles in this part deal with genetic algorithms. Berard and Bienvenüe consider a simplified context with finite population but develop a rigorous mathematical treatment, which leads to a detailed understanding of the effects of selection. They obtain the convergence in law of the normalized population and a large deviation principle. These results confirm numerical simulations and biological experiments. Mazza and Piau consider the case of infinite population; the genetic algorithm is modeled as a discrete-time measure valued dynamical system. This Markov chain approach gives the asymptotic distribution and a large deviation principle. The article of Dror, Fortin and Roucairol tackles the deterministic problem of transporting nondedicated commodities from a set of suppliers to a set of customers with one vehicle of limited capacity. The authors give a VRP-like formulation, the dimension of associated polytope and complex analysis of practical complexity. The last article by Métivier, Saheb and Zemmari introduces and analyses a randomized algorithm to get rendez-vous in a graph.

Performance evaluation.
Ben Mamoun and Pekergin introduce a class of Markov chains on finite state-space, which have a closed form solution to compute steady-state distribution; they also provide an algorithm to construct a bounding Markov chain in this class. Dayar produces an efficient iterative aggregation-disaggregation algorithm to compute the stationary vector of discrete-time stochastic automata networks that are lumpable. Delcoigne and De La Fortelle state large deviation principles for polling systems. The paper of Fayolle and Lasgouttes deals with a symmetrical star-shaped network comprising N links and such that all routes are of lenght 2; they are mainly concerned with policies that can be used to share the bandwidth of the links between active connections. A functional analysis approach is used to characterize the behaviour of the network.

Other topics.
This part contains articles on stochastic subjects, which either are relevant to several previous parts (branching processes, generating functions, walk generations) or introduce new mathematical topics related to computer science (Brownian excursion, random sentences). The first three articles deal with branching processes. Geiger gives a new proof of Yaglom's theorem for critical Galton-Watson branching process. Liu presents an interesting survey on recent results concerning the branching measure, the exact Hausdorff measure and the exact packing measure, defined on the boundary of the Galton-Waltson tree. Löcherbach's article concerns statistical models for branching particle systems; an explicit version of the likelihood ratio process and local asymptotic normality are derived. Louchard and Rocques use tools from combinatorics, probability and singularity analysis to achieve a complete asymptotic analysis of the cost of a Schröder walk generation algorithm; five different probability distributions are observed in the study. Malyshev introduces

a generalization of Gibbs distributions when the space (lattice, graph) is random. These generalized distributions are well suited to study local probability structures on graphs with random topology; this is a new connection between mathematical physics and computer science. Pemantle produces a nice asymptotic study for meromorphic generating functions; deep tools and nontrivial technics from analytic geometry, sheaf cohomology and Stein spaces theory are used. In what he calls a "speculative report", Spencer presents approaches to obtain asymptotic formula for the ultrahigh moments for Brownian excursion; the paper also contains a conjecture and many indications for further research. Ycart and Rousset consider probability distributions on a set of sentences; they show a zero-one law for random sentences under these distributions.

All the articles in the present book were also presented as talks at the colloquium on MCS in Versailles, September 2000. Some of the speakers of the colloquium are unfortunately not represented here.

Although the content of this book is of high level, it also has a pedagogical interest; it is intended for a large public, including graduate students, in Mathematics and in Computer Science.

We hope this book will help to deepen the connections between Mathematics and Computer Science and will be followed by many others on MCS.

D. Gardy, A. Mokkadem

Part I

Trees and Analysis of Algorithms

Trends in Mathematics, © 2000 Birkhäuser Verlag Basel/Switzerland

Rawa Trees

ANDRAS ANTOS *Institute of Communication Electronics, Technical University of Budapest 1111 Stoczek u., Budapest XI, Hungary H-1521.* antos@inf.bme.hu
LUC DEVROYE *School of Computer Science, McGill University, 3480 University Street, Montreal, Canada H3A 2K6.* luc@cs.mcgill.ca

Abstract. *A rawa tree is a binary search tree for an ordinary* ***ra****ndom* ***wa****lk* $0, S_1, S_2, S_3, \ldots$*, where* $S_n = \sum_{i=1}^n X_i$ *and the* X_i*'s are i.i.d. distributed as* X. *We study the height* H_n *of the rawa tree, and show that if* X *is absolutely continuous with bounded symmetric density, if* X *has finite variance, and if the density of* X *is bounded away from zero near the origin, then* $H_n/\sqrt{2n}$ *tends to the Erdős-Kac-Rényi distribution.*

Key words. *Random binary search tree, probabilistic analysis, random walk, Catalan constant, limit distributions.*

Introduction

Binary search trees are the most common data structures for storing information. Given a sequence S of real numbers and a real number x, let $L(S,x)$ denote the subsequence of S consisting of all numbers less than x and let $R(S,x)$ be the subsequence consisting of numbers greater than x. In particular, if $S = (S_0, S_1, \ldots, S_n)$ are real numbers then the binary search tree for these data is recursively defined as follows: it is a binary tree with root S_0, with left subtree the binary search tree for $L(S, S_0)$ and with right subtree the binary search tree for $R(S, S_0)$. Binary search trees permit searching in time bounded by the height of the tree, where the height H_n is the maximal path distance from any node to the root (Knuth, 1973; Cormen, Leiserson and Rivest, 1990).

The standard random binary search tree is based on an i.i.d. sequence

$$(S_0, S_1, \ldots, S_n) \,.$$

In that case, it is well-known (Robson, 1979, Devroye, 1986, 1987) that if the common distribution is absolutely continuous, then

$$H_n \sim 4.33107\ldots \log n$$

almost surely as $n \to \infty$. Rather little is known about H_n when the defining sequence is not i.i.d. We define a running average model for a generic random variable X and an averaging parameter $p \in [0, \infty)$ as follows: let $X_1, \ldots, X_n$ be i.i.d., distributed as X, and define $S_0 = 0$ and

$$S_n = \frac{X_1 + \cdots + X_n}{n^p} \,, n \geq 1.$$

The running average random binary search tree is based on $(S_0, S_1, \ldots, S_n)$. Within this model, there are only two choices of practical interest, $p = 1/2$ and $p = 0$: for

$p = 0$, S_n is just a partial sum of all X_i's with $i \le n$, and we obtain a binary search for an ordinary random walk. We define a rawa tree $\mathcal{T}_n$ (or $\mathcal{T}_n(X)$) as a binary search tree based on $(S_0, S_1, \ldots, S_n)$ where $S_0 = 0$, $S_n = \sum_{i=1}^n X_i$ and $X_1, \ldots, X_n$ are i.i.d. distributed as X. We study H_n for a large class of rawa trees. If X has nonzero mean, the random walk will drift off, and the binary search tree has an uninteresting shape and height $H_n = \Theta(n)$. For this reason, and technical reasons that will be encountered later, we restrict ourselves to nice random variables, which are random variables with the following properties:

- A. X has a density, is symmetric about zero, and has a finite variance σ^2.
- B. The density f of X is bounded: $\|f\|_\infty < \infty$.
- C. f is bounded away from 0 in a neighborhood of the origin: $\liminf_{x \downarrow 0} f(x) > 0$.

Random variables satisfying A only will be called simple.

While the shape of $\mathcal{T}_n(X)$ indeed depends heavily on the distribution of X, it is quite interesting that for all nice random variables, the limit behavior for H_n is essentially identical.

THEOREM. *Let X be a nice random variable, and let $\mathcal{T}_n$ be a rawa tree. Then, for all $x > 0$,*

$$\lim_{n\to\infty} \mathbf{P}\left\{ \frac{H_n}{\sqrt{2n}} < x \right\} = \mathcal{L}(x) ,$$

where $\mathcal{L}$ is the Erdős-Kac-Rényi distribution function

$$\mathcal{L}(x) = \begin{cases} \frac{4}{\pi} \sum_{n=0}^{\infty} \frac{(-1)^n}{2n+1} e^{-\frac{(2n+1)^2\pi^2}{8x^2}} & \textit{if } x > 0 \\ 0 & \textit{otherwise.} \end{cases}$$

Note that the limit law does not depend upon the distribution of X, yet the shape of the rawa tree depends upon the distribution in a substantial manner. Also, the Theorem does not exclude the possibility that there are symmetric random variables with different limit laws. Other questions of a more universal nature may be asked: for example, is $\mathbf{E}H_n \ge \Omega(\sqrt{n})$ for all X with a density? Is $\mathbf{E}H_n = O(\sqrt{n})$ for all X with a symmetric density? Binary search trees extract a lot of fine detail from the underlying sequence in terms of permutations and other global properties. They thus help in the understanding of the behavior of sequences.

In the figure below, we show six rawa trees to indicate possible pathways for the proof.

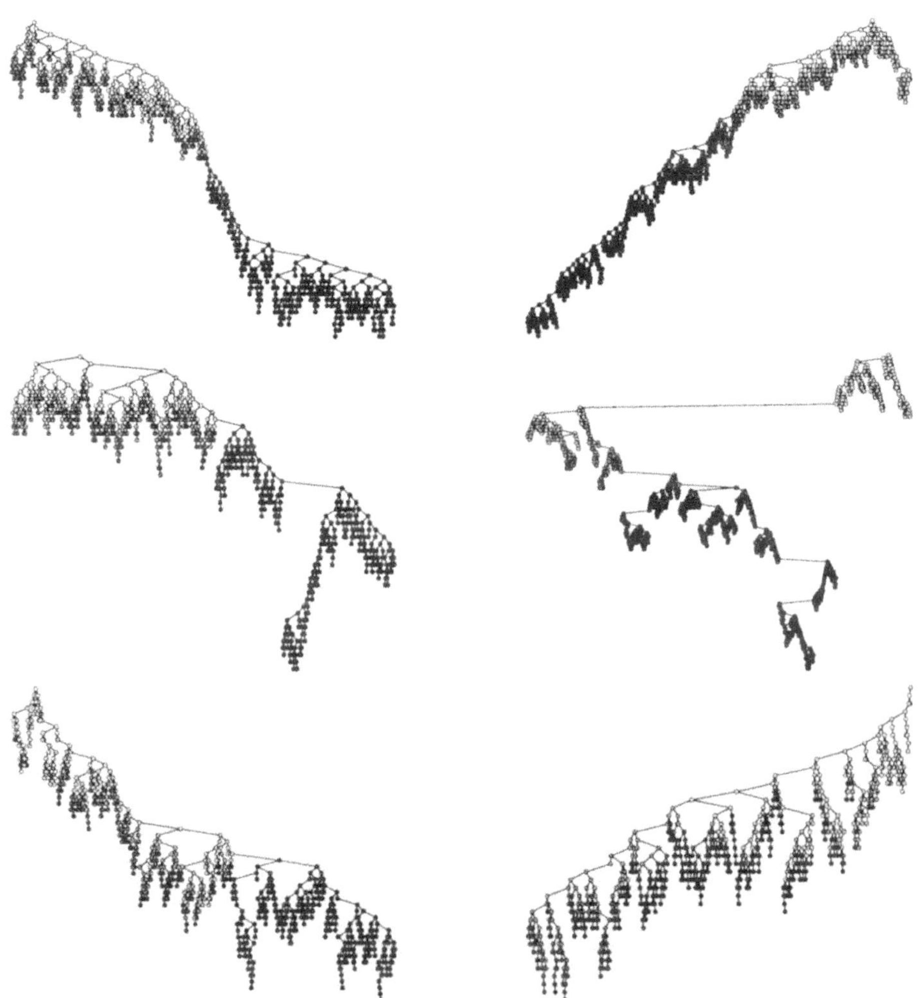

Figure 1. Six rawa trees are shown, from left to right, top to bottom: the random variables X are normal, Laplace, $s/\sqrt{|U_2|}$, $1/U_1$, $U_1U_2U_3U_4$ and $s(|U_1| + 5Z)$, where s is a random sign, Z is Bernoulli $(1/3)$ and the U_i's are uniform $[-1, 1]$.

Relationship between height and record sequences

The proof uses records in sequences. In a sequence $s = (s_0, s_1, s_2, \ldots)$, we say that s_n is an up-record if $s_n = \max\{s_0, s_1, \ldots, s_n\}$ and that it is a down-record if $s_n = \min\{s_0, s_1, \ldots, s_n\}$. Let $U(s)$ and $D(s)$ be the number of up-records and down-records in the sequence s. It is well-known (Devroye, 1988) that the path distance from the node for S_n to the root in a binary search tree is

$$D_n = U(L(S_{<n}, S_n), S_n) + D(R(S_{<n}, S_n), S_n) - 2$$

where $S_{<n} = (S_0, S_1, \ldots, S_{n-1})$. Thus, the height of the binary search tree is

$$H_n = \max_{1\le i\le n} D_i = \max_{1\le i\le n} (U(L(S_{<i}, S_i), S_i) + D(R(S_{<i}, S_i), S_i)) - 2 \, .$$

It is easy to see that

$$H_n \ge H'_n \stackrel{\text{def}}{=} \max(U(S_{<n+1}), D(S_{<n+1})) - 1 \, .$$

The proof is based on the following two lemmas, proved in the remainder of the paper.

LEMMA 1. *Let X be a simple random variable. Then, for all $x > 0$,*

$$\lim_{n\to\infty} \mathbf{P}\left\{ \frac{H'_n}{\sqrt{2n}} < x \right\} = \mathcal{L}(x) \, ,$$

where $\mathcal{L}$ is the Erdős-Kac-Rényi distribution function.

LEMMA 2. *Let X be a nice random variable. Then for all $\epsilon > 0$,*

$$\lim_{n\to\infty} \mathbf{P}\{H_n - H'_n > \epsilon\sqrt{n}\} = 0 \, .$$

Records and ladder heights for random walks

We first consider the increasing ladder epochs and ladder heights for

$$S_0, S_1, \ldots, S_n,$$

where $S_0 = 0$. The (increasing) ladder epochs are at $0 = T_0 < T_1 < T_2 < \cdots$, where

$$T_i = \inf\{j > T_{i-1} : S_j > S_{T_{i-1}}\} \, .$$

We say that S_{T_i} is an up-record and call $S_{T_i} - S_{T_{i-1}} = \mathcal{H}_{T_i}$ an ascending ladder height. Similarly, we have decreasing ladder epochs for consecutive minima. The epochs are denoted by $0 = T'_0 < T'_1 < T'_2 < \ldots$. We say that $S_{T'_i}$ is a down-record and call $S_{T'_i} - S_{T'_{i-1}}$ a descending ladder height. Define $M_n = \max_{i\le n} S_i$,

$M'_n = -\min_{i \le n} S_i$. Let $R_n = \max\{k : T_k \le n\}$ and $R'_n = \max\{k : T'_k \le n\}$. The R_n ascending ladder heights are denoted by $\mathcal{H}_1, \ldots, \mathcal{H}_{R_n}$. The absolute values of the R'_n descending ladder heights are denoted by $\mathcal{H}'_1, \ldots, \mathcal{H}'_{R'_n}$. Note that $R_n = U(S_0, S_1, \ldots, S_n) - 1$, $R'_n = D(S_0, S_1, \ldots, S_n) - 1$ and $H'_n = \max(R_n, R'_n)$. We define the distribution function

$$\mathcal{G}(x) = \begin{cases} 2\Phi(x) - 1 & \text{if } x > 0 \\ 0 & \text{otherwise,} \end{cases}$$

where Φ is the standard normal distribution function. Thus, $\mathcal{G}$ is the distribution function of the absolute value of a standard normal random variable. Most of the properties of this section are available in standard references such as Feller (1971, chapter 12) or Spitzer (1976). The first half of proposition 2 is due to Erdős and Kac (1946). For more on $\mathcal{L}$, see Rényi (1963). Feller (1971) and Spitzer (1976) may be consulted for proposition 1 and for more properties of ladder heights and ladder epochs in random walks.

PROPOSITION 1. *Let X be a simple random variable. Then $\mathcal{H}_1, \mathcal{H}_2, \ldots$ are independent identically distributed with mean $\sigma/\sqrt{2}$. The same is true for $\mathcal{H}'_1, \mathcal{H}'_2, \ldots$.*

PROPOSITION 2. *Let X be a simple random variable. Then*

$$\frac{M_n}{\sigma\sqrt{n}} \xrightarrow{\mathcal{L}} \mathcal{G} \; ; \; \frac{M'_n}{\sigma\sqrt{n}} \xrightarrow{\mathcal{L}} \mathcal{G} \; ; \text{ and } \frac{\max(M_n, M'_n)}{\sigma\sqrt{n}} \xrightarrow{\mathcal{L}} \mathcal{L} .$$

Furthermore,

$$\frac{R_n}{\sqrt{2n}} \xrightarrow{\mathcal{L}} \mathcal{G} \text{ and } \frac{R'_n}{\sqrt{2n}} \xrightarrow{\mathcal{L}} \mathcal{G} .$$

REMARK: EXPECTED VALUES. As $\mathcal{G}$ has expected value $\sqrt{2/\pi}$, and $\mathcal{L}$ has expected value $\kappa \stackrel{\text{def}}{=} \gamma\sqrt{32/\pi^3} = 0.930527\ldots$, where $\gamma = 0.915965\ldots$ is Catalan's constant ($\sum_{n=0}^{\infty} \frac{(-1)^n}{(2n+1)^2}$), it is possible to show that for simple random variables,

$$\mathbf{E}M_n \sim \sigma\sqrt{2n/\pi} \, , \, \mathbf{E}M'_n \sim \sigma\sqrt{2n/\pi}$$

and

$$\mathbf{E}\{\max(M_n, M'_n)\} \sim \kappa\sigma\sqrt{n} \, .$$

PROPOSITION 3. *Let X be a simple random variable.* $\lim_{n\to\infty} R_n = \lim_{n\to\infty} R'_n = \infty$ *almost surely and* $\mathbf{E}R_n = \Theta(\sqrt{n})$, $\mathbf{E}R'_n = \Theta(\sqrt{n})$.

The proof of the following routine fact is omitted.

LEMMA 3. *If $X_1, X_2, \ldots$ are i.i.d. random variables with finite mean m, and if $R_n \to \infty$ almost surely, then $\sum_{i=1}^{R_n} X_i/R_n \to m$ almost surely. In particular, with M_n and R_n as in proposition 2, we have $M_n/R_n \to \sigma/\sqrt{2}$ almost surely.*

PROPOSITION 4. *Let X be a simple random variable. Then, for all x,*

$$\mathbf{P}\left\{\frac{R_n}{\sqrt{2n}} < x\right\} - \mathbf{P}\left\{\frac{M_n}{\sigma\sqrt{n}} < x\right\} \to 0\ ,$$

$$\mathbf{P}\left\{\frac{R'_n}{\sqrt{2n}} < x\right\} - \mathbf{P}\left\{\frac{M'_n}{\sigma\sqrt{n}} < x\right\} \to 0\ ,$$

and

$$\mathbf{P}\left\{\frac{\max(R_n, R'_n)}{\sqrt{2n}} < x\right\} - \mathbf{P}\left\{\frac{\max(M_n, M'_n)}{\sigma\sqrt{n}} < x\right\} \to 0\ .$$

PROOF. We only show the proof for the last statement. We may assume $x > 0$. For arbitrary small $\epsilon > 0$

$$\begin{aligned}
\mathbf{P} & \ \{\max(R_n, R'_n) < \sqrt{2n}x\} \\
= & \ \mathbf{P}\{\max(M_n, M'_n) < (\sigma+\epsilon)\sqrt{n}x, \max(R_n, R'_n) < \sqrt{2n}x\} + \\
& + \mathbf{P}\{\max(M_n, M'_n) \geq (\sigma+\epsilon)\sqrt{n}x, \max(R_n, R'_n) < \sqrt{2n}x\} \\
\leq & \ \mathbf{P}\{\max(M_n, M'_n) < (\sigma+\epsilon)\sqrt{n}x\} + \\
& + \mathbf{P}\left\{\frac{\mathcal{H}_1 + \ldots + \mathcal{H}_{R_n}}{R_n} > \frac{\sigma+\epsilon}{\sqrt{2}}\right\} + \mathbf{P}\left\{\frac{\mathcal{H}'_1 + \ldots + \mathcal{H}'_{R'_n}}{R'_n} > \frac{\sigma+\epsilon}{\sqrt{2}}\right\}
\end{aligned}$$

By Lemma 3, the last two terms tend to zero, so

$$\limsup_{n\to\infty} \mathbf{P}\{\max(R_n, R'_n) < \sqrt{2n}x\} \leq \limsup_{n\to\infty} \mathbf{P}\{\max(M_n, M'_n) < (\sigma+\epsilon)\sqrt{n}x\},$$

and by the continuity of the limit distribution of $\max(M_n, M'_n)$ in Proposition 2

$$\limsup_{n\to\infty} \mathbf{P}\{\max(R_n, R'_n) < \sqrt{2n}x\} \leq \limsup_{n\to\infty} \mathbf{P}\{\max(M_n, M'_n) < \sigma\sqrt{n}x\} = \mathcal{L}(x).$$

On the other hand for arbitrary small $\epsilon > 0$

$$\begin{aligned} \mathbf{P} & \{\max(M_n, M_n') < (\sigma - \epsilon)\sqrt{nx}\} \\ = & \ \mathbf{P}\{\max(R_n, R_n') < \sqrt{2nx}, \max(M_n, M_n') < (\sigma - \epsilon)\sqrt{nx}\} + \\ & + \mathbf{P}\{\max(R_n, R_n') \geq \sqrt{2nx}, \max(M_n, M_n') < (\sigma - \epsilon)\sqrt{nx}\} \\ \leq & \ \mathbf{P}\{\max(R_n, R_n') < \sqrt{2nx}\} + \\ & + \mathbf{P}\left\{\frac{\mathcal{H}_1 + \ldots + \mathcal{H}_{R_n}}{R_n} < \frac{\sigma - \epsilon}{\sqrt{2}}\right\} + \mathbf{P}\left\{\frac{\mathcal{H}_1' + \ldots + \mathcal{H}_{R_n'}'}{R_n'} < \frac{\sigma - \epsilon}{\sqrt{2}}\right\} \end{aligned}$$

and by Lemma 3 the last two terms tend to zero, so

$$\liminf_{n\to\infty} \mathbf{P}\{\max(R_n, R_n') < \sqrt{2nx}\} \geq \liminf_{n\to\infty} \mathbf{P}\{\max(M_n, M_n') < (\sigma - \epsilon)\sqrt{nx}\} .$$

By the continuity of $\mathcal{L}$,

$$\liminf_{n\to\infty} \mathbf{P}\{\max(R_n, R_n') < \sqrt{2nx}\} \geq \liminf_{n\to\infty} \mathbf{P}\{\max(M_n, M_n') < \sigma\sqrt{nx}\} = \mathcal{L}(x)$$

gives the result. □

COROLLARY. Let X be a simple random variable. For all $x > 0$

$$\lim_{n\to\infty} \mathbf{P}\{\max(R_n, R_n')/\sqrt{2n} < x\} = \mathcal{L}(x) .$$

Note that this implies Lemma 1.

Concentration results for random walks

In this section, we study upper tail bounds for an empirical concentration function for the random walk,

$$Q_n(\ell) = \sup_x \sum_{i=1}^n \mathbf{1}_{[S_i \in [x, x+\ell]]}$$

for a particular range of interval sizes ℓ roughly between $1/\sqrt{n}$ and $1/n^{1/3}$. Classical concentration inequalities for S_n are nicely described by Petrov (1995, section 2.4). For example, Petrov (1995, (2.71)) shows that there exists a positive constant A (referred to below as Petrov's constant) such that uniformly over all ℓ,

$$\sup_x \mathbf{P}\{S_n \in [x, x+\ell]\} \leq \frac{A \sup_x \mathbf{P}\{X \in [x, x+\ell]\}}{\sqrt{n(1 - \sup_x \mathbf{P}\{X \in [x, x+\ell]\})}} ,$$

where X is the generic summand in the random walk. In particular, if X has a density f bounded by $\|f\|_\infty$, and $\ell\|f\|_\infty \leq 3/4$, then

$$\sup_x \mathbf{P}\{S_n \in [x, x+\ell]\} \leq \frac{2A\ell\|f\|_\infty}{\sqrt{n}} .$$

We do not expect $Q_n(\ell)$ to be substantially larger than n times the latter bound, i.e., $Q_n(\ell)$ should be $O(\ell\sqrt{n})$. This result, even without a tight constant, will suffice for the present paper. Lemma 4 takes care of this.

LEMMA 4. *Let X have density f, and let f be symmetric and bounded by $\|f\|_\infty$. Let $\ell > 0$ be so small that $\ell\|f\|_\infty \le 3/4$, and let A be Petrov's constant. Then, if $Y_i = \mathbf{1}_{[S_i \in [x, x+\ell]]}$, we have*

$$\mathbf{E}\left\{\left(\sum_{i=1}^{n} Y_i\right)^4\right\} \le 543 \max\left(16, \lambda^4 n^2\right)$$

where $\lambda = 2A\|f\|_\infty \ell$.

PROOF. Note that Y_i is increasing in ℓ. If $\lambda\sqrt{n} < 2$, then we increase ℓ to make $\lambda\sqrt{n} = 2$. Thus, without loss of generality, we can assume that $\lambda\sqrt{n} \ge 2$, and we need only prove the inequality with $543\lambda^4 n^2$ on the right-hand-side. We proceed by repeated use of Petrov's inequality. The following simple summation bounds are easy to verify:

$$\sum_{i=1}^{n} \frac{1}{\sqrt{i}} \le 2\sqrt{n} \ ,$$

$$\sum_{1 \le i_1 < i_2 \le n} \frac{1}{\sqrt{i_1}\sqrt{i_2 - i_1}} \le (2\sqrt{n})^2 \ ,$$

$$\sum_{1 \le i_1 < i_2 < i_3 \le n} \frac{1}{\sqrt{i_1}\sqrt{i_2 - i_1}\sqrt{i_3 - i_2}} \le (2\sqrt{n})^3 \ ,$$

$$\sum_{1 \le i_1 < i_2 < i_3 < i_4 \le n} \frac{1}{\sqrt{i_1}\sqrt{i_2 - i_1}\sqrt{i_3 - i_2}\sqrt{i_4 - i_3}} \le (2\sqrt{n})^4 \ .$$

Thus,

$$\begin{aligned}
&\mathbf{E}\left\{\left(\sum_{i=1}^{n} Y_i\right)^4\right\} \\
&= \sum_{1 \le i_1, i_2, i_3, i_4 \le n} \mathbf{E}\{Y_{i_1} Y_{i_2} Y_{i_3} Y_{i_4}\} \\
&= 24 \sum_{1 \le i_1 < i_2 < i_3 < i_4 \le n} \mathbf{E}\{Y_{i_1} Y_{i_2} Y_{i_3} Y_{i_4}\} + 36 \sum_{1 \le i_1 < i_2 < i_3 \le n} \mathbf{E}\{Y_{i_1} Y_{i_2} Y_{i_3}\} \\
&\quad + 14 \sum_{1 \le i_1 < i_2 \le n} \mathbf{E}\{Y_{i_1} Y_{i_2}\} + \sum_{1 \le i_1 \le n} \mathbf{E}\{Y_{i_1}\}
\end{aligned}$$

$$
\begin{aligned}
&\leq 24 \sum_{1 \leq i_1 < i_2 < i_3 < i_4 \leq n} \frac{\lambda^4}{\sqrt{i_1(i_2-i_1)(i_3-i_2)(i_4-i_3)}} \\
&\quad +36 \sum_{1 \leq i_1 < i_2 < i_3 \leq n} \frac{\lambda^3}{\sqrt{i_1(i_2-i_1)(i_3-i_2)}} \\
&\quad +14 \sum_{1 \leq i_1 < i_2 \leq n} \frac{\lambda^2}{\sqrt{i_1(i_2-i_1)}} + \sum_{1 \leq i_1 \leq n} \frac{\lambda}{\sqrt{i_1}} \\
&\leq 24(2\lambda\sqrt{n})^4 + 36(2\lambda\sqrt{n})^3 + 14(2\lambda\sqrt{n})^2 + (2\lambda\sqrt{n}) \\
&\leq (2\lambda\sqrt{n})^4(24 + 36/4 + 14/16 + 1/64) \\
&\leq 543(\lambda\sqrt{n})^4 \, . \square
\end{aligned}
$$

LEMMA 5. *Let f be a symmetric density bounded by $\|f\|_\infty$, and assume that f has finite variance σ^2. Let ℓ be a sequence of interval sizes depending upon n such that $n\ell^3 = o(1)$. Then, for every $\epsilon > 0$,*

$$
\lim_{n \to \infty} \mathbf{P}\{Q_n(\ell) \geq \epsilon\sqrt{n}\} = 0 \, .
$$

PROOF. We may assume without loss of generality that $n\ell^2 \to \infty$. If not, we increase ℓ artificially, which by virtue of the monotonicity of $Q_n(\ell)$ with respect to ℓ is allowable. We partition $\mathbf{R}$ into intervals of length 2ℓ each and let N_i denote the number of S_j's that land in the i-th interval, $1 \leq j \leq n$. Clearly, $Q_n(\ell) \leq 2\max_i N_i$. Let B denote the set of interval indices for intervals that intersect $[-n, n]$, and note that $|B| \leq 2 + n/\ell$. Let N denote the number of S_j's that fall outside $[-n, n]$. Let n be so large that $2\ell\|f\|_\infty \leq 3/4$. Define $k = \epsilon\sqrt{n}/2$. We have, if X denotes the generic summand with density f,

$$
\begin{aligned}
\mathbf{P}\{N \geq k\} &\leq \mathbf{E}\{N\}/k = \sum_{j=1}^{n} \mathbf{P}\{|S_j| \geq n\}/k \\
&\leq \frac{\sum_{j=1}^{n} \mathbf{E}\{S_j^2\}}{kn^2} = \frac{\sum_{j=1}^{n} j\mathbf{E}\{X^2\}}{kn^2} \\
&= \frac{(n+1)\sigma^2}{2kn} \leq \frac{\sigma^2}{k} \, .
\end{aligned}
$$

Let A be Petrov's constant. By Lemma 4, if n is so large that $4A\ell\|f\|_\infty\sqrt{n} \geq 2$,

$$
\begin{aligned}
\mathbf{P}\{\max_{i \in B} N_i \geq k\} &\leq |B| \max_i \mathbf{P}\{N_i \geq k\} \\
&\leq \frac{|B|\mathbf{E}\{N_i^4\}}{k^4} \\
&\leq \frac{543|B|(4A\|f\|_\infty\ell)^4 n^2}{k^4} \, .
\end{aligned}
$$

Thus,

$$\begin{aligned}\mathbf{P}\{Q_n(\ell) \geq 2k\} &\leq \mathbf{P}\{N \geq k\} + \mathbf{P}\{\max_{i\in B} N_i \geq k\} \\ &= O(1/k) + O\left(n\ell^3\right) \\ &= o(1) \; . \square\end{aligned}$$

The height of a rawa tree: proof of Lemma 2

Throughout this section, we set $\ell = 1/n^{4/10}$. Observe that the condition of Lemma 5, $n\ell^3 \to 0$, holds. We call the depth $D(x)$ of x the path distance from x to the root for the binary search tree defined by $S_0, S_1, \ldots, S_n, x$. Consider the collection $C = \{i\ell : i \text{ integer}\}$. Recalling the definition of $Q_n(\ell)$ from the previous section, we note that the height H_n of the rawa tree satisfies

$$H_n \leq Q_n(\ell) + \max_{x\in C} D(x) \; .$$

Fix $x > 0$. Let $1 \leq T_1 < T_2 < \cdots$ be the epochs at which the random walk reaches a maximum: so, T_{i+1} is the first index greater than T_i for which $S_{T_{i+1}} > S_{T_i}$. We define $A(x)$ and $B(x)$ as follows: let i be the unique index for which $S_{T_i} \leq x < S_{T_{i+1}}$, where $S_{T_{i+1}}$ is replaced by ∞ if $T_{i+1} > n$ (to take care of the rightmost interval); then set $A(x) = S_{T_i}$ and $B(x) = S_{T_{i+1}}$. A similar symmetric definition is used for $x < 0$ with possibly $A(x) = -\infty$. Observe that

$$H_n \leq Q_n(\ell) + \max_{x\in C:|A(x)|+|B(x)|<\infty} D(x) \; .$$

For $x > 0$, there are no S_j's strictly in $(A(x), B(x))$ with $j \leq T_{i+1}$. We condition on the history of the random walk up to T_{i+1} and call it $\mathcal{F}$. Note that $D(x)$ is bounded by the sum of $i+2$ and the number of local left records, plus the number of local right records. A local left record is a record value (maximum) among those S_j's that fall in $(A(x), x)$. Clearly, each value of j must exceed T_{i+1}. A local right record is a record value (minimum) among those S_j's that fall in $(x, B(x))$. Observe that $i + 1 \leq H'_n$ if $|A(x)| + |B(x)| < \infty$. Denoting by $L(x)$ and $R(x)$ the number of local left records and local right records for x respectively, we see that $D(x) \leq i + 2 + L(x) + R(x)$, and that

$$H_n - H'_n \leq Q_n(\ell) + \max_{x\in C:|A(x)|+|B(x)|<\infty} L(x) + \max_{x\in C:|A(x)|+|B(x)|<\infty} R(x) \; .$$

The proof is complete if we can show that for each $\epsilon > 0$,

$$\lim_{n\to\infty} \mathbf{P}\left\{Q_n(\ell) > \epsilon\sqrt{n}\right\} = 0 \; ,$$

$$\lim_{n\to\infty} \mathbf{P}\left\{\max_{x\in C:|A(x)|+|B(x)|<\infty} L(x) > \epsilon\sqrt{n}\right\} = 0 \; ,$$

and

$$\lim_{n\to\infty} \mathbf{P}\left\{\max_{x\in C:|A(x)|+|B(x)|<\infty} R(x) > \epsilon\sqrt{n}\right\} = 0 \; .$$

The first part was shown in Lemma 5. We will show the second part involving $L(x)$, as the third part is similar.

Given $\mathcal{F}$, we may have large values for $L(x)$ if $B(x) - A(x)$ is large. So, let G be the event

$$\left[\max_{x \in C: |A(x)|+|B(x)|<\infty} |B(x) - A(x)| \leq \delta\sqrt{n}\right],$$

where $\delta > 0$ is to be picked as a function of ϵ. Let G^c denote its complement. Clearly,

$$\begin{aligned}
\mathbf{P}\{G^c\} &\leq \mathbf{P}\left\{\max_i |X_i| > \delta\sqrt{n}\right\} \\
&\leq n\mathbf{P}\left\{|X_1| > \delta\sqrt{n}\right\} \\
&\leq n\frac{\mathbf{E}\left\{X_1^2 \mathbf{1}_{[|X_1|>\delta\sqrt{n}]}\right\}}{\delta^2 n} \\
&= o(1)
\end{aligned}$$

if $\mathbf{E}\left\{X_1^2\right\} < \infty$. Thus, by Lemma 5,

$$\begin{aligned}
&\mathbf{P}\left\{\max_{x \in C: |A(x)|+|B(x)|<\infty} L(x) > \epsilon\sqrt{n}\right\} \\
\leq\ & \mathbf{P}\{G^c\} + \mathbf{P}\{Q_n(\ell) > (\epsilon/3)\sqrt{n}\} \\
& +\mathbf{P}\left\{G, Q_n(\ell) \leq (\epsilon/3)\sqrt{n}, \max_{x \in C: |A(x)|+|B(x)|<\infty} L(x) > \epsilon\sqrt{n}\right\} \\
\leq\ & o(1) \\
& + \mathbf{P}\left\{\bigcup_{\substack{x \in C: \\ |A(x)|+|B(x)|<\infty}} \left[Q_n(\ell) \leq (\epsilon/3)\sqrt{n}, |B(x) - A(x)| \leq \delta\sqrt{n}, L(x) > \epsilon\sqrt{n}\right]\right\} \\
\leq\ & o(1) \\
& + \mathbf{E}\left\{\sum_{\substack{x \in C: \\ |A(x)|+|B(x)|<\infty}} \mathbf{1}_{[|B(x)-A(x)|\leq\delta\sqrt{n}]} \mathbf{P}\left\{Q_n(\ell) \leq (\epsilon/3)\sqrt{n}, L(x) > \epsilon\sqrt{n} \middle| \mathcal{F}\right\}\right\}.
\end{aligned}$$

We show that we can find $\delta > 0$ so that the last term is $o(1)$.

Using the chain of inequalities for N in the proof of Lemma 5, we see that the probability that there is one occurrence of $|S_j| > n^2$ is bounded by the expected number of such occurrences, which by Chebyshev's inequality does not exceed σ^2/n^2. Thus, with probability at least $1 - O(1/n^2)$, the number of $x \in C$ with $|A(x)| + |B(x)| < \infty$ is not more than $2 + 2n^2/\ell = O(n^{24/10})$. By trivial bounding then, it would suffice if we can show that for an appropriate choice of $\delta > 0$, and uniformly over all x, and all histories $\mathcal{F}$ with $|B(x) - A(x)| \leq \delta\sqrt{n}$,

$$\mathbf{P}\left\{Q_n(\ell) \leq (\epsilon/3)\sqrt{n}, L(x) > \epsilon\sqrt{n} \middle| \mathcal{F}\right\} \leq \exp\left(-cn^{1/10}\right)$$

for some $c > 0$. We drop the conditioning in the notation. By condition (C) on f (see the definition of nice random variables), we find a pair $b > 0$ and $r > 0$ such that $\inf_{|x| \leq r} f(x) \geq b > 0$. Any such pair will do. Then we partition $[A(x), x]$ into intervals of length $r/2$ starting from $A(x)$, taking care not to cover x (thus leaving an interval of length less than r that reaches x; call that interval I). Let J be the rightmost interval. Replace J by $J \cup I - [x - \ell, x]$ and I by $[x - \ell, x]$. Thus, $r - \ell \geq |J| \geq r/2 - \ell \geq r/3$ for n large enough. Consider the following process: a local left record S_j arrives in one of the intervals (I and J excluded). At that time, given that we are in an interval to the left of J, there is a probability of at least $(r/3)b$ that S_{j+1} hits one of the intervals to the right, J included. Picture this as a success. Let the number of intervals up to and including J be k, and note that $k \leq 1 + 2\delta\sqrt{n}/r$. Let N^* be the number of local left records in the intervals to the left of J. In what follows, we set $m = \lfloor (\epsilon/3)\sqrt{n} \rfloor$. $[N^* > m]$ implies that in m such trials we had less than k successes. In other words,

$$\mathbf{P}\{N^* > m\} \leq \mathbf{P}\{\text{binomial}(m, br/3) < k\} .$$

If we set $k < mbr/7$, then this probability is $\exp(-\Omega(\sqrt{n}))$, by Hoeffding's inequality (Hoeffding, 1963). This condition is satisfied for n large enough when $\delta = \epsilon br^2/43$, which is the choice we will adopt. Let N' be the number of local left records in J. Here, we move to I with probability at least $(1/2)b\ell$, so that by the same argument,

$$\begin{aligned} \mathbf{P}\{N' > m\} &\leq \mathbf{P}\{\text{binomial}(m, b\ell/2) = 0\} = (1 - b\ell/2)^m \\ &\leq \exp(-mb\ell/2) = \exp\left(-\Omega\left(n^{1/10}\right)\right) . \end{aligned}$$

Finally, let N'' be the number of local left records in I. This is clearly bounded by $Q_n(\ell)$, uniformly over all x and all $A(x), B(x)$. As $L(x) \leq N^* + N' + Q_n(\ell)$, we see that

$$\begin{aligned} \mathbf{P}\{Q_n(\ell) \leq (\epsilon/3)\sqrt{n}, L(x) > \epsilon\sqrt{n}\} &\leq \mathbf{P}\{N^* > m\} + \mathbf{P}\{N' > m\} \\ &= \exp\left(-\Omega\left(n^{1/10}\right)\right) . \end{aligned}$$

This concludes the proof of Lemma 2. □

References

T. H. Cormen, C. E. Leiserson, and R. L. Rivest, *Introduction to Algorithms*, MIT Press, Boston, 1990.

L. Devroye, "A note on the height of binary search trees," *Journal of the ACM*, vol. 33, pp. 489–498, 1986b.

L. Devroye, "Branching processes in the analysis of the heights of trees," *Acta Informatica*, vol. 24, pp. 277–298, 1987.

L. Devroye, "Applications of the theory of records in the study of random trees," *Acta Informatica*, vol. 26, pp. 123–130, 1988.

P. Erdős and M. Kac, "On certain limit theorems of the theory of probability," *Bulletin of the American Mathematical Society*, vol. 52, pp. 292–302, 1946.

W. Feller, *An Introduction to Probability Theory and its Applications, Volume 2*, John Wiley, New York, 1971.

I. S. Gradshteyn and I. M. Ryzhik, *Table of Integrals, Series and Products*, Academic Press, New York, 1980.

W. Hoeffding, "Probability inequalities for sums of bounded random variables," *Journal of the American Statistical Association*, vol. 58, pp. 13–30, 1963.

D. E. Knuth, *The Art of Computer Programming, Vol. 3 : Sorting and Searching*, Addison-Wesley, Reading, MA, 1973.

H. M. Mahmoud, *Evolution of Random Search Trees*, John Wiley, New York, 1992.

V. V. Petrov, *Limit Theorems of Probability Theory*, Oxford University Press, 1995.

J. M. Robson, "The height of binary search trees," *The Australian Computer Journal*, vol. 11, pp. 151–153, 1979.

A. Rényi, "On the distribution function $L(z)$," *Selected Translations in Mathematical Statistics and Probability*, vol. 4, pp. 219–224, 1963.

F. Spitzer, *Principles of Random Walk (2nd ed)*, Springer-Verlag, New York, 1976.

R. L. Wheeden and A. Zygmund, *Measure and Integral*, Marcel Dekker, New York, 1977.

Trends in Mathematics, © 2000 Birkhäuser Verlag Basel/Switzerland

The height and width of simple trees

P. CHASSAING, J.F. MARCKERT *Laboratoire de Mathématiques, IECN, 54 506 Vandoeuvre lès Nancy Cedex.* [chassain,marckert]@iecn.u-nancy.fr
M. YOR *Université Paris VI, Laboratoire de Probabilités et modèles aléatoires, Tour 56-4, place Jussieu - 75252 Paris Cedex 05.*

Abstract. *The limit law of the couple height-width for simple trees can be seen as a consequence of deep results of Aldous, Drmota and Gittenberger, and Jeulin. We give here an elementary proof in the case of binary trees.*

1 Introduction

Let $Z_i(t)$ denote the number of nodes at distance i from the root of a rooted tree t. The *profile* of the tree t is the sequence $(Z_i(t))_{i\geq 0}$. The *width* $w(t)$ and *height* $h(t)$ of the tree t are defined by:

$$\begin{aligned} w(t) &= \max_i \{Z_i(t)\}, \\ h(t) &= \max\{i | Z_i(t) > 0\}. \end{aligned}$$

Let $T_B^{(n)}$ denote the set of *binary trees* with n leaves ($2n-1$ nodes), endowed with the uniform probability, and let $H_B^{(n)}$ (resp. $W_B^{(n)}$) be the restriction of h (resp. w) to $T_B^{(n)}$. One can also see $H_B^{(n)}$ and $W_B^{(n)}$ as the height and width of a Galton-Watson tree with offspring distribution 0 or 2 with probability 1/2, conditioned to have total progeny $2n-1$ (see [1, pp. 27-28]). Then, the limit law of the height [15, 23] and of the width [7, 13, 25] are given by:

$$\frac{H_B^{(n)}}{\sqrt{2n}} \xrightarrow[n\to+\infty]{law} 2V, \tag{1.1}$$

$$\frac{W_B^{(n)}}{\sqrt{2n}} \xrightarrow[n\to+\infty]{law} V, \tag{1.2}$$

where:

$$\Pr(V \leq x) = \sum_{-\infty<k<+\infty} (1-4k^2x^2)\exp\left(-2k^2x^2\right). \tag{1.3}$$

Connections between the distribution of V on one hand, the Brownian motion and Jacobi's Theta function on the other hand, are discussed in [5, 9, 20]. For instance, let $\left(e(s)\right)_{0\leq s\leq 1}$ denote a standard normalized Brownian excursion (see Subsection 3.1). Then, the random variables

$$H = \int_0^1 \frac{ds}{e(s)}$$

and

$$W = \max_{0\leq s\leq 1} e(s)$$

satisfy

$$V \stackrel{law}{=} W \stackrel{law}{=} \frac{H}{2}. \tag{1.4}$$

The first identity is due to Chung [9], the second was first stressed in [4, p. 69]. The aim of this paper is to give a simple proof of the following theorem:

Theorem 1.1

$$\left(\frac{H_B^{(n)}}{\sqrt{2n}}, \frac{W_B^{(n)}}{\sqrt{2n}} \right) \underset{n \to +\infty}{\stackrel{law}{\longrightarrow}} (H, W).$$

Note that the obvious negative correlation between height and width of a tree with given size n, is reflected in the dependence between $\int_0^1 \frac{ds}{e(s)}$ and $\max_{0 \le s \le 1} e(s)$. Previous results [15, 23] about height and width of simple trees belongs to the foundations of computer science. Surprisingly, Theorem 1.1 does not seem to be stated anywhere, though it can be deduced easily from deep results of Aldous on one hand (about the continuum random tree [1, 2]) and on the other hand of Drmota & Gittenberger [12], using a clever idea due to Aldous [3, Th. 3] again. We felt that this consequence of [3, Th. 3] deserved to be pointed out, and that the reader would welcome an 'elementary' and direct proof.

Let $\Phi(\alpha, \gamma, z)$ denote the confluent hypergeometric function, defined, for $|z| < +\infty, \gamma \neq 0, -1, -2, \cdots$, by:

$$\Phi(\alpha, \gamma, z) = \sum_{k=0}^{+\infty} \frac{(\alpha)_k \, z^k}{(\gamma)_k \, k!}$$

where $(\lambda)_k = (\lambda)(\lambda+1) \ldots (\lambda+k-1)$. The joint law of (H, W) has been investigated recently by Catherine Donati-Martin [11]. With the help of the agreement formula for the Itô measure (see [4, 22] and [26]), she obtains the following results:

Theorem 1.2 *For* $\lambda \ge 0, \alpha \ge 0$,

$$E\Big(W \exp\Big(- \frac{\lambda^2}{2W^2} - \frac{\alpha^2 H}{2W} \Big)\Big) = \sqrt{\frac{\pi}{2}} \frac{\exp(2\lambda)}{\Phi^2(1 + \alpha^2/(2\lambda), 2, 2\lambda)}.$$

As a consequence, for $Re(s) > 1, Re(t) < 0$ *and* $Re(s+t) > 1$*:*

$$E(W^s H^t) = \sqrt{\frac{\pi}{2}} \frac{2^{\frac{5+t-s}{2}}}{\Gamma(-t)\Gamma(\frac{s+t-1}{2})} \int_0^{+\infty} \int_0^{+\infty} \frac{\lambda^{s+t-2} \alpha^{-(1+2t)} \exp(2\lambda)}{\Phi^2(1 + \alpha^2/(2\lambda), 2, 2\lambda)} \, d\alpha \, d\lambda.$$

2 First proof of Theorem 1.1

Aldous [3, Th. 3] proves that, suitably rescaled, the depth-first walk and the profile of a random rooted labeled tree with n nodes converges jointly to $(2e, l/2)$, where l is the *local time* of the normalized Brownian excursion e, defined by:

$$\int_0^a l(x) \, dx = \int_0^1 I_{[0,a]}(e(s)) ds.$$

Let $H_L^{(n)}$ (resp. $W_L^{(n)}$) denote the restriction of h (resp. w) to the set of rooted labeled trees with n nodes, endowed with the uniform probability. Invariance principle yields at once that:

$$\frac{1}{\sqrt{n}}\left(H_L^{(n)}, W_L^{(n)}\right) \xrightarrow{law} \left(2 \max_{0\leq t\leq 1} e(t), \frac{1}{2}\max_{x\geq 0} l(x)\right).$$

For a general class of simple trees with n leaves, the proof of [3, Th. 3] is still valid (see [2, Th. 23], and for binary trees, [17]), the limit being now $\left(\frac{2e}{\sigma}, \frac{\sigma l}{2}\right)$. Here σ^2 denotes the variance of the offspring distribution of the corresponding critical Galton-Watson tree (see [1, p. 28, formula (8)] for the meaning of σ in term of simple trees). In the special case of binary trees with $n-1$ internal nodes and n leaves, it yields:

$$\frac{1}{\sqrt{2n}}\left(H_B^{(n)}, W_B^{(n)}\right) \xrightarrow{law} \left(2 \max_{0\leq t\leq 1} e(t), \frac{1}{2}\max_{x\geq 0} l(x)\right). \tag{2.5}$$

Theorem 1.1 is deduced from (2.5) through Jeulin's description of the local time of Brownian excursion. Let $\left(e(s)\right)_{0\leq s\leq 1}$ be a normalized Brownian excursion with local time $\left(l(x)\right)_{x\geq 0}$. Define

$$L(y) = \int_0^y l(x)\, dx$$

and

$$\psi(t) = L^{-1}(t) = \sup\left\{y \middle| \int_0^y l(x)dx < t\right\}.$$

Jeulin [18] proved that the process $\left(\tilde{e}(s)\right)_{0\leq s\leq 1}$ defined by:

$$\tilde{e}(s) = \frac{1}{2} l(\psi(s)) \tag{2.6}$$

is itself a normalized Brownian excursion (see also [4, p. 70] and interesting heuristic arguments [1, pp. 47-48]). Taking the derivative in $\psi(t) = L^{-1}(t)$, we obtain $\psi' = \frac{1}{L'\circ\psi} = \frac{1}{2\,\tilde{e}}$ and

$$\psi(t) = \int_0^t \frac{du}{2\,\tilde{e}(u)},$$

so Jeulin's representation can be rewritten:

$$l\left(\int_0^s \frac{du}{2\,\tilde{e}(u)}\right) = 2\,\tilde{e}(s). \tag{2.7}$$

A direct consequence is the identity:

$$\left(2 \max_{0\leq t\leq 1} e(t), \frac{1}{2}\max_{x\geq 0} l(x)\right) = \left(\int_0^1 \frac{ds}{\tilde{e}(s)}, \max_{0\leq s\leq 1} \tilde{e}(s)\right). \tag{2.8}$$

The equality between first components of (2.8) follows from (2.7) because

$$\int_0^1 \frac{ds}{2\,\tilde{e}(s)} = \psi(1) = \max_{0\leq s\leq 1} e(s)\,, \tag{2.9}$$

while the equality between second components follows by taking the maximum on each side of (2.7). Thus, (2.5) is equivalent to Theorem 1.1. □

Thus Theorem 1.1 is a direct consequence of Jeulin's representation [18], and of [3, Theorem 3] which relies itself on two deep, but technical, papers [2, 12]. The line of the second proof of Theorem 1.1 is close to that of [7, 25]: the profile of the tree is seen as the breadth-first search random walk, changed of time, giving a discrete converse of Jeulin's representation. That the change of time has precisely the form given by Jeulin, follows, in the discrete case, from a counting principle due to Odlyzko [8, 21].

3 Second proof of Theorem 1.1

3.1 Brownian excursion and Bernoulli excursion

Let us call *Bernoulli excursion* of size $2n$, any $2n$-steps random walk $\omega = \left(S_k(\omega)\right)_{k=0,\cdots,2n}$ that satisfy:

$$S_0(\omega) = 0\,, \quad S_{2n}(\omega) = 0, \quad S_{k+1}(\omega) = S_k(\omega) \pm 1$$

and

$$S_k(\omega) > 0 \text{ for } k \in \{1, \cdots, 2n-1\}.$$

Let $Es(2n)$ denote the set of Bernoulli excursions of size $2n$, endowed with the uniform probability. It is well known that

$$\#Es(2n) = \#T_B^{(n)} = \binom{2n-1}{n-1}\frac{1}{2n-1},$$

is the $n-1^{th}$ Catalan number: C_{n-1} (see [24, pp.220-221, and 256-257]). Note that there is an obvious one-to-one correspondence between Bernoulli excursions and Dyck paths.

Any Bernoulli excursion ω defines a random element

$$e_n(t) = \frac{S_{\lfloor 2nt \rfloor}}{\sqrt{2n}}, \quad 0 \leq t \leq 1,$$

of the set $D([0,1])$ of right continuous left limit functions, endowed with the Skorohod topology. The weak limit of e_n is called the *normalized Brownian excursion* (see [16]). The normalized Brownian excursion e is usually defined by the following path transformation of the standard linear Brownian motion $\mathcal{B} = (\mathcal{B}_t)_{t\geq 0}$: let g (resp. d) be the last zero of $\mathcal{B}$ before 1 (resp. after 1), and set

$$e(t) = \frac{\mathcal{B}_{g+t(d-g)}}{\sqrt{d-g}}, \qquad 0 \leq t \leq 1.$$

3.2 Breadth-first search correspondence

Let S_k be the height of the queue at the kth step of the breadth-first search of a rooted binary tree $t \in T_B^{(n)}$ (see [10, Section 23.2], and Figure (3.1) for an example). Then $\omega = (S_k)_{k=0,\cdots,2n}$ belongs to $Es(2n)$, and this is a one-to-one correspondence (for instance, one can adapt [24, p. 256, 6.19.d]). We explain below how to obtain an expression of $(h(t), w(t))$ in term of functionals of the corresponding Bernoulli excursion ω.

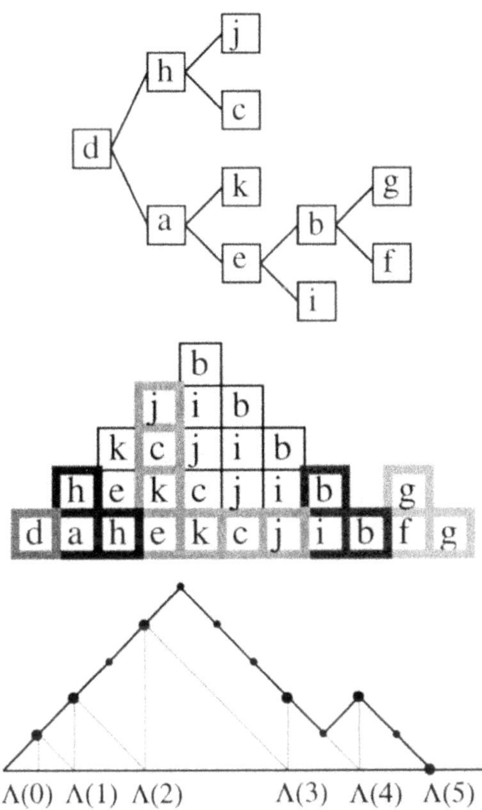

Figure 3.1 : Excursion - Binary tree

The width

As already noted in [7, 19, 25], the profile of t can be read on ω: assuming $S_k(\omega) = 0$ for $k \geq 2n+1$, we have

$$\begin{aligned} Z_0(t) &= S_1(\omega) = 1 \\ Z_1(t) &= S_{1+Z_0(t)}(\omega) \\ Z_2(t) &= S_{1+Z_0(t)+Z_1(t)}(\omega) \\ &\dots \\ Z_{k+1}(t) &= S_{1+Z_0(t)+\cdots+Z_k(t)}(\omega). \end{aligned}$$

Set

$$\begin{aligned} \Lambda(k) &= 1 + Z_0(t) + \cdots + Z_{k-1}(t), \\ M_{2n}(\omega) &= \max_k S_k(\omega). \end{aligned}$$

The triplet (S, Z, Λ) can be seen as the discrete version of $(\tilde{e}, l, L)$ appearing in Jeulin's representation. Since

$$W_B^{(n)}(t) = \max_k S_{\Lambda(k)}(\omega),$$

we obtain:

$$M_{2n}(\omega) \geq W_B^{(n)}(\omega),$$

but, actually, moderate variation of S_k (see Lemma 3.3) yields that:

Lemma 3.1

$$E\Big[|W_B^{(n)}(t) - M_{2n}(\omega)|\Big] = O(n^{1/4}\sqrt{\log n}).$$

The height

Set

$$\begin{aligned} \Psi(k) &= \sum_{j=1}^{2k-1} \frac{1}{S_j(\omega)}, \\ \tilde{\Psi}(k) &= -1 + \inf\{j \mid \Lambda(j) = 2k\}. \end{aligned}$$

We see easily that

$$H_B^{(n)}(t) = \tilde{\Psi}(n). \tag{3.10}$$

The following Lemma can be seen as the discrete version of (2.9):

Lemma 3.2

$$E\left[|H_B^{(n)}(t) - \Psi(n)|\right] = o(\sqrt{n}).$$

Remark. Obviously, if the speed of a traveller at point y of the line is $s(y)$, then the duration t of the journey from point 0 to point x satisfies:

$$t = \int_0^x \frac{dy}{s(y)}. \tag{3.11}$$

Lemma 3.2 can be seen as a stochastic analog of relation (3.11), as $H_B^{(n)}(t)$ is the time needed to go from point 0 to point $2n$, doing one step (from $\Lambda(k)$ to by $\Lambda(k+1)$) by time unit, so the speed at point $\Lambda(k)$ is $\Lambda(k+1) - \Lambda(k) = S_{\Lambda(k)}$. This counting principle was used in [8, Section 2] and [21] in order to study the average cost of some search algorithms.

3.3 Proofs of Lemmata 3.1 and 3.2

The proofs of Lemmata 3.1 and 3.2 rely on a property of moderate variation of Bernoulli excursions, inherited from the simple symmetric random walk.
Let $\tilde{\Omega}_c(2n)$ denote the set of Bernoulli excursions $\omega = (S_k(\omega))_{k=0,\cdots,2n}$ such that for any l, k in the set $\{0, 1, \cdots, 2n\}$,

$$|S_k(\omega) - S_l(\omega)| \leq c\sqrt{|k-l|\log n}.$$

Lemma 3.3 *For every $\beta > 0$ there exist $c > 0$ such that, for n sufficiently large:*

$$\Pr(\tilde{\Omega}_c(2n)) \geq 1 - n^{-\beta}.$$

Proof : The lemma is easily proved for a simple symmetric random walk $\omega = (S_k(\omega))_{k=0,\cdots,2n}$, using Chernoff bounds:

$$\forall x \geq 0,\ \forall k,\ \Pr(|S_k(\omega)| > x) \leq 2\exp(-\frac{x^2}{2k})$$

(see for instance [8]). But $\Pr(\tilde{\Omega}_c(2n))$ in Lemma 3.3 is just $\Pr(\omega \in \tilde{\Omega}_c(2n) | \omega \in Es(2n))$, and in the other hand the probability that a simple symmetric random walk ω belongs to $Es(2n))$ is $\Theta(n^{-\frac{3}{2}})$. Finally, choose $A = \complement\tilde{\Omega}_c(2n)$ in:

$$\Pr(\omega \in A \,|\, \omega \in Es(2n)) \leq \frac{\Pr(\omega \in A)}{\Pr(\omega \in Es(2n))} \leq c_1 n^{\frac{3}{2}} \Pr(\omega \in A). \quad \square$$

Proof of Lemma 3.1: We have

$$0 \leq E\big(M_{2n}(\omega) - W_B^{(n)}(t)\big) = E\big(M_{2n}(\omega) - \max_k S_{\Lambda(k)}(\omega)\big).$$

We consider an index $K(\omega)$ such that $M_{2n}(\omega) = S_{K(\omega)}(\omega)$. There exists an integer $i(\omega)$ such that

$$\Lambda(i) \leq K \leq \Lambda(i+1).$$

Then,

$$\begin{aligned} E(M_{2n} - \max_k S_{l(k)}) &\leq E(S_K - S_{\Lambda(i)}) \\ &\leq E\left(\mathbb{I}_{\tilde{\Omega}_c} c\sqrt{(K-\Lambda(i))\log(2n)} + n\,\mathbb{I}_{\complement\tilde{\Omega}_c}\right) \\ &\leq E\left(c\sqrt{M_{2n}\log(2n)}\right) + n\Pr(\complement\tilde{\Omega}_c) \\ &= O\left(n^{1/4}(\log n))^{3/4}\right), \end{aligned}$$

for c large enough. □

Proof of Lemma 3.2. For any positive integers l, k, such that $k \leq l \leq 2n-k$, we have

$$\Pr(S_k = l) = \frac{n\,l^2}{k(2n-k)} \frac{\binom{k}{\frac{k-l}{2}}\binom{2n-k}{\frac{2n-k-l}{2}}}{\binom{2n-2}{n-1}} \mathbb{I}_{k\equiv l[2]}, \tag{3.12}$$

since $\frac{l}{k}\binom{k}{\frac{k-l}{2}}$ is the number of positive paths from $(0,0)$ to (k,l), and $\frac{l}{2n-k}\binom{2n-k}{\frac{2n-k-l}{2}}$ is the number of positive paths from (k,l) to $(2n,0)$. We have

$$H_B^{(n)} = \tilde{\Psi}(n) = \sum_{i=0}^{\tilde{\Psi}(n)-1} \sum_{h=\Lambda(i)}^{\Lambda(i+1)-1} \frac{1}{Z(i)}.$$

Let α be a real number in $]0,1/2[$. Then

$$\left| \sum_{i=0}^{\tilde{\Psi}(n)-1} \sum_{h=\Lambda(i)}^{\Lambda(i+1)-1} \frac{1}{Z(i)} - \sum_{h=1}^{2n-1} \frac{1}{S_h} \right| \leq 2n^{1/2-\alpha} + A_n + B_n$$

where

$$\begin{aligned} A_n &\leq \sum_{i=0}^{\tilde{\Psi}(n)-1} \sum_{h=\Lambda(i)}^{\Lambda(i+1)-1} \left| \frac{1}{Z(i)} - \frac{1}{S_h} \right| \mathbb{I}_{Z(i)\geq(\log^{1+\epsilon} n)}\, \mathbb{I}_{[n^{1/2-\alpha}, 2n-n^{1/2-\alpha}]}(h), \\ B_n &\leq \sum_{i=0}^{\tilde{\Psi}(n)-1} \sum_{h=\Lambda(i)}^{\Lambda(i+1)-1} \left| \frac{1}{Z(i)} - \frac{1}{S_h} \right| \mathbb{I}_{Z(i)\leq(\log n)^{1+\epsilon}}\, \mathbb{I}_{[n^{1/2-\alpha}, 2n-n^{1/2-\alpha}]}(h). \end{aligned}$$

First,

$$\begin{aligned} E(B_n) &\leq E\big(\#\{h|h \in [n^{1/2-\alpha}, 2n - n^{1/2-a}], S_h \leq (\log n)^{1+\varepsilon}\}\big) \\ &\leq \sum_{h=n^{1/2-\alpha}}^{2n-n^{1/2-\alpha}} \sum_{l=1}^{[\log^{1+\varepsilon} n]} \Pr(S_h = l) \\ &= O(\log^{3+3\varepsilon} n) \end{aligned}$$

where the last equality follows from (3.12), Stirling formula and $\binom{k}{k/2} \geq \binom{k}{(k-l)/2}$. Clearly,

$$\mathbb{I}_{\complement\tilde{\Omega}_c(2n)}(\omega) A_n \leq n I_{\complement\tilde{\Omega}_c(2n)}(\omega).$$

Finally, using the moderate variation property to bound

$$\left|\frac{1}{Z(i)} - \frac{1}{S_h}\right|,$$

we obtain:

$$\mathbb{I}_{\tilde{\Omega}_c(2n)}(\omega) A_n \leq (\log^{-\varepsilon/2} n)\ \Psi(n).$$

Lemma 3.2 follows, for c large enough. □

3.4 Convergence of $(\Psi(n), M_{2n})$

Lemmata 3.1 and 3.2 together yields that:

$$\frac{1}{\sqrt{n}} \left\| (H_B^{(n)}, W_B^{(n)}) - (\Psi(n), M_{2n}) \right\|_1 = o(1).$$

Thus, the proof of Theorem 1.1 reduces to the proof of

Proposition 3.4

$$\frac{1}{\sqrt{2n}} (\Psi(n), M_{2n}) \overset{law}{\longrightarrow} \left(\int_0^1 \frac{1}{e(s)}\, ds, \max_{0 \leq s \leq 1} e(s) \right).$$

We use the following Lemma [6, Th.4.2 p.25]:

Lemma 3.5 *Let $(X_n)_n$ and $(X_n^{(a)})_{n,a}$ be two families of $\mathbb{R}^2$ valued r.v., defined on the same probability space, such that:*

$$(X_n^{(a)})_n \underset{n \longrightarrow +\infty}{\overset{law}{\longrightarrow}} X^{(a)}$$

and

$$X^{(a)} \underset{a \longrightarrow 0}{\overset{law}{\longrightarrow}} X.$$

Assume that

$$\lim_{a \longrightarrow 0} \left[\limsup_{n \longrightarrow +\infty} P(\|X_n - X_n^{(a)}\|_1 \geq \epsilon) \right] = 0$$

for each positive ϵ. *Then*

$$X_n \xrightarrow{law} X.$$

Proof of Proposition 3.4 : We have

$$M_{2n}/\sqrt{2n} = \max_{0 \leq t \leq 1} e_n(t)$$

and

$$\frac{\Psi(n)}{\sqrt{2n}} = \frac{1}{2n} \sum_{k=1}^{2n-1} \frac{1}{e_n(k/2n)}.$$

Define $\Psi^{(a)}(n)$ by

$$\Psi^{(a)}(n) = \frac{1}{2n} \sum_{k=1}^{2n-1} \frac{\mathbb{I}_{\{e_n \geq a\}}}{e_n(k/2n)}.$$

Set

$$\begin{aligned}
X_n &= \frac{1}{\sqrt{2n}} \big(\Psi(n), M_{2n} \big), \\
X_n^{(a)} &= \frac{1}{\sqrt{2n}} \big(\Psi^{(a)}(n), M_{2n} \big), \\
X &= \Big(\int_0^1 \frac{1}{e(s)}\, ds, \max_{0 \leq s \leq 1} e(s) \Big), \\
X^{(a)} &= \Big(\int_0^1 \frac{\mathbb{I}_{\{e(s) \geq a\}}}{e(s)}\, ds, \max_{0 \leq s \leq 1} e(s) \Big).
\end{aligned}$$

Proposition 3.4 is equivalent to

$$X_n \xrightarrow{law} X.$$

The convergence of $X_n^{(a)}$ to $X^{(a)}$ when n goes to ∞ results from the continuity of the functional. To conclude, it suffices to prove the two following lemmas:

Lemma 3.6 *There exists a positive constant* C_1 *such that, for any* $a > 0$,

$$\|X_n - X_n^{(a)}\|_1 = \frac{1}{\sqrt{2n}} E\Big(\sum_{k=1}^{2n-1} \frac{I_{S_k \leq a\sqrt{2n}}}{S_k} \Big) \leq C_1\, a.$$

Lemma 3.7 *There exists a positive constant C_2 such that, for any $a > 0$,*

$$\|X - X^{(a)}\|_1 = E\Big(\int_0^1 \frac{\mathbb{I}_{e(s)\le a}}{e(s)}\, ds\Big) \le C_2\, a.$$

Proof of Lemma 3.6 : Using Formula (3.12), we have

$$\begin{aligned}
\frac{1}{\sqrt{2n}} E\Big(\sum_{k=1}^{2n-1} \frac{I_{S_k \le a\sqrt{2n}}}{S_k}\Big) &= \frac{1}{\sqrt{2n}} \sum_{k=1}^{2n-1} \sum_{l=1}^{a\sqrt{2n}} \frac{\tilde{\mathbb{P}}(S_k = l)}{l} \\
&\le c_1 . \frac{n}{\sqrt{n}\binom{2n-2}{n-1}} \sum_{k=1}^{n} \sum_{l=1}^{a\sqrt{2n}} \frac{l\binom{k}{\frac{k-l}{2}}\binom{2n-k}{\frac{2n-k}{2}}}{k(2n-k)} \mathbb{I}_{k\equiv l[2]} \\
&\le c_2 \frac{1}{\sqrt{n}} \sum_{l=1}^{a\sqrt{2n}} \sum_{k=1}^{n} \frac{l\binom{k}{\frac{k-l}{2}}}{2^k k} \mathbb{I}_{k\equiv l[2]}.
\end{aligned}$$

Note that

$$\begin{aligned}
\frac{l\binom{k}{\frac{k-l}{2}}}{2^k k} \mathbb{I}_{k\equiv l[2]} &= \mathbb{P}(S_k = l)\frac{l}{k} \\
\mathbb{P}(S_k = l)\frac{l}{k} &= \frac{l}{k}\frac{1}{2\pi}\int_0^{2\pi} \cos^k(t)\cos(lt)dt \\
&= \frac{2}{\pi}\int_0^{\pi/2} \sin(lt)\cos^{k-1}(t)\sin t\; dt \text{ for } k \ge 1.
\end{aligned}$$

Thus,

$$\begin{aligned}
\frac{1}{\sqrt{2n}} E\Big(\sum_{k=1}^{2n-1} \frac{I_{S_k \le a\sqrt{2n}}}{S_k}\Big) &\le \frac{c_3}{\sqrt{n}} \sum_{l=1}^{a\sqrt{2n}} \int_0^{\pi/2} \sin(lt)\sin t\, \frac{1-\cos^n t}{1-\cos t} dt \\
&= \frac{c_3}{\sqrt{n}} \sum_{l=1}^{a\sqrt{2n}} \int_0^{\pi/2} \sin(lt)\frac{(1+\cos t)(1-\cos^n t)}{\sin t} dt \qquad (3.13)
\end{aligned}$$

Let us expand this sum and bound its terms. Set

$$\begin{aligned}
I_l &= \int_0^{\pi/2} \frac{\sin(lt)\cos t}{\sin t} dt, \\
J_l &= \int_0^{\pi/2} \frac{\sin(lt)}{\sin t} dt.
\end{aligned}$$

We notice that

$$J_l = I_{l-1} + \frac{\sin((l-1)\frac{\pi}{2})}{l-1} \text{ and } I_l = J_{l-1} + \frac{\sin(\frac{\pi l}{2})}{l}.$$

So J_l and I_l are uniformly bounded. We have

$$\begin{aligned}\left|\int_0^{\pi/2} \frac{\sin(lt)}{\sin t}\cos^n t\, dt\right| &\leq \int_0^{\pi/2} \left|\frac{\sin(lt)}{\sin t}\right| \cos^n t\, dt \\ &\leq l \int_0^{\pi/2} \cos^n t\, dt \\ &= \frac{\sqrt{\pi}}{2} \frac{l\,\Gamma(\frac{n+1}{2})}{\Gamma(n/2+1)}\end{aligned}$$

Due to Stirling formula, this last term is uniformly bounded for $l \in \{1, \cdots, a\sqrt{2n}\}$, so the terms of the sum in (3.13) are uniformly bounded and the proof is complete. □

Proof of Lemma 3.7 : According to [14, Prop. 3.4],

$$E\Big(\int_0^1 \mathbb{I}_{e(s)\leq a}\, ds\Big) = 1 - 2e^{-2a^2},$$

we have:

$$E\Big(\int_0^1 \frac{\mathbb{I}_{e(s)\leq a}}{e(s)}\, ds\Big) = \int_0^a 8e^{-2a^2}\, da \leq 8\,a. \quad \square$$

References

[1] D. Aldous, (1991) *The continuum random tree II: An overview,* Stochastic analysis, Proc. Symp., Durham. UK 1990, Lond. Math. Soc. Lect. Note Ser. 167, 23-70.

[2] D. Aldous, (1993) *The continuum random tree III,* Ann. of Probab. 21, No.1, 248-289.

[3] D. Aldous, (1998) *Brownian excursion conditionned on its local time*, Elect. Comm. in Probab., **3** , 79-90.

[4] P. Biane, M. Yor, (1987) *Valeurs principales associées aux temps locaux browniens*, Bull. Sci. Maths 111, 23-101.

[5] P. Biane, J. Pitman, M. Yor, (1999) *Probability laws related to the Jacobi theta and Riemann zeta functions, and Brownian excursions*, (see http://www.stat.berkeley.edu/~pitman/)

[6] P. Billingsley, (1968) *Convergence of Probability Measures.* John Wiley & Sons, Inc., New York-London-Sydney.

[7] P. Chassaing, J.F. Marckert, (1999) *Parking functions, empirical processes and the width of rooted labeled trees*, preprint Elie Cartan, Université Nancy I.

[8] P. Chassaing, J.F. Marckert, M. Yor, (1999) *A Stochastically Quasi-Optimal Search Algorithm for the Maximum of the Simple Random Walk*, to appear in Ann. of App. Prob.

[9] K.L. Chung, (1976) *Excursions in Brownian motion*, Ark. för Math., **14**, 155-177.

[10] T. H. Cormen, C. E. Leiserson, & R. L. Rivest, (1990) *Introduction to algorithms*, MIT Press, Cambridge, MA; McGraw-Hill Book Co., New York.

[11] C. Donati-Martin, (1999) *Some remarks about the identity in law for the Bessel bridge* $\int_0^1 \frac{ds}{r(s)} \overset{(law)}{=} 2 \sup_{s\leq 1} r(s)$, prepub. Toulouse.

[12] M. Drmota, B. Gittenberger, (1997) *On the profile of random trees*, Random Structures Algorithms 10, no. 4, 421–451.

[13] M. Drmota, B. Gittenberger, (2000) *The width of Galton-Watson trees*, preprint.

[14] R.T. Durrett, D.L. Iglehart, (1977) *Functionals of Brownian meander and Brownian excursion*, Ann. Probab. 5, 130-135.

[15] P. Flajolet, A. Odlyzko, (1982) *The average height of binary trees and other simple trees*, J. Comp. and Sys. Sci., Vol. 25, No.2.

[16] I.I. Gikhman, A.V.Skorohod, (1969) *Introduction to the theory of random processes*, W.B. Saunders, Philadelphia.

[17] W. Gutjahr, G. Ch. Pflug, (1992) *The asymptotic contour process of a binary tree is Brownian excursion*, Stochastic Proc. Appl. 41, 69-90.

[18] Th. Jeulin, (1980) *Semi-martingales et grossissement d'une filtration*, Lecture Notes in Mathematics, 833, Springer-Verlag.

[19] D.G. Kendall, (1951) *Some problems in the theory of queues*, J. of the Roy. Stat. Soc. **B 13**, 151-185.

[20] G. Louchard, (1984) *Kac's formula, Levy's local time and Brownian excursion*, J. Appl. Prob. **21** , 479-499.

[21] A.M. Odlyzko, (1995) *Search for the maximum of a random walk*, Ran. Struct. Alg., Vol. 6, p. 275-295.

[22] J. Pitman, M. Yor, (1996) *Decomposition at the maximum for excursions and bridges of one-dimensional diffusions*, Ito's stochastic calculus and probability theory. Springer. 293-310.

[23] A. Renyi, G. Szekeres, (1967) *On the height of trees*, J. Aust. Math. Soc. 7, 497-507.

[24] R. P. Stanley, (1999) *Enumerative combinatorics. Vol. 2.*, Cambridge Studies in Advanced Mathematics, 62. Cambridge University Press, Cambridge.

[25] L. Takács, (1993) *Limit distributions for queues and random rooted trees*, J. Appl. Math. Stoch. Ana., **6** , No.3, p.189 - 216.

[26] D. Williams, (1990) *Brownian motion and the Riemann zeta-function*, Disorder in physical systems, Clarendon press, Oxford, 361-372.

Trends in Mathematics, © 2000 Birkhäuser Verlag Basel/Switzerland

On the node structure of binary search trees

F.M. DEKKING AND S. DE GRAAF AND L.E. MEESTER *Thomas Stieltjes Institute for Mathematics and Delft University of Technology*
F.M.Dekking@its.tudelft.nl

Abstract. *The external nodes of a binary search tree are of two types: arm nodes whose parents have degree 2, and foot nodes whose parents have degree 1. We study the positioning of these two types on the tree. We prove that the conditional distribution of the insertion depth of a key given that it is inserted in a foot node equals that of the conditional distribution given that it is inserted in an arm node shifted by 1. We further prove that the normalized path length of the arm nodes converges almost surely to $\frac{1}{3}$ times the limit distribution of the normalized path length of all external nodes.*

1 Introduction

Let $\mathcal{T}_n$ be the binary search tree generated by a sequence of keys $(K_1, \dots, K_n)$. As usual we assume that all $n!$ permutations of the keys have the same probability. Let $\mathcal{E}_n$ be the set of external nodes of $\mathcal{T}_n$, i.e., the elements of $\mathcal{E}_n$ are the children of the leaves of $\mathcal{T}_n$, and one of these nodes will receive the next key K_{n+1}. The set $\mathcal{E}_n$ can be split in a natural way into a set $\mathcal{E}_n^A$ containing the arm nodes, external nodes whose parents have degree 2, and the set $\mathcal{E}_n^F$ of foot nodes whose parents have degree 1. It has been known for a long time [2] that the expected values of the cardinality of $\mathcal{E}_n^A$, respectively $\mathcal{E}_n^F$, are $\frac{1}{3}(n+1)$ and $\frac{2}{3}(n+1)$. In our work we study the question of how these foot and arm nodes are distributed over the tree. Clearly, foot nodes will occur further from the root of the tree than arm nodes, since foot nodes cannot occur at level 1 (except in $\mathcal{T}_1$), and arm nodes cannot occur at the highest level. Rather surprisingly, this property is in some sense the only restriction that prevents the foot and arm nodes from being identically distributed.

Let U_n be the insertion depth of key K_{n+1}. It is well-known [3] that U_n is distributed as $B_0 + \dots + B_{n-1}$, where the B_k are independent Bernoulli random variables with parameter $p_k = \frac{2}{k+2}$. We consider the insertion depth of K_{n+1} given that K_{n+1} is inserted in an arm node, and prove

$$[U_n \mid K_{n+1} \in \mathcal{E}_n^A] \stackrel{d}{=} 1 + B_2 + \dots B_{n-1},$$

where as before the B_k are independent Bernoulli random variables with parameter p_k. We also prove that

$$[U_n \mid K_{n+1} \in \mathcal{E}_n^F] \stackrel{d}{=} 2 + B_2 + \dots B_{n-1}.$$

To prove these results we define $U_n^A = U_n 1_{[K_{n+1} \in \mathcal{E}_n^A]}$ and $U_n^F = U_n 1_{[K_{n+1} \in \mathcal{E}_n^F]}$, and determine in Section 2 the joint probability generating function:

$$\mathrm{E} s^{U_n^A} t^{U_n^F} = \frac{1}{3} s \prod_{k=2}^{n-1} \frac{k+2s}{k+2} + \frac{2}{3} t^2 \prod_{k=2}^{n-1} \frac{k+2t}{k+2}.$$

In sections 3 and 4 we consider almost sure properties of the distribution of the arm and foot nodes over $\mathcal{T}_n$ by studying the total path length X_n^A of the arm nodes and X_n^F of the foot nodes. Well-studied is $X_n = X_n^A + X_n^F$, the total path length of all external nodes, which equals the number of comparisons needed to quicksort n different numbers. It was shown by [5] (see also [6]) that $Z_n = \frac{1}{n+1}(X_n - \mathrm{E}X_n)$ is an L^2-bounded martingale, which hence converges almost surely to a limiting random variable Z, the 'quicksort distribution'. The corresponding normalized arm node path length $Z_n^A = \frac{1}{n+1}(X_n^A - \mathrm{E}X_n^A)$ and foot node path length $Z_n^F = \frac{1}{n+1}(X_n^F - \mathrm{E}X_n^F)$ are not martingales. In fact we show that the only sequences $a = (a_n)_{n=3}^{\infty}$ and $b = (b_n)_{n=3}^{\infty}$ which turn

$$M_n = a_n Z_n^A + b_n Z_n^F \quad n = 3, 4, \ldots$$

into a martingale are linear combinations of the constant sequences $\left(\begin{smallmatrix} a \\ b \end{smallmatrix}\right) = \left(\left(\begin{smallmatrix} 1 \\ 1 \end{smallmatrix}\right)\right)$ and the sequences $\left(\begin{smallmatrix} a \\ b \end{smallmatrix}\right) = \left(p_n \left(\begin{smallmatrix} 2 \\ -1 \end{smallmatrix}\right)\right)$, where $p_n = n^3 - n$. However, the martingales corresponding to the second solution have unbounded variances. This follows immediately from the asymptotics for the variances: as $n \to \infty$

$$\begin{aligned} \mathrm{Var}(Z_n^A) - \frac{7}{9} + \frac{2\pi^2}{27} &\sim \frac{32}{45}\frac{\log^2 n}{n}, \\ \mathrm{Var}(Z_n^F) - \frac{28}{9} + \frac{8\pi^2}{27} &\sim \frac{32}{45}\frac{\log^2 n}{n}. \end{aligned}$$

We furthermore show that the path lengths Z_n^A and Z_n^F are negatively correlated for $n \leq 168$ and positively correlated for $n \geq 169$. In fact the correlation between $2Z_n^A$ and Z_n^F tends to 1 as $n \to \infty$. This is related to our final result, which gives a self-similarity result for binary search trees.

Theorem Almost surely $\lim_{n\to\infty} Z_n^A = \frac{1}{3}Z$, $\lim_{n\to\infty} Z_n^F = \frac{2}{3}Z$, where Z is the quicksort distribution.

2 Where are the arm nodes? Distributional results.

We will determine the joint probability generating function

$$G_n(s,t) = \mathrm{E}s^{U_n^A} t^{U_n^F} = \sum_{k,l=0}^{\infty} s^k t^l \mathrm{P}(U_n^A = k, U_n^F = l).$$

The definition of U_n^A and U_n^F implies that

$$\begin{aligned} \mathrm{P}(U_n^A = k, U_n^F = l, K_{n+1} \in \mathcal{E}_n^A) &= 0 \text{ when } l \neq 0, \text{ and} \\ \mathrm{P}(U_n^A = k, U_n^F = l, K_{n+1} \in \mathcal{E}_n^F) &= 0 \text{ when } k \neq 0. \end{aligned}$$

Since also clearly $U_n^A \le n$ and $U_n^F \le n$, we can simplify G_n to

$$G_n(s,t) = \sum_{k=1}^{n} \left[s^k \mathrm{P}(U_n^A = k) + t^k \mathrm{P}(U_n^F = k) \right].$$

Let $\mathcal{F}_n$ be the σ-field generated by the first n keys, let $X_{n,k}^A$ be the number of arm nodes, and let $X_{n,k}^F$ be the number of foot nodes of $\mathcal{T}_n$ at level k. We then have for $k \ge 1$

$$\mathrm{P}(U_n^A = k \mid \mathcal{F}_n) = \mathrm{P}(U_n = k, K_{n+1} \in \mathcal{E}_n^A \mid \mathcal{F}_n) = \frac{X_{n,k}^A}{n+1}, \tag{1}$$

$$\mathrm{P}(U_n^F = k \mid \mathcal{F}_n) = \mathrm{P}(U_n = k, K_{n+1} \in \mathcal{E}_n^F \mid \mathcal{F}_n) = \frac{X_{n,k}^F}{n+1}, \tag{2}$$

and find

$$\mathrm{P}(U_n^A = k) = \mathrm{E}\frac{X_{n,k}^A}{n+1} \text{ and } \mathrm{P}(U_n^F = k) = \mathrm{E}\frac{X_{n,k}^F}{n+1}. \tag{3}$$

Hence we obtain

$$G_n(s,t) = \sum_{k=1}^{n} \left[s^k \mathrm{E}\frac{X_{n,k}^A}{n+1} + t^k \mathrm{E}\frac{X_{n,k}^F}{n+1} \right] = \mathrm{E}\sum_{k=1}^{n} \left[s^k \frac{X_{n,k}^A}{n+1} + t^k \frac{X_{n,k}^F}{n+1} \right].$$

Writing $\Lambda_n(s,t)$ for the simultaneous generating function of $\left(\frac{X_{n,k}^A}{n+1}\right)$ and $\left(\frac{X_{n,k}^F}{n+1}\right)$, the previous equation reduces to

$$G_n(s,t) = \mathrm{E}\Lambda_n(s,t).$$

We will derive a recursion for $G_n(s,t)$. Observe that

$$\begin{aligned} \mathrm{E}(\Lambda_{n+1}(s,t) \mid \mathcal{F}_n) &= \mathrm{E}\left(\sum_{k=1}^{n+1} \left[s^k \frac{X_{n+1,k}^A}{n+2} + t^k \frac{X_{n+1,k}^F}{n+2} \right] \mid \mathcal{F}_n \right) \\ &= \sum_{k=1}^{n+1} \left[s^k \frac{\mathrm{E}(X_{n+1,k}^A \mid \mathcal{F}_n)}{n+2} + t^k \frac{\mathrm{E}(X_{n+1,k}^F \mid \mathcal{F}_n)}{n+2} \right]. \end{aligned} \tag{4}$$

Using

$$\mathrm{E}(X_{n+1,k}^A 1_{[U_n^A = l]} \mid \mathcal{F}_n) = \begin{cases} X_{n,k}^A \cdot \mathrm{P}(U_n^A = l \mid \mathcal{F}_n) & \text{if } l \ne k, \\ (X_{n,k}^A - 1) \cdot \mathrm{P}(U_n^A = k \mid \mathcal{F}_n) & \text{if } l = k, \end{cases}$$

and

$$\mathrm{E}(X_{n+1,k}^A 1_{[U_n^F = l]} \mid \mathcal{F}_n) = \begin{cases} X_{n,k}^A \cdot \mathrm{P}(U_n^F = l \mid \mathcal{F}_n) & \text{if } l \ne k, \\ (X_{n,k}^A + 1) \cdot \mathrm{P}(U_n^F = k \mid \mathcal{F}_n) & \text{if } l = k, \end{cases}$$

we obtain

$$\begin{aligned}\mathrm{E}(X^A_{n+1,k} \mid \mathcal{F}_n) &= \sum_{l\neq k}^{n} X^A_{n,k}\cdot \mathrm{P}(U^A_n = l \mid \mathcal{F}_n) + (X^A_{n,k}-1)\mathrm{P}(U^A_n = k \mid \mathcal{F}_n)\\ &\quad + \sum_{l\neq k}^{n} X^A_{n,k}\cdot \mathrm{P}(U^F_n = l \mid \mathcal{F}_n) + (X^A_{n,k}+1)\mathrm{P}(U^F_n = k \mid \mathcal{F}_n)\\ &= X^A_{n,k} - \mathrm{P}(U^A_n = k \mid \mathcal{F}_n) + \mathrm{P}(U^F_n = k \mid \mathcal{F}_n).\end{aligned}$$

Applying equations (1) and (2) this yields

$$\mathrm{E}(X^A_{n+1,k} \mid \mathcal{F}_n) = \frac{nX^A_{n,k}}{n+1} + \frac{X^F_{n,k}}{n+1}. \tag{5}$$

In a similar fashion one derives

$$\mathrm{E}(X^F_{n+1,k} \mid \mathcal{F}_n) = \frac{(n-1)X^F_{n,k}}{n+1} + 2\frac{X^A_{n,k-1}}{n+1} + 2\frac{X^F_{n,k-1}}{n+1}. \tag{6}$$

Let $X_{n,k} = X^A_{n,k} + X^F_{n,k}$ be the number of external nodes at level k in $\mathcal{T}_n$. Substituting equation (5) and (6) in equation (4) and writing $X^A_{n,k-1} + X^F_{n,k-1} = X_{n,k-1}$ yields

$$\begin{aligned}&(n+2)\mathrm{E}(\Lambda_{n+1}(s,t) \mid \mathcal{F}_n)\\ &\qquad = \sum_{k=1}^{n+1}\left[s^k\left(\frac{nX^A_{n,k}}{n+1} + \frac{X^F_{n,k}}{n+1}\right) + t^k\left(\frac{(n-1)X^F_{n,k}}{n+1} + 2\frac{X_{n,k-1}}{n+1}\right)\right].\end{aligned}$$

Taking the terms together and shifting the index of the last term, and writing $\Gamma_n(s)$ for the generating function of $(\frac{X_{n,k}}{n+1})$, we obtain

$$\begin{aligned}&(n+2)\mathrm{E}(\Lambda_{n+1}(s,t) \mid \mathcal{F}_n)\\ &= \sum_{k=1}^{n+1}(n-1)\left[s^k\frac{X^A_{n,k}}{n+1} + t^k\frac{X^F_{n,k}}{n+1}\right] + \sum_{k=1}^{n+1}\left[s^k\frac{X_{n,k}}{n+1}\right] + \sum_{k=1}^{n}\left[2t^{k+1}\frac{X_{n,k}}{n+1}\right]\\ &= (n-1)\Lambda_n(s,t) + \Gamma_n(s) + 2t\Gamma_n(t).\end{aligned}$$

Taking expectations we arrive at the following recursion

$$(n+2)G_{n+1}(s,t) = (n-1)G_n(s,t) + \mathrm{E}\Gamma_n(s) + 2t\mathrm{E}\Gamma_n(t).$$

It is well-known [3] that $\mathrm{E}\Gamma_n(s) = \prod_{k=0}^{n-1}\frac{k+2s}{k+2}$ for $n \geq 1$. Note that $G_1(s,t) = t$. The solution to the recursion is, for $n \geq 2$

$$G_n(s,t) = \frac{1}{3}s\prod_{k=2}^{n-1}\frac{k+2s}{k+2} + \frac{2}{3}t^2\prod_{k=2}^{n-1}\frac{k+2t}{k+2}. \tag{7}$$

Directly from this expression we obtain

$$\left[U_n \mid K_{n+1} \in \mathcal{E}_n^A\right] \stackrel{d}{=} 1 + B_2 + \dots B_{n-1},$$

where the B_k are independent Bernoulli variables with parameter $p_k = \frac{2}{k+2}$. Similarly, for the foot nodes we obtain

$$\left[U_n \mid K_{n+1} \in \mathcal{E}_n^F\right] \stackrel{d}{=} 2 + B_2 + \dots B_{n-1}.$$

3 Variances and covariances of total path lengths.

In order to obtain further characterisations of the positions of the arm and foot nodes we introduce the total path length X_n^A of arm nodes and X_n^F of foot nodes, defined by $X_n^A = \sum_{k=1}^n kX_{n,k}^A$ and $X_n^F = \sum_{k=1}^n kX_{n,k}^F$. In this section we determine the covariance matrix of these random variables.

If the next key is inserted in an arm node, we loose an arm node at depth U_n, and get two foot nodes at depth $U_n + 1$. If the next key is inserted in a foot node, we get a new arm node and loose two foot nodes at level U_n, and get two new foot nodes at level $U_n + 1$. Therefore

$$X_{n+1}^A - X_n^A = \begin{cases} -U_n, & \text{if } K_{n+1} \in \mathcal{E}_n^A, \\ U_n, & \text{if } K_{n+1} \in \mathcal{E}_n^F, \end{cases} \tag{8}$$

and

$$X_{n+1}^F - X_n^F = \begin{cases} 2U_n + 2, & \text{if } K_{n+1} \in \mathcal{E}_n^A, \\ 2, & \text{if } K_{n+1} \in \mathcal{E}_n^F. \end{cases} \tag{9}$$

Let $\vec{U}_n = \begin{pmatrix} U_n^A \\ U_n^F \end{pmatrix}$ and $\vec{X}_n = \begin{pmatrix} X_n^A \\ X_n^F \end{pmatrix}$. We have

$$\vec{X}_{n+1} - \vec{X}_n = \begin{pmatrix} -1 & 1 \\ 2 & 0 \end{pmatrix} \vec{U}_n + \begin{pmatrix} 0 \\ 2 \end{pmatrix}. \tag{10}$$

It is convenient to normalize the total path lengths. Let $Z_n^A = \frac{1}{n+1}(X_n^A - \mathrm{E}X_n^A)$, $Z_n^F = \frac{1}{n+1}(X_n^F - \mathrm{E}X_n^F)$, $\vec{Z}_n = \begin{pmatrix} Z_n^A \\ Z_n^F \end{pmatrix}$, and $\Sigma_n = \mathrm{Var}((n+1)\vec{Z}_n)$ be the covariance matrix of $(n+1)\vec{Z}_n$. In terms of $\vec{Z}_n$ equation (10) becomes:

$$(n+2)\vec{Z}_{n+1} = (n+1)\vec{Z}_n + C(\vec{U}_n - \mathrm{E}\,\vec{U}_n),$$

where $C = \begin{pmatrix} -1 & 1 \\ 2 & 0 \end{pmatrix}$. In order to use this relation to express Σ_{n+1} in terms of Σ_n and $\mathrm{Var}(\vec{U}_n)$ we note that the mixed terms in $\mathrm{Var}((n+1)\vec{Z}_n + C(\vec{U}_n - \mathrm{E}\,\vec{U}_n))$ simplify as follows:

$$\begin{aligned} \mathrm{E}(n+1)\vec{Z}_n(\vec{U}_n - \mathrm{E}\,\vec{U}_n)^T C^T &= (n+1)\mathrm{E}(\mathrm{E}(\vec{Z}_n(\vec{U}_n - \mathrm{E}\,\vec{U}_n)^T \mid \mathcal{F}_n))C^T \\ &= (n+1)\mathrm{E}(\vec{Z}_n\,\mathrm{E}(\vec{U}_n - \mathrm{E}\,\vec{U}_n \mid \mathcal{F}_n))C^T \\ &= (n+1)\mathrm{E}(\vec{Z}_n\vec{Z}_n{}^T)C^T = \frac{1}{n+1}\Sigma_n C^T. \end{aligned}$$

Hence, we find

$$\Sigma_{n+1} = \Sigma_n + \frac{1}{n+1}(\Sigma_n C^T + C\Sigma_n) + C\mathrm{Var}(\vec{U}_n)C^T.$$

We can rewrite this covariance matrix recursion to three recursions: one for $V_n^A = \mathrm{Var}(Z_n^A)$, one for $V_n^F = \mathrm{Var}(Z_n^F)$, and one for $C_n = \mathrm{Cov}(Z_n^A, Z_n^F)$. We obtain

$$(n+2)^2 V_{n+1}^A = (n^2-1)V_n^A + 2(n+1)C_n + \mathrm{Var}(U_n^F - U_n^A), \tag{11}$$
$$(n+2)^2 V_{n+1}^F = (n+1)^2 V_n^F) + 4(n+1)C_n + 4\mathrm{Var}(U_n^A), \tag{12}$$
$$\begin{aligned}(n+2)^2 C_{n+1} &= n(n+1)C_n + 2(n+1)V_n^A \\ &\quad + (n+1)V_n^F - 2\mathrm{Var}(U_n^A) + 2\mathrm{Cov}(U_n^A, U_n^F).\end{aligned} \tag{13}$$

This system of recursions, equations (11), (12), and (13), can be solved. We start by creating a new recursion from the first two recursions, equations (11) and (12): take twice equation (11), and add equation (12). We have

$$\begin{aligned}&(n+2)^2(2V_{n+1}^A + V_{n+1}^F) \\ &\quad = 2(n^2-1)V_n^A + 4(n+1)C_n + 2\mathrm{Var}(U_n^F - U_n^A) \\ &\qquad + (n+1)^2 V_n^F + 4(n+1)C_n + 4\mathrm{Var}(U_n^A).\end{aligned}$$

Remembering that $V_n = V_n^A + 2C_n + V_n^F$, or

$$8(n+1)C_n = 4(n+1)V_n - 4(n+1)V_n^A - 4(n+1)V_n^F,$$

we can use this to eliminate the covariance term from the recursion:

$$\begin{aligned}&(n+2)^2(2V_{n+1}^A + V_{n+1}^F) \\ &\quad = (2(n^2-1) - 4(n+1))V_n^A + 2\mathrm{Var}(U_n^F - U_n^A) \\ &\qquad + ((n+1)^2 - 4(n+1))V_n^F + 4\mathrm{Var}(U_n^A) + 4(n+1)V_n.\end{aligned}$$

This expression simplifies to

$$\begin{aligned}&(n+2)^2(2V_{n+1}^A + V_{n+1}^F) \\ &\quad = 2(n^2-2n-3)V_n^A + (n^2-2n-3)V_n^F \\ &\qquad + 2\mathrm{Var}(U_n^F - U_n^A) + 4\mathrm{Var}(U_n^A) + 4(n+1)V_n \\ &\quad = (n+1)(n-3)(2V_n^A + V_n^F) + 4(n+1)V_n + 2\mathrm{Var}(U_n^F - U_n^A) + 4\mathrm{Var}(U_n^A).\end{aligned}$$

Defining $R_n = 4(n+1)V_n + 2\mathrm{Var}(U_n^F - U_n^A) + 4\mathrm{Var}(U_n^A)$, we can rewrite this as

$$(n+2)^2(2V_{n+1}^A + V_{n+1}^F) = (n+1)(n-3)(2V_n^A + V_n^F) + R_n. \tag{14}$$

We can solve this recursion. Recalling that $\mathrm{Var}(X) = G_X''(1) + G_X'(1) - (G_X'(1))^2$, we can use equation (7) to find, for $n > 2$,

$$\mathrm{Var}(U_n^A) = \frac{176}{27} - \frac{4}{3}H_{n+1}^{(2)} - \frac{46}{27}H_{n+1} + \frac{8}{9}(H_{n+1})^2, \text{ and}$$
$$\mathrm{Var}(U_n^F) = \frac{146}{81} - \frac{8}{3}H_{n+1}^{(2)} - \frac{4}{27}H_{n+1} + \frac{8}{9}(H_{n+1})^2,$$

where we used the notation $H_n = \sum_{k=1}^{n} \frac{1}{k}$ and $H_n^{(2)} = \sum_{k=1}^{n} \frac{1}{k^2}$. From (7) we can also obtain

$$\mathrm{Cov}(U_n^A, U_n^F) = -\frac{80}{81} + \frac{52}{27} H_{n+1} - \frac{8}{9}(H_{n+1})^2.$$

This leads to the conclusion that

$$R_n = (28n + \frac{1204}{27}) - (16n + \frac{88}{3}) H_{n+1}^{(2)} - \frac{236}{9} H_{n+1} + \frac{32}{3}(H_{n+1})^2.$$

Using Maple[1] we find that the solution to (14) is, for $n > 2$,

$$\begin{aligned} 2V_n^A + V_n^F = {} & \frac{2}{3375}\frac{7875n + 14858}{n+1} - \frac{8}{15}\frac{5n+9}{n+1} H_{n+1}^{(2)} \\ & - \frac{1372}{225(n+1)} H_{n+1} + \frac{32}{15(n+1)}(H_{n+1})^2. \end{aligned} \tag{15}$$

Using this result, we can simplify the recursion for C_n, equation (13). This yields

$$(n+2)^2 C_{n+1} = n(n+1) C_n + S_n, \tag{16}$$

where

$$\begin{aligned} S_n &= (n+1)(2V_n^A + V_n^F) - 2\mathrm{Var}(U_n^A) + 2\mathrm{Cov}(U_n^A, U_n^F) \\ &= (\frac{14}{3}n + \frac{25148}{10125}) - (\frac{8}{3}n + \frac{32}{15}) H_{n+1}^{(2)} + \frac{784}{675} H_{n+1} - \frac{64}{45}(H_{n+1})^2. \end{aligned}$$

Again, using Maple, we find the solution to recursion (16), for $n > 2$:

$$\begin{aligned} C_n = {} & \frac{15750n^3 + 10534n^2 - 5216n - 1500}{10125n(n+1)^2} - \frac{8}{45}\frac{5n+1}{n+1} H_{n+1}^{(2)} \\ & + \frac{872}{675(n+1)} H_{n+1} - \frac{32}{45(n+1)}(H_{n+1})^2. \end{aligned} \tag{17}$$

It is interesting to note that $C_n < 0$ for $n \leq 168$ and $C_n > 0$ for $n \geq 169$. The quickest way to calculate V_n^A and V_n^F is now to use (14), (17), the relation $V_n = V_n^A + 2C_n + V_n^F$, and the known expression for $V_n = \frac{7n+6}{n+1} - 4H_{n+1}^{(2)} - \frac{2H_{n+1}}{n+1}$ (see [3], page 90). We obtain

$$\begin{aligned} V_n^A = {} & \frac{7875n^3 + 25841n^2 + 17966n - 3000}{10125n(n+1)^2} - \frac{20n+52}{45(n+1)} H_{n+1}^{(2)} \\ & - \frac{1022}{675(n+1)} H_{n+1} + \frac{32}{45(n+1)}(H_{n+1})^2, \end{aligned} \tag{18}$$

and

$$\begin{aligned} V_n^F = {} & \frac{31500n^3 + 84716n^2 + 53216n - 6000}{10125n(n+1)^2} - \frac{80n+112}{45(n+1)} H_{n+1}^{(2)} \\ & - \frac{2072}{675(n+1)} H_{n+1} + \frac{32}{45(n+1)}(H_{n+1})^2. \end{aligned} \tag{19}$$

[1] We have verified these results by hand.

We thus find that as $n \to \infty$

$$\begin{aligned} \mathrm{Var}(Z_n^A) - \frac{7}{9} + \frac{2\pi^2}{27} &\sim \frac{32}{45}\frac{\log^2 n}{n}, \\ \mathrm{Var}(Z_n^F) - \frac{28}{9} + \frac{8\pi^2}{27} &\sim \frac{32}{45}\frac{\log^2 n}{n}, \\ \mathrm{Cov}(Z_n^A, Z_n^F) - \frac{14}{9} + \frac{4\pi^2}{27} &\sim -\frac{32}{45}\frac{\log^2 n}{n}. \end{aligned}$$

4 Where are the arm nodes? Almost sure results.

The normalized total path length Z_n has an interesting asymptotic behaviour described by the following result.

Theorem [5, 6] The random variables $Z_n = \frac{1}{n+1}(X_n - \mathrm{E}X_n)$ form a martingale with zero mean and second moment[2] $\mathrm{E}Z_n^2 = 7 - \frac{2\pi^2}{3} + O(\frac{\log n}{n})$. Hence $Z_n \to Z$ in L^2 and almost surely for some random variable Z (which we call the quicksort distribution).

This raises the question whether a linear combination of Z_n^A and Z_n^F can be a martingale, i.e., whether there exist sequences of real numbers $a = (a_n)_{n=3}^\infty$ and $b = (b_n)_{n=3}^\infty$, such that the random variables $M_n = a_n Z_n^A + b_n Z_n^F$ form a martingale. Because $Z_1^A \equiv Z_2^A \equiv Z_1^F \equiv Z_2^F \equiv 0$, we can let the sequences start at $n = 3$. Now, a direct computation shows that (M_n) is a martingale iff, for $n \geq 3$,

$$\begin{pmatrix} a_n \\ b_n \end{pmatrix} = c_1 \begin{pmatrix} 1 \\ 1 \end{pmatrix} + c_2(n^3 - n) \begin{pmatrix} 2 \\ -1 \end{pmatrix}.$$

where c_1 and c_2 can be chosen freely. Defining $p_n = n^3 - n$, we thus find that, for $n \geq 3$,

$$M_n = c_1 Z_n + c_2 p_n (2Z_n^A - Z_n^F)$$

is a martingale. We see that M_n is a linear combination of two martingales: Z_n and $p_n(2Z_n^A - Z_n^F)$. However, as follows from the results of the previous section, the martingales corresponding to the second solution have unbounded variance.

As we cannot use the L^2-martingale convergence theorem, we will use another approach to prove the theorem stated in Section 1. Since $Z_n^A + Z_n^F = Z_n$ and $Z_n \to Z$ a.s. according to the theorem above, it suffices to show that, as $n \to \infty$, $W_n = 2Z_n^A - Z_n^F \to 0$ a.s. If we can prove that

$$\sum_{n=1}^{\infty} \mathrm{Var}(W_n) < \infty,$$

then $W_n \to 0$ almost surely by the Borel-Cantelli lemma.

[2] Régnier made a small mistake. The order is not, as she wrote, $O(\frac{1}{n})$, but $O(\frac{\log n}{n})$.

Using the results from the previous section, we find

$$\mathrm{Var}(W_n) = 4\mathrm{Var}(Z_n^A) - 4\mathrm{Cov}(Z_n^A, Z_n^F) + \mathrm{Var}(Z_n^F) \sim \frac{32}{5}\frac{\log^2 n}{n}.$$

The sum of the variances diverges. However, we see that $\sum_{n=1}^{\infty} \mathrm{Var}(W_{n^2})$ is finite, so $W_{n^2} \to 0$ almost surely, as $n \to \infty$.
To prove that $W_n \stackrel{a.s.}{\to} 0$, we will show that the intermediate terms behave well: writing $W_{n^2+j} = W_{n^2} + T_{n,j}$, we will prove that $\sup_{0\le j\le 2n} |T_{n,j}| \stackrel{a.s.}{\to} 0$.
For $0 \le j \le 2n$,

$$\begin{aligned} T_{n,j} &= W_{n^2+j} - W_{n^2} = 2Z_{n^2+j}^A - Z_{n^2+j}^F - (2Z_{n^2}^A - Z_{n^2}^F) \\ &= 2(Z_{n^2+j}^A - Z_{n^2}^A) - (Z_{n^2+j}^F - Z_{n^2}^F), \end{aligned}$$

so that

$$\sup_{0\le j\le 2n} |T_{n,j}| \le 2 \sup_{0\le j\le 2n} |Z_{n^2+j}^A - Z_{n^2}^A| + \sup_{0\le j\le 2n} |Z_{n^2+j}^F - Z_{n^2}^F|. \tag{20}$$

We consider the arm node term in the right hand side of (20)

$$\begin{aligned} \sup_{0\le j\le 2n} |Z_{n^2+j}^A - Z_{n^2}^A| &= \sup_{0\le j\le 2n} \left| \frac{X_{n^2+j}^A - \mathrm{E}X_{n^2+j}^A}{n^2+j+1} - \frac{X_{n^2}^A - \mathrm{E}X_{n^2}^A}{n^2+1} \right| \\ &\le \sup_{0\le j\le 2n} \left| \frac{X_{n^2+j}^A}{n^2+j+1} - \frac{X_{n^2}^A}{n^2+1} \right| + \sup_{0\le j\le 2n} \left| \frac{\mathrm{E}X_{n^2+j}^A}{n^2+j+1} - \frac{\mathrm{E}X_{n^2}^A}{n^2+1} \right|. \end{aligned} \tag{21}$$

The part with the expectations can be calculated. From equation (3) it follows that $\mathrm{E}X_n^A = \frac{1}{n+1}\mathrm{E}U_n^A$, and from (7) we obtain that $\mathrm{E}U_n^A = \frac{1}{3}\sum_{k=1}^{n-1}\frac{2}{k+2}$. So we obtain, for $n \ge 2$,

$$\begin{aligned} \sup_{0\le j\le 2n} \left| \frac{\mathrm{E}X_{n^2+j}^A}{n^2+j+1} - \frac{\mathrm{E}X_{n^2}^A}{n^2+1} \right| &= \sup_{0\le j\le 2n} |\mathrm{E}U_{n^2+j}^A - \mathrm{E}U_{n^2}^A| \\ &= \sup_{0\le j\le 2n} \frac{1}{3} \sum_{k=n^2}^{n^2+j-1} \frac{2}{k+2} \le \frac{4n}{3(n^2+2)}. \end{aligned}$$

The first part of the right hand side of (21) can be simplified by noting that

$$\left| \frac{X_{n^2+j}^A}{n^2+j+1} - \frac{X_{n^2}^A}{n^2+1} \right| \le \left| \frac{X_{n^2+j}^A - X_{n^2}^A}{n^2+j+1} \right| + \frac{jX_{n^2}^A}{(n^2+j+1)(n^2+1)}. \tag{22}$$

We denote the height of $\mathcal{T}_n$ by D_n. Clearly $X_{n^2}^A \le X_{n^2} \le (n^2+1)D_{n^2}$, hence for the right most term in (22) we obtain

$$\sup_{0\le j\le 2n} \frac{jX_{n^2}^A}{(n^2+j+1)(n^2+1)} \le \frac{2nX_{n^2}^A}{n^2+1} \le \frac{2n}{n^2+1} D_{n^2}.$$

To bound the middle term in (22), we note that by equation (8),

$$|X_{n+1}^A - X_n^A| = U_n \leq D_n,$$

and thus, using that (D_n) is increasing, for $0 \leq j \leq 2n$,

$$|X_{n^2+j}^A - X_{n^2}^A| \leq \sum_{k=n^2}^{n^2+j-1} |X_{k+1}^A - X_k^A| \leq \sum_{k=n^2}^{n^2+j-1} D_k \leq 2nD_{(n+1)^2}.$$

Substituting these three bounds in (22) and (21) we obtain

$$\sup_{0\leq j\leq 2n} |Z_{n^2+j}^A - Z_{n^2}^A| \leq \frac{2nD_{(n+1)^2}}{n^2+1} + \frac{2nD_{n^2}}{n^2+1} + \frac{4n}{3(n^2+2)}.$$

Combining this inequality with the well-known convergence result $\frac{D_n}{\log n} \to \lambda$ almost surely [4, 1], leads to

$$\sup_{0\leq j\leq 2n} |Z_{n^2+j}^A - Z_{n^2}^A| \xrightarrow{a.s.} 0.$$

For the foot nodes we proceed similarly, and find that the foot node term in (20) also tends to 0 almost surely. This finishes the proof of $\sup_{0\leq j\leq 2n} |T_{n,j}| \to 0$ almost surely, hence we have $W_n \to 0$, and thus

$$Z_n^A \to \tfrac{1}{3}Z, \text{ and } Z_n^F \to \tfrac{2}{3}Z \text{ almost surely.}$$

References

[1] Luc Devroye. A note on the height of binary search trees. *J. Assoc. Comput. Mach.*, 33(3):489–498, 1986.

[2] H. M. Mahmoud. The expected distribution of degrees in random binary search trees. *Comput. J.*, 29(1):36–37, 1986.

[3] Hosam M. Mahmoud. *Evolution of random search trees.* John Wiley & Sons Inc., New York, 1992. A Wiley-Interscience Publication.

[4] B. Pittel. Asymptotical growth of a class of random trees. *Ann. Probab.*, 13(2):414–427, 1985.

[5] Mireille Régnier. A limiting distribution for quicksort. *RAIRO Inform. Théor. Appl.*, 23(3):335–343, 1989.

[6] Uwe Rösler. A limit theorem for "Quicksort". *RAIRO Inform. Théor. Appl.*, 25(1):85–100, 1991.

Trends in Mathematics, © 2000 Birkhäuser Verlag Basel/Switzerland

The Saturation Level in Binary Search Tree

MICHAEL DRMOTA *Institut für Geometrie, TU Wien, Wiedner Hauptstrasse 8-10/118, A-1040 Wien, Austria.* drmota@tuwien.ac.at

Abstract. *Let H'_n denote the saturation level in binary search trees. It is shown that* $\mathbf{E}H'_n = c' \log n + \mathcal{O}\left((\log n)^{1/2} \log\log n\right)$, *where $c' = 0.373\ldots < 2$ is the solution of the equation $\left(\frac{2e}{c'}\right)^{c'} = e$. The methods used rely on the analysis of a retarded differential equation of the form $\Phi'(u) = -\alpha^{-2}\Phi(u/\alpha)^2$ with $\alpha > 1$.*

1 Introduction

Let us consider the usual probabilistic model for binary search trees (see Mahmoud [7] for a description and an overview of the state of the art). The saturation level H'_n of a binary search tree is defined to be maximal level h' such that for all levels h up to h' there are no external nodes, i.e. the binary search has 2^h (internal) nodes for all levels $h \le h'$ but less than $2^{h'+1}$ (internal) nodes at level $h'+1$. For example, if follows from Biggins [1] that

$$\frac{H'_n}{c' \log n} \to 1 \quad a.s.,$$

where $c' = 0.373\ldots < 2$ is the solution of the equation $\left(\frac{2e}{c'}\right)^{c'} = e$. (It can also be shown that $\mathbf{E}H'_n \sim c' \log n$, compare with Mahmoud [7].) Quite recently, the author [6] could prove the following property for $\mathbf{E}H'_n$ and for $\mathbf{E}|H'_n - \mathbf{E}H'_n|^k$.

Theorem 1 *Let $z_h(x)$ $(h \ge 0)$ be recursively defined by $z_0(x) \equiv 1$ and by*

$$z_{h+1}(x) := 1 + \int_0^x z_h(t)\left(\frac{2}{1-t} - z_h(t)\right) dt \qquad (h \ge 0).$$

Then the expected value $\mathbf{E}H'_n$ of the saturation level of binary search trees is given by

$$\mathbf{E}H'_n = \max\{h : z_h(1-n^{-1}) \le n/2\} + \mathcal{O}(1) \qquad (n \to \infty) \tag{1}$$

and all centralized moments of H'_n are bounded:

$$\mathbf{E}|H'_n - \mathbf{E}H'_n|^k = \mathcal{O}(1) \qquad (n \to \infty). \tag{2}$$

Note that (1) is quite implicit. It does not reprove the limiting relation $\mathbf{E}H'_n \sim c' \log n$. For this purpose we have to discuss the functions $z_h(x)$ in more detail. (Interestingly, it is not necessary to know the asymptotic behaviour of $\mathbf{E}H'_n$ to prove boundedness of the variance etc., compare with [6].) The purpose of this paper is to extend the methods of [6] to obtain proper bounds for $z_h(x)$ and to prove the following quantified limiting relation.

Theorem 2 *The expected value $\mathbf{E}H'_n$ of the saturation level of binary search trees is given by*

$$\mathbf{E}H'_n = c' \log n + \mathcal{O}\left((\log n)^{1/2} \log\log n\right). \tag{3}$$

It is very likely that the error term $\mathcal{O}\left((\log n)^{1/2}\log\log n\right)$ in (3) is not optimal. It might be expected that there is a constant d' such that

$$\mathbf{E}H'_n = c'\log n + d'\log\log n + \mathcal{O}(1).$$

The method presented below is adapted from that used in [5] and [6] to treat the height of binary search trees. (For results and references concerning the height of binary search trees we refer to [8, 2, 3, 10, 5, 9].)

Before we start with the proof we want to mention that the functions $z_h(x)$ introduced in Theorem 1 are just the generating functions of the probabilities $\mathbf{P}[H'_n \le n]$:

$$z_h(x) = \sum_{n\ge 0} \mathbf{P}[H'_n \le n]\, x^n.$$

The paper is organized in the following way. In section 2 we prove an upper bound for $z_h(x)$ yielding an lower bound for $\mathbf{E}H'_n$. Sections 3 and 4 are devoted to the analysis of an integral equation. Those results will be then applied in section 5 to provide a lower bound for $z_h(x)$ which finally yields an upper bound for $\mathbf{E}H'_n$

2 A lower bound for $\mathbf{E}H'_n$.

Interestingly, a lower bound for $\mathbf{E}H'_n$ is much easier to obtain than an upper bound. We can use the following lemma.

Lemma 1 *For $0 \le x < 1$ we have*

$$z_h(x) \le 2^h \sum_{k=0}^{h} \frac{\left(\log\frac{1}{1-x}\right)^k}{k!}.$$

Proof. Obviously, we have equality for $h = 0$. Now we use the inequality

$$z'_{h+1}(x) = z_h(x)\left(\frac{2}{1-x} - z_h(x)\right) \le \frac{2}{1-x} z_h(x) \tag{4}$$

and proceed by induction. ◊

Corollary *Let $h = c'\log n + \frac{c'}{2(1-c')}\log\log n + r$. Then*

$$z_h\left(1-\frac{1}{n}\right) \ll n\left(\frac{2}{c'}\right)^r$$

uniformly for $r = \mathcal{O}(1)$.

Thus, it follows from Theorem 1 that

$$\mathbf{E}H'_n \ge c'\log n + \frac{c'}{2(1-c')}\log\log n - C_1 \tag{5}$$

for some constant $C_1 > 0$.

3 An Integral Equation

Our aim is to find proper solutions $\Phi(u)$ of the (retarded) differential equation

$$\Phi'(u) = -\frac{1}{\alpha^2}\Phi\left(\frac{u}{\alpha}\right)^2 \tag{6}$$

with

$$\alpha \geq \alpha_0 = e^{1/c'} = 14.59\ldots$$

Interestingly there is an explicit solution

$$\Phi(u) = \frac{1+u^{1/4}}{u}e^{-u^{1/4}} \tag{7}$$

with $\alpha = 16$. This function was used to prove Theorem 1 in [6]. However, if we want to get more then we need solutions for every $\alpha > \alpha_0$.

Instead of solving (6) directly we consider a related integral equation. In fact, we will prove the following relations.

Lemma 2 *Let $\alpha > \alpha_0 = e^{1/c'}$ and let $\beta > c' - 1$ be the solution of the equation*

$$2\alpha^\beta = \beta + 1.$$

Then there exists a function $\Psi(y)$, $y \geq 0$ with the following properties:

1. $\Psi(y) = 1 - y^\beta + \mathcal{O}((1-(\alpha/\alpha_0)^{c'-1})^{-1}y^{c'-1})$ *as $y \to \infty$ with an $\mathcal{O}$-constant independent of α.*
2. $\Psi(y) \leq e^{-C(\alpha)y^{-\gamma}}$ *for $\gamma = \log 2/(\log(\alpha/2))$ and $y \leq y_0(\alpha)$, where $C(\alpha)$ and $y_0(\alpha)$ can be chosen in a way that* $C(\alpha) = \exp(-c_1(\alpha-\alpha_0)^{-1/2}\log((\alpha-\alpha_0)^{-1}))$ *and* $y_0(\alpha) = \exp(c_2(\alpha-\alpha_0)^{-1/2}\log((\alpha-\alpha_0)^{-1}))$ *for some constants* $c_1, c_2 > 0$.
3. $0 \leq \Psi(y) \leq 1$, $0 \leq y < \infty$, *is strictly increasing.*
4. $\int_0^y \Psi(z)\Psi(y-z)\,dz = y\Psi(y/\alpha)$, $(0 \leq y < \infty)$.

Proof. We first observe that for $\alpha > \alpha_0 = e^{1/c'}$ the equation

$$2\alpha^\beta = \beta + 1 \tag{8}$$

has two negative solutions β_1, β_2 with $\beta_1 < \beta_0 = c' - 1 = -0.626\cdots < \beta_2$. The solution we are interested in is $\beta = \beta_2 > \beta_0$. For $\alpha = \alpha_0$ there is only one solution $\beta = \beta_0 < 0$ and for $\alpha < \alpha_0$ there are no negative solutions. It is also an easy exercise to show that for $\alpha > \alpha_0$ we have

$$\beta = \beta_0 + \sqrt{\frac{2(c')^2(1-c')}{\alpha_0}}\sqrt{\alpha-\alpha_0} + \mathcal{O}(\alpha-\alpha_0).$$

Now let $\mathcal{F}$ denote the set of functions $\Psi(y)$, $y > 0$, with the following properties:

1. $\Psi(y) = 1 - y^\beta + \mathcal{O}(y^{\beta_0})$ as $y \to \infty$.

2. $\Psi(0) = 0$.

3. $0 \le \Psi(y) \le 1$, $0 \le y < \infty$.

4. $\Psi(y)$, $0 \le y < \infty$, is strictly decreasing.

It is clear that $\mathcal{F}$ with the distance

$$d(\Psi_1, \Psi_2) := \sup_{y>0} |(\Psi_1(y) - \Psi_2(y))y^{-\beta_0}|$$

is a complete metric space. Now we show that the operator I, defined by

$$(I\Psi)(y) := \frac{1}{\alpha y}\int_0^{\alpha y} \Psi(z)\Psi(\alpha y - z)\, dz$$

is a contraction on $\mathcal{F}$.

Firstly, we prove that $I\Psi \in \mathcal{F}$ for all $\Psi \in \mathcal{F}$. Suppose that $\Psi \in \mathcal{F}$. Then

$$\Psi(z)\Psi(\alpha y - z) = 1 - (z^\beta + (\alpha y - z)^\beta) + \mathcal{O}(z^{\beta_0}) + \mathcal{O}((\alpha y - z)^{\beta_0}).$$

Since

$$\int_0^{\alpha y} z^\beta\, dz = \frac{(\alpha y)^{\beta+1}}{\beta+1}$$

it immediately follows that

$$\begin{aligned}(I\Psi)(y) &= \frac{1}{\alpha y}\int_0^{\alpha y} \Psi(z)\Psi(\alpha y - z)\, dz \\ &= 1 - 2\frac{(\alpha y)^{\beta+1}}{\alpha y(\beta+1)} + \mathcal{O}(y^{\beta_0}) \\ &= 1 - y^\beta + \mathcal{O}(y^{\beta_0}).\end{aligned}$$

Furthermore, it is clear that $(I\Psi)(0) = 0$ and $0 \le (I\Psi)(y) \le 1$. Finally, by using the representation

$$(I\Psi)(y) = \int_0^1 \Psi(\alpha y x)\Psi(\alpha y(1-x))\, dx$$

it is also clear that $(I\Psi)(y)$ is increasing (if $\Psi(y)$ is increasing).

Now suppose that $\Psi_1, \Psi_2 \in \mathcal{F}$ with $d(\Psi_1, \Psi_2) = \delta$. Then it follows from $0 \le \Psi_j(y) \le 1$ that

$$\begin{aligned}&|\Psi_1(z)\Psi_1(\alpha y - z) - \Psi_2(z)\Psi_2(\alpha y - z)| \\ &\le |\Psi_1(z) - \Psi_2(z)| + |\Psi_1(\alpha y - z) - \Psi_2(\alpha y - z)| \\ &\le \delta\left(z^{\beta_0} + (\alpha y - z)^{\beta_0}\right)\end{aligned}$$

and consequently

$$\begin{aligned} |(I\Psi_1)(y) - (I\Psi_2)(y)| &\le \delta \frac{2}{\alpha y} \int_0^{\alpha y} z^{\beta_0}\, dz \\ &= \delta \frac{2\alpha^{\beta_0}}{\beta_0 + 1} y^{\beta_0} \end{aligned}$$

which implies

$$d(I\Psi_1, I\Psi_2) \le L\, d(\Psi_1, \Psi_2)$$

with

$$L = \frac{2\alpha^{\beta_0}}{\beta_0 + 1} = \left(\frac{\alpha}{\alpha_0}\right)^{\beta_0} < 1.$$

Thus, $I : \mathcal{F} \to \mathcal{F}$ is as contraction.

By Banach's fixed point theorem there exists a unique fixed point $\Psi \in \mathcal{F}$. By definition, this fixed points satisfies properties 1., 3., and 4. of Lemma 2.

If we start with

$$\Psi_0(y) = \max\{1 - y^\beta, 0\}$$

and define iteratively $\Psi_{k+1} := I\Psi_k$ then the fixed point is the limit $\Psi = \lim\limits_{k\to\infty} \Psi_k$ and we have

$$d(\Psi_0, \Psi) = \mathcal{O}\left(\frac{1}{1-L}\right),$$

which directly translates to the precise estimate in 1.

In order to prove 2. we set

$$\overline{\Psi}(y) := e^{-Cy^{-\gamma}}$$

with $\gamma = \log 2/\log(\alpha/2)$. Since $\gamma > 1$ we surely have

$$\begin{aligned} \frac{1}{y}\int_0^y \overline{\Psi}(z)\overline{\Psi}(y-z)\, dz &\le \frac{1}{y}\int_0^y \overline{\Psi}(y/2)^2\, dz \\ &= \overline{\Psi}(y/2)^2 \\ &= \overline{\Psi}(y/\alpha). \end{aligned}$$

Thus, if we know that $\Psi_k(y) \le \overline{\Psi}(y)$ for $y \le y_0$ then it follows that $(I\Psi_k)(y) = \Psi_{k+1}(y) \le \overline{\Psi}(y)$ for $y \le y_0/\alpha$. However, there is an a-priori bound for all Ψ_k of the form

$$\Psi_k(y) \le 1 - y^\beta + C'(1 - (\alpha/\alpha_0)^{c'-1})^{-1} y^{\beta_0}.$$

Hence, if we choose C and y_0 appropriately (i.e., as proposed) then we can assure that $\Psi_0(y) \le \overline{\Psi}(y)$ for $y \le y_0$ and (a-priori) $\Psi_k(y) \le \overline{\Psi}(y)$ for $y_0/\alpha \le y \le y_0$. Thus, 2. follows by induction. ◇

Corollary *Let $\Psi(y)$ be as in Lemma 2 then the Laplace transform*

$$\Phi(u) = \int_0^\infty \Psi(y) e^{-uy}\, dy$$

is an analytic function for $\Re u > 0$ and satisfies the following properties.

1. *$\Phi(u)$ and and $u\Phi(u)$ are decreasing for real $u > 0$.*

2. $1 - u\Phi(u) = \Gamma(\beta+1)u^{-\beta} + \mathcal{O}\left((1-(\alpha/\alpha_0)^{c'-1})^{-1}u^{1-c'}\right)$ *as $u \to 0+$.*

3. *$\Phi(u) = \mathcal{O}(e^{-C(\alpha)u^{\gamma'}})$ as $u \to \infty$, where $\gamma' > 0$ and $C(\alpha)$ is of the form stated in Lemma 2.*

4. $\Phi'(u) = -\frac{1}{\alpha^2}\Phi\left(\frac{u}{\alpha}\right)^2.$

Proof. It is clear that the Laplace transform $\Phi(u)$ is an analytic function for $\Re u > 0$. Moreover, since $\Phi(y)$ is non-negative, it follows by definition that $\Phi(u)$ is decreasing.

By partial integration we get for any $u > 0$

$$u\Phi(u) = -\int_0^\infty \Psi'(y)e^{-uy}\,dy.$$

Since $\Psi'(y) > 0$ for $y > 0$ it also follows that $u\Phi(u)$ is decreasing for $u > 0$, too.

Next, the expansion $\Phi(y) = 1 - y^\beta + \mathcal{O}(y^{\beta_0})$ as $y \to \infty$ directly translates to

$$\Phi(u) = \frac{1}{u} - \frac{\Gamma(\beta+1)}{u^{\beta+1}} + \mathcal{O}\left((1-(\alpha/\alpha_0)^{c'-1})^{-1}\frac{1}{u^{\beta_0+1}}\right)$$

as $u \to 0+$.

Finally the integral equation for $\Psi(y)$ induces the proposed differential equation for $\Phi(u)$. ◇

4 Auxiliary Functions

We will now work with the auxiliary functions

$$\tilde{z}_h(x) := \frac{1}{1-x}\left(1 - \alpha^h(1-x)\Phi(\alpha^h(1-x))\right), \tag{9}$$

where h is a real parameter.

The properties of $\Phi(u)$ can be translated to corresponding properties of $\tilde{z}_h(x)$. The proof is immediate. The idea behind is that theses functions imitate the original functions $z_h(x)$.

Lemma 3 *The functions $\tilde{z}_h(x)$, $h \geq 0$, $0 \leq x < 1$, defined by (9) satisfy*

1. $0 < \tilde{z}_h(0) < 1$.

2. *$0 \leq \tilde{z}_h(x) \leq 1/(1-x)$ for $0 \leq x < 1$.*

3. *$\tilde{z}_h(x)$ is strictly increasing for $0 \leq x < 1$.*

4. $\tilde{z}'_{h+1}(x) = \tilde{z}_h(x)\left(\frac{2}{1-x} - \tilde{z}_h(x)\right).$

We will also make use of the following property of the functions $z_h(x)$.

Lemma 4 *For every non-negative integer h and for every (real) D there exists $0 < \overline{x}_{h,D} < 1$ such that*

$$\tilde{z}_{h+D}(x) < z_h(x) \qquad (0 \le x < \overline{x}_{h,D}) \tag{10}$$

and

$$z_h(x) < \tilde{z}_{h+D}(x) \qquad (\overline{x}_{h,D} < x < 1). \tag{11}$$

Furthermore

$$\overline{x}_{h+1,D} > \overline{x}_{h,D}. \tag{12}$$

Proof. We proceed by induction. Since $\tilde{z}_D(x)$ is strictly increasing and satisfies $0 < \tilde{z}_D(0) < 1$ and $\lim_{x\to 1-} \tilde{z}_D(x) = \infty$ the assertion is surely true for $h = 0$. Now suppose that (10) and (11) are satisfied for some $h \ge 0$, i.e. the difference

$$\delta_{h,D}(x) := z_h(x) - \tilde{z}_{h+D}(x)$$

has a unique zero $\overline{x}_{h,D} > 0$ such that $\delta_{h,D}(x) > 0$ for $0 \le x < x_{h,D}$ and $\delta_{h,D}(x) < 0$ for $x > x_{h,D}$. Now we have

$$\begin{aligned}
\delta'_{h+1,D}(x) &= z'_{h+1}(x) - \tilde{z}'_{h+1+D}(x) \\
&= z_h(x)\left(\frac{2}{1-x} - z_h(x)\right) - \tilde{z}_{h+D}(x)\left(\frac{2}{1-x} - \tilde{z}_{h+D}(x)\right) \\
&= \delta_{h,D}(x)\left(\frac{2}{1-x} - z_h(x) - \tilde{z}_{h+D}(x)\right).
\end{aligned}$$

Hence, $\delta_{h+1,D}(x)$ is increasing for $0 \le x < \overline{x}_{h,D}$ and decreasing for $x > \overline{x}_{h,D}$. Since $\delta_{h+1,D}(0) > 0$ and $\lim_{x\to\infty} \delta_{h+1,D}(x) = -\infty$ there exists a unique zero $\overline{x}_{h+1,D} > \overline{x}_{h,D}$ of $\delta_{h+1,D}(x)$ such that $\delta_{h+1,D}(x) > 0$ for $0 \le x < \overline{x}_{h+1,D}$ and $\delta_{h+1,D}(x) < 0$ for $x > \overline{x}_{h+1,D}$. $\Diamond$

Finally, we can provide a uniform upper bound for $\tilde{z}_h(x)$.

Lemma 5 *Let $\alpha > \alpha_0$ and $z_h(x)$ be defined by (9). Then we have*

$$\begin{aligned}
\tilde{z}_h(x) &\le \Gamma(\beta+1)\alpha^{\beta d}\sum_{k\ge 0}\frac{(\beta+1)^k}{k!}\left(\log\frac{1}{1-x}\right)^k \\
&+ \mathcal{O}\left(\frac{1}{1-(\alpha/\alpha_0)^{c'-1}}\alpha^{(c'-1)h}\sum_{k\ge 0}\frac{(c')^k}{k!}\left(\log\frac{1}{1-x}\right)^k\right)
\end{aligned}$$

uniformly for all real h and all $0 \le x < 1$.

Proof. Firstly we note that the inequality

$$1 - u\Phi(u) \le \Gamma(\beta+1)u^{-\beta} + \mathcal{O}\left((1-(\alpha/\alpha_0)^{c'-1})^{-1}u^{1-c'}\right)$$

is valid for all $u \geq 0$. Thus, by definition

$$z_h(x) \leq \frac{\Gamma(\beta+1)}{(1-x)^{\beta+1}\alpha^{\beta h}} + \mathcal{O}\left(\frac{(1-(\alpha/\alpha_0)^{c'-1})^{-1}}{(1-x)^{c'}\alpha^{(c'-1)h}}\right),$$

which proves the lemma. ◇

5 An upper bound for $\mathbf{E}H_n'$.

For every integer $h \geq 0$ we will now use a special chosen

$$\alpha = \alpha_h := \alpha_0 + \frac{1}{h}.$$

This α is associated with

$$\beta = \beta_h = c' - 1 + \frac{d_0}{\sqrt{h}} + \mathcal{O}\left(\frac{1}{h}\right)$$

(for some $d_0 > 0$), the solution $> c' - 1$ of the equation $2\alpha^\beta = \beta + 1$. Hence, for every integer $h \geq 0$ there is a function $\Psi(y) = \Psi_h(y)$ (guaranteed by Lemma 2) satisfying

$$\Psi_h(y) = 1 - y^\beta + \mathcal{O}\left(hy^{c'-1}\right) \tag{13}$$

as $y \to \infty$ and and its Laplace transform $\Phi(u) = \Phi_h(u)$ with

$$\Phi_h(u) = 1 - \Gamma(\beta+1)u^{-\beta} + \mathcal{O}\left(hu^{c'-1}\right) \tag{14}$$

as $u \to 0+$ and $\mathcal{O}$-constants which are uniform in h. Thus, (13) and (14) are only significant for

$$y \geq e^{d_1\sqrt{h}\log h}$$

resp. for

$$u < e^{-d_2\sqrt{h}\log h},$$

where $d_1, d_2 > 0$ are appropriately chosen constants.

The key lemma for the proof of an lower bound of $z_h(x)$ is the following one.

Lemma 6 *For every real number d and for every integer $h \geq 0$ we have*

$$\begin{aligned}\tilde{z}_{h+d}(x) - z_h(x) \quad &\leq \quad \Gamma(\beta_h+1)\left(\frac{2}{\beta_h+1}\right)^{h+d}\sum_{k\geq h}\frac{(\beta_h+1)^k}{k!}\left(\log\frac{1}{1-x}\right)^k \\ &+ \quad \mathcal{O}\left(h\left(\frac{2}{c'}\right)^h\alpha_h^{(1-c')d}\sum_{k\geq h}\frac{(c')^k}{k!}\left(\log\frac{1}{1-x}\right)^k\right)\end{aligned}$$

for all $0 \leq x < 1$.

Proof. For $h = 0$ this inequality is just Lemma 5. Now we can proceed by induction. We only have to observe that $\tilde{z}_{h+d}(0) < 1 = z_h(0)$ and that

$$\tilde{z}'_{h+1+d}(x) - z'_{h+1}(x) = \left(\frac{2}{1-x} - z_{h+d}(x) - z_h(x)\right)(\tilde{z}_{h+d}(x) - z_h(x))$$

and that

$$\frac{2}{1-x} - z_{h+d}(x) - z_h(x) \le \frac{2}{1-x}.$$

(The proof is quite similar to that of Lemma 1). ◇

The next lemma uses this estimate to get a lower bound for $z_h(x)$ at a specific value.

Lemma 7 *There exists a constant $D_0 > 0$ such that*

$$z_h\left(1 - e^{-h/(\beta_h+1)}\right) \ge \frac{\Gamma(c')}{3} e^{h/(\beta_h+1)} \left(\frac{\beta_h+1}{2}\right)^{D_0\sqrt{h}\log h}$$

Proof. We use Lemma 6 with $x = 1 - e^{-h/(\beta_h+1)}$ and

$$d = -h\left(1 - \frac{1}{(\beta_h+1)\log\alpha_h}\right) - D_0\sqrt{h}\log h,$$

where $D_0 > 0$ will be chosen in the sequel. We directly get (after some algebra)

$$\left(\frac{2}{\beta_h+1}\right)^{h+d} \sum_{k\ge h} \frac{(\beta_h+1)^k}{k!}\left(\log\frac{1}{1-x}\right)^k$$
$$\sim \frac{1}{2} e^{h/(\beta_h+1)} \left(\frac{\beta_h+1}{2}\right)^{D_0\sqrt{h}\log h}$$

and similarly

$$h\left(\frac{2}{c'}\right)^h \alpha_h^{(1-c')d} \sum_{k\ge h} \frac{(c')^k}{k!}\left(\log\frac{1}{1-x}\right)^k$$
$$\ll e^{h/(\beta_h+1)} h \left(\frac{\beta_h+1}{2}\right)^{\frac{c'-1}{\beta_h}(D_0\sqrt{h}\log h)}$$

Since

$$\frac{c'-1}{\beta_h} \ge 1 + \frac{d_4}{\sqrt{h}}$$

(for some constant $c_4 > 0$) it follows that there exists $D_0 > 0$ such that

$$h\left(\frac{\beta_h+1}{2}\right)^{\frac{c'-1}{\beta_h}(D_0\sqrt{h}\log h)} = o\left(\left(\frac{\beta_h+1}{2}\right)^{D_0\sqrt{h}\log h}\right)$$

as $h \to \infty$.

In a similar way we can discuss $\tilde{z}_{h+d}(x)$. By definition (and after some algebra) we get

$$\begin{aligned}\tilde{z}_{h+d}(x) &= \Gamma(\beta_h+1)e^h\left(\frac{2}{\beta_h+1}\right)^{h+d} + \mathcal{O}\left(he^{hc'/(\beta_h+1)}\alpha_h^{(1-c')(h+d)}\right)\\ &= \Gamma(\beta_h+1)e^{h/(\beta_h+1)}\left(\frac{\beta_h+1}{2}\right)^{D_0\sqrt{h}\log h}(1+o(1))\end{aligned}$$

if D_0 is sufficiently large. Consequently

$$\begin{aligned}\tilde{z}_h(x) &\geq \Gamma(\beta_h+1)\left(1-\frac{1}{2}\right)e^{h/(\beta_h+1)}\left(\frac{\beta_h+1}{2}\right)^{D_0\sqrt{h}\log h}(1+o(1))\\ &\geq \frac{\Gamma(c')}{3}e^{h/(\beta_h+1)}\left(\frac{\beta_h+1}{2}\right)^{D_0\sqrt{h}\log h}\end{aligned}$$

as proposed. ◊

Corollary *Set*

$$h' = \frac{h}{(\beta_h+1)\log\alpha_h} - D_0\sqrt{h}\log h - 1.$$

Then we have

$$z_h(x) \geq \tilde{z}_{h'}(x)$$

for $x \leq 1 - e^{-h/(\beta_h+1)}$.

Proof. As in the proof of Lemma 7 it follows that

$$\begin{aligned}\tilde{z}_{h'}\left(1-e^{-h/(\beta_h+1)}\right) &= \frac{\beta_h+1}{2}\Gamma(\beta_h+1)e^{h/(\beta_h+1)}\left(\frac{\beta_h+1}{2}\right)^{D_0\sqrt{h}\log h}(1+o(1))\\ &\leq \frac{\Gamma(\beta_h+1)}{3}e^{h/(\beta_h+1)}\left(\frac{\beta_h+1}{2}\right)^{D_0\sqrt{h}\log h}.\end{aligned}$$

Thus,

$$\tilde{z}_{h'}\left(1-e^{-h/(\beta_h+1)}\right) \leq z_h\left(1-e^{-h/(\beta_h+1)}\right)$$

and consequently, by Lemma 4

$$\tilde{z}_{h'}(x) \leq z_h(x)$$

for $x \leq 1 - e^{-h/(\beta_h+1)}$. ◊

Now we can complete the proof of Theorem 2. For given $n \geq 1$ choose a constant $D_1 > 0$ such that

$$h := \lfloor c'\log n + D_1(\log n)^{1/2}\log\log n\rfloor$$

is large enough that

$$u\Phi_h(u) \le \frac{1}{4}$$

for

$$u = \frac{1}{n}\alpha_h^{h/((\beta_h+1)\log \alpha_h) - D_0\sqrt{h}\log h}.$$

(Here we use 3. of the Corollary of Lemma 2.) Hence it follows that for $h' = h/((\beta_h+1)\log\alpha_h) - D_0\sqrt{h}\,\log h$

$$z_h(1-n^{-1}) \ge z_{h'}(1-n^{-1}) = n(1-u\Phi_h(u)) \ge \frac{3}{4}n.$$

Thus, by Theorem 1 we get

$$\mathbf{E}H_n' \le c'\log n + D_1(\log n)^{1/2}\log\log n.$$

This completes the proof of Theorem 2.

References

[1] J. D. BIGGINS, *How fast does a general branching random walk spread ?*, in: IMA Vol. Math. Appl. **84** (1997), 19–39.

[2] L. DEVROYE, *A note on the height of binary search trees*, J. Assoc. Comput. Mach. **33** (1986), 489–498.

[3] L. DEVROYE AND B. REED, *On the variance of the height of random binary search trees*, SIAM J. Comput. **24** (1995), 1157–1162.

[4] M. DRMOTA, *An Analytic Approach to the Height of Binary Search Trees*, Algorithmica, to appear.

[5] M. DRMOTA, *The Variance of the Height of Binary Search Trees*, Theoret. Comput. Sci., submitted.

[6] M. DRMOTA, *An Analytic Approach to the Height of Binary Search Trees. II*, manuscript.

[7] H. M. MAHMOUD, *Evolution of Random Search Trees*, John Wiley & Sons, New York, 1992.

[8] B. PITTEL, *On growing random binary trees*, J. Math. Anal. Appl. **103** (1984), 461–480.

[9] B. REED, *How tall is a tree*, Proceedings of STOC 2000.

[10] B. REED, *The height of a random binary search tree*, manuscript.

Trends in Mathematics, © 2000 Birkhäuser Verlag Basel/Switzerland

Smoothness and Decay Properties of the Limiting Quicksort Density Function

JAMES ALLEN FILL[1] *Department of Mathematical Sciences, The Johns Hopkins University, U.S.A.* jimfill@jhu.edu ; http://www.mts.jhu.edu/~fill/
SVANTE JANSON *Department of Mathematics, Uppsala University, Sweden.* svante.janson@math.uu.se ; http://www.math.uu.se/~svante/

Abstract. *Using Fourier analysis, we prove that the limiting distribution of the standardized random number of comparisons used by* Quicksort *to sort an array of n numbers has an everywhere positive and infinitely differentiable density f, and that each derivative $f^{(k)}$ enjoys superpolynomial decay at $\pm\infty$. In particular, each $f^{(k)}$ is bounded. Our method is sufficiently computational to prove, for example, that f is bounded by* 16.

Key words. Quicksort, *density, characteristic function, sorting algorithm, Fourier analysis, rapidly decreasing C^∞ function, tempered distribution, integral equation.*

AMS 2000 subject classification. *Primary 68W40; secondary 68P10, 60E05, 60E10.*

1 Introduction and summary

The Quicksort algorithm of Hoare [7] is "one of the fastest, the best-known, the most generalized, the most completely analyzed, and the most widely used algorithms for sorting an array of numbers" [2]. Quicksort is the standard sorting procedure in Unix systems, and Philippe Flajolet, a leader in the field of analysis of algorithms, has noted that it is among "some of the most basic algorithms—the ones that do deserve deep investigation" [4]. Our goal in this introductory section is to review briefly some of what is known about the analysis of Quicksort and to summarize how this paper advances that analysis.

The Quicksort algorithm for sorting an array of n numbers is extremely simple to describe. If $n = 0$ or $n = 1$, there is nothing to do. If $n \geq 2$, pick a number uniformly at random from the given array. Compare the other numbers to it to partition the remaining numbers into two subarrays. Then recursively invoke Quicksort on each of the two subarrays.

Let X_n denote the (random) number of comparisons required (so that $X_0 = 0$). Then X_n satisfies the distributional recurrence relation

$$X_n \overset{\mathcal{L}}{=} X_{U_n-1} + X^*_{n-U_n} + n - 1, \qquad n \geq 1,$$

where $\overset{\mathcal{L}}{=}$ denotes equality in law (i.e., in distribution), and where, on the right, U_n is distributed uniformly on the set $\{1, \ldots, n\}$, $X^*_j \overset{\mathcal{L}}{=} X_j$, and

$$U_n;\ X_0, \ldots, X_{n-1};\ X^*_0, \ldots, X^*_{n-1}$$

are all independent.

[1]Research supported by NSF grant DMS–9803780, and by the Acheson J. Duncan Fund for the Advancement of Research in Statistics.

As is well known and quite easily established, for $n \geq 0$ we have

$$\mu_n := \mathbf{E}X_n = 2(n+1)H_n - 4n \sim 2n \ln n,$$

where $H_n := \sum_{k=1}^{n} k^{-1}$ is the nth harmonic number and $\sim$ denotes asymptotic equivalence. It is also routine to compute explicitly the standard deviation of X_n (see Exercise 6.2.2-8 in [9]), which turns out to be $\sim n\sqrt{7 - \frac{2}{3}\pi^2}$.

Consider the standardized variate

$$Y_n := (X_n - \mu_n)/n, \qquad n \geq 1.$$

Régnier [11] showed using martingale arguments that $Y_n \to Y$ in distribution, with Y satisfying the distributional identity

$$Y \stackrel{\mathcal{L}}{=} UY + (1-U)Z + g(U) =: h_{Y,Z}(U), \tag{1.1}$$

where

$$g(u) := 2u \ln u + 2(1-u)\ln(1-u) + 1, \tag{1.2}$$

and where, on the right of $\stackrel{\mathcal{L}}{=}$ in (1.1), U, Y, and Z are independent, with $Z \stackrel{\mathcal{L}}{=} Y$ and $U \sim \text{unif}(0,1)$. Rösler [12] showed that (1.1) characterizes the limiting law $\mathcal{L}(Y)$, in the precise sense that $F := \mathcal{L}(Y)$ is the *unique* fixed point of the operator

$$G = \mathcal{L}(V) \mapsto SG := \mathcal{L}(UV + (1-U)V^* + g(U))$$

(in what should now be obvious notation) subject to

$$\mathbf{E}V = 0, \qquad \mathbf{Var}V < \infty.$$

Thus it is clear that fundamental (asymptotic) probabilistic understanding of `Quicksort`'s behavior relies on fundamental understanding of the limiting distribution F. In this regard, Rösler [12] showed that

the moment generating function (mgf) of Y is everywhere finite, (1.3)

and Hennequin [5] [6] and Rösler showed how all the moments of Y can be pumped out one at a time, though there is no known expression for the mgf nor for the general pth moment in terms of p. Tan and Hadjicostas [15] proved that F has a density f which is almost everywhere positive, but their proof does not even show whether f is continuous.

The main goal of this paper is to prove that F has a density f which is infinitely differentiable, and that each derivative $f^{(k)}(y)$ decays as $y \to \pm\infty$ more rapidly than any power of $|y|^{-1}$: this is our main Theorem 3.1. In particular, it follows that each $f^{(k)}$ is bounded (cf. Theorem 3.3).

Our main tool will be Fourier analysis. We begin in Section 2 by showing (see Theorem 2.9) that the characteristic function ϕ for F has rapidly decaying derivatives of every order. Standard arguments reviewed briefly at the outset of Section 3 then immediately carry this result over from ϕ to f. Finally, in Section 4 we will use the boundedness and continuity of f to establish an integral equation for f (Theorem 4.1). As a corollary, f is everywhere positive (Corollary 4.2).

Remark 1.1. (a) Our method is sufficiently computational that we will prove, for example, that f is bounded by 16. This is not sharp numerically, as Figure 4 of [15] strongly suggests that the maximum value of f is about 2/3. However, in future work we will rigorously justify (and discuss how to obtain bounds on the error in) the numerical computations used to obtain that figure, and the rather crude bounds on f and its derivatives obtained in the present paper are needed as a starting point for that more refined work.

(b) Very little is known rigorously about f. For example, the figure discussed in (a) indicates that f is unimodal. Can this be proved? Is f in fact *strongly* unimodal (i.e., log-concave)? What can one say about changes of signs for the derivatives of f?

(c) Knessl and Szpankowski [8] purport to prove very sharp estimates of the rates of decay of $f(y)$ as $y \to -\infty$ and as $y \to \infty$. Roughly put, they assert that the left tail of f decays doubly exponentially (like the tail of an extreme-value density) and that the right tail decays exponentially. But their results rely on several unproven assumptions (as noted in their paper). Among these, for example, is their assumption (59) that

$$\mathbf{E}e^{-\lambda Y} \sim \exp(\alpha\lambda \ln \lambda + \beta\lambda + \gamma \ln \lambda + \delta) \qquad \text{as } \lambda \to \infty$$

for some constants $\alpha(> 0), \beta, \gamma, \delta$. (Having assumed this, they derive the values of α, γ, and δ exactly, and the value of β numerically.)

2 Bounds on the limiting Quicksort characteristic function

We will in this section prove the following result on superpolynomial decay of the characteristic function of the limit variable Y.

Theorem 2.1. *For every real* $p \geq 0$ *there is a smallest constant* $0 < c_p < \infty$ *such that the characteristic function* $\phi(t) :\equiv \mathbf{E}e^{itY}$ *satisfies*

$$|\phi(t)| \leq c_p |t|^{-p} \quad \text{for all } t \in \mathbf{R}. \tag{2.1}$$

These best possible constants c_p *satisfy* $c_0 = 1$, $c_{1/2} \leq 2$, $c_{3/4} \leq \sqrt{8\pi}$, $c_1 \leq 4\pi$, $c_{3/2} < 187$, $c_{5/2} < 103215$, $c_{7/2} < 197102280$, *and the relations*

$$c_{p_1}^{1/p_1} \leq c_{p_2}^{1/p_2}, \qquad 0 < p_1 \leq p_2; \tag{2.2}$$

$$c_{p+1} \leq 2^{p+1} c_p^{1+(1/p)} p/(p-1), \qquad p > 1; \tag{2.3}$$

$$c_p \leq 2^{p^2+6p}, \qquad p > 0. \tag{2.4}$$

[The numerical bounds are not sharp (except in the trivial case of c_0); they are the best that we can get without too much work, but we expect that substantial improvements are possible.]

Proof. The basic approach is to use the fundamental relation (1.1). We will first show, using a method of van der Corput [1, 10], that the characteristic function of $h_{y,z}(U)$ is bounded by $2|t|^{-1/2}$ for each y, z. Mixing, this yields Theorem 2.1 for $p = 1/2$. Then we will use another consequence of (1.1), namely, the functional equation

$$\phi(t) = \int_{u=0}^{1} \phi(ut)\,\phi((1-u)t)\,e^{itg(u)}\,du, \quad t \in \mathbf{R}, \tag{2.5}$$

or rather its consequence

$$|\phi(t)| \leq \int_{u=0}^{1} |\phi(ut)|\;|\phi((1-u)t)|\,du, \tag{2.6}$$

and obtain successive improvements in the exponent p.

We give the details as a series of lemmas, beginning with a standard calculus estimate [10]. Note that it suffices to consider $t > 0$ in the proofs because $\phi(-t) = \overline{\phi(t)}$ and thus $|\phi(-t)| = |\phi(t)|$. Note also that the best constants satisfy $c_p = \sup_{t>0} t^p|\phi(t)|$ (although we do not know in advance of proving Theorem 2.1 that these are finite), and thus $c_p^{1/p} = \sup_{t>0} t|\phi(t)|^{1/p}$, which clearly satisfies (2.2) because $|\phi(t)| \leq 1$.

Lemma 2.2. *Suppose that a function h is twice continuously differentiable on an open interval (a, b) with*

$$h'(x) \geq c > 0 \quad \text{and} \quad h''(x) \geq 0 \quad \text{for } x \in (a,b).$$

Then

$$\left| \int_{x=a}^{b} e^{ith(x)}\,dx \right| \leq \frac{2}{ct} \quad \text{for all } t > 0.$$

Proof. By considering subintervals $(a+\varepsilon, b-\varepsilon)$ and letting $\varepsilon \to 0$, we may without loss of generality assume that h is defined and twice differentiable at the endpoints, too. Then, using integration by parts, we calculate

$$\begin{aligned}\int_{x=a}^{b} e^{ith(x)}\,dx &= \frac{1}{it}\int_{x=a}^{b} \left[\frac{d}{dx}e^{ith(x)}\right]\frac{dx}{h'(x)} \\ &= \frac{1}{it}\left\{ \frac{e^{ith(x)}}{h'(x)}\bigg|_{x=a}^{b} - \int_{x=a}^{b} e^{ith(x)}\,d\left(\frac{1}{h'(x)}\right)\right\}.\end{aligned}$$

So

$$\left|\int_{x=a}^{b} e^{ith(x)}\,dx\right| \le \frac{1}{t}\left\{\left(\frac{1}{h'(b)}+\frac{1}{h'(a)}\right)+\int_{x=a}^{b}\left|d\left(\frac{1}{h'(x)}\right)\right|\,dx\right\}$$
$$=\frac{1}{t}\left\{\left(\frac{1}{h'(b)}+\frac{1}{h'(a)}\right)+\int_{x=a}^{b}\left[-d\left(\frac{1}{h'(x)}\right)\right]\,dx\right\}$$
$$=\frac{1}{t}\left\{\left(\frac{1}{h'(b)}+\frac{1}{h'(a)}\right)+\left(\frac{1}{h'(a)}-\frac{1}{h'(b)}\right)\right\}$$
$$=\frac{2}{th'(a)}\le\frac{2}{ct}.$$

□

Lemma 2.3. *For any real numbers y and z, the random variable $h_{y,z}(U)$ defined by (1.1) satisfies*

$$|\mathbf{E}e^{ith_{y,z}(U)}| \le 2|t|^{-1/2}.$$

Proof. We will apply Lemma 2.2, taking h to be $h_{y,z}$. Observe that

$$h''_{y,z}(u) = 2\left(\frac{1}{u}+\frac{1}{1-u}\right)=\frac{2}{u(1-u)}\ge 8 \quad \text{for } u\in(0,1)$$

and that

$$h'_{y,z}(u)=0 \quad \text{if and only if} \quad u=\alpha_{y,z}:=\frac{1}{1+\exp\left(\frac{1}{2}(y-z)\right)}\in(0,1).$$

Let $t>0$ and $\gamma>0$. If in Lemma 2.2 we take $a:=\alpha_{y,z}+\gamma t^{-1/2}$ and $b:=1$, and assume that $a<b$, then note

$$h'(u)=h'_{y,z}(u)=\int_{x=\alpha_{y,z}}^{u} h''_{y,z}(x)\,dx \ge 8(u-\alpha_{y,z})\ge 8\gamma t^{-1/2} \quad \text{for all } u\in(a,b).$$

So, by Lemma 2.2,

$$\left|\int_{u=\alpha_{y,z}+\gamma t^{-1/2}}^{1} e^{ith_{y,z}(u)}\,du\right| \le \frac{2}{t}[8\gamma t^{-1/2}]^{-1}=\frac{1}{4\gamma}t^{-1/2}.$$

Trivially,

$$\left|\int_{u=\alpha_{y,z}}^{\alpha_{y,z}+\gamma t^{-1/2}} e^{ith_{y,z}(u)}\,du\right|\le \gamma t^{-1/2},$$

so we can conclude

$$\left|\int_{u=\alpha_{y,z}}^{1} e^{ith_{y,z}(u)}\,du\right|\le [(4\gamma)^{-1}+\gamma]t^{-1/2}.$$

This result is trivially also true when $a = \alpha_{y,z} + \gamma t^{-1/2} \geq b = 1$, so it holds for all $t, \gamma > 0$. The optimal choice of γ here is $\gamma_{\text{opt}} = 1/2$, which yields

$$\left| \int_{u=\alpha_{y,z}}^{1} e^{ith_{y,z}(u)}\, du \right| \leq t^{-1/2} \qquad \text{for all } t > 0.$$

Similarly, for example by considering $u \mapsto h(1-u)$,

$$\left| \int_{0}^{\alpha_{y,z}} e^{ith_{y,z}(u)}\, du \right| \leq t^{-1/2} \qquad \text{for all } t > 0,$$

and we conclude that the lemma holds for all $t > 0$, and thus for all real t. □

Lemma 2.4. *For any real t, $|\phi(t)| \leq 2|t|^{-1/2}$.*

Proof. Lemma 2.3 shows that

$$\left| \mathbf{E}\left(e^{ith_{Y,Z}(U)} \;\middle|\; Y, Z \right) \right| \leq 2|t|^{-1/2}$$

and thus

$$|\phi(t)| = \left| \mathbf{E} e^{ith_{Y,Z}(U)} \right| \leq \mathbf{E} \left| \mathbf{E}\left(e^{ith_{Y,Z}(U)} \;\middle|\; Y, Z \right) \right| \leq 2|t|^{-1/2}. \qquad \square$$

The preceding lemma is the case $p = 1/2$ of Theorem 2.1. We now improve the exponent.

Lemma 2.5. *Let $0 < p < 1$. Then*

$$c_{2p} \leq \frac{\left[\Gamma(1-p)\right]^2}{\Gamma(2-2p)} c_p^2.$$

Proof. By (2.6) and the definition of c_p,

$$|\phi(t)| \leq \int_{u=0}^{1} c_p^2 |ut|^{-p} |(1-u)t|^{-p}\, du = c_p^2 |t|^{-2p} \int_{u=0}^{1} u^{-p}(1-u)^{-p}\, du,$$

and the result follows by evaluating the beta integral. □

In particular, recalling $\Gamma(1/2) = \sqrt{\pi}$, Lemmas 2.4 and 2.5 yield

$$|\phi(t)| \leq \frac{4\pi}{|t|}. \tag{2.7}$$

This proves (2.1) for $p = 1$, with $c_1 \leq 4\pi$, and thus by (2.2) for every $p \leq 1$ with $c_p \leq (4\pi)^p$; applying Lemma 2.5 again, we obtain the finiteness of c_p in (2.1) for all $p < 2$. Somewhat better numerical bounds are obtained for $1/2 < p < 1$ by taking a geometric average between the cases $p = 1/2$ and $p = 1$: the inequality

$$|\phi(t)| \leq (2t^{-1/2})^{2-2p} (4\pi t^{-1})^{2p-1} = 2^{2p} \pi^{2p-1} t^{-p}, \qquad t > 0,$$

shows that $c_p \leq 2^{2p}\pi^{2p-1}$, $1/2 \leq p \leq 1$. In particular, we have $c_{3/4} \leq \sqrt{8\pi}$, and thus, by Lemma 2.5, $c_{3/2} \leq 8\pi^{1/2}\left[\Gamma(1/4)\right]^2 < 186.4 < 187$.

Lemma 2.6. *Let $p > 1$. Then*

$$c_{p+1} \le 2^{p+1} c_p^{1+(1/p)} p/(p-1).$$

Proof. Assume that $t \ge 2c_p^{1/p}$. Then, again using (2.6),

$$\begin{aligned}
|\phi(t)| &\le \int_{u=0}^{1} \min\left(\frac{c_p}{(ut)^p}, 1\right) \min\left(\frac{c_p}{[(1-u)t]^p}, 1\right) du \\
&= 2\int_{u=0}^{c_p^{1/p}t^{-1}} \frac{c_p}{[(1-u)t]^p}\, du + \int_{u=c_p^{1/p}t^{-1}}^{1-c_p^{1/p}t^{-1}} \frac{c_p^2}{[u(1-u)t^2]^p}\, du \\
&\le \frac{2}{\left[1-c_p^{1/p}t^{-1}\right]^p} \frac{c_p^{1+(1/p)}}{t^{p+1}} + 2\frac{c_p^2}{t^{2p}} \int_{u=c_p^{1/p}t^{-1}}^{1/2} \frac{du}{[u(1-u)]^p} \\
&\le \frac{2}{(1/2)^p} c_p^{1+(1/p)} t^{-(p+1)} + \frac{2}{(1/2)^p} \frac{c_p^2}{t^{2p}} \int_{u=c_p^{1/p}t^{-1}}^{1/2} u^{-p}\, du \\
&\le 2^{p+1}\left\{c_p^{1+(1/p)} t^{-(p+1)} + \frac{1}{p-1} c_p^2 t^{-2p} \left[c_p^{1/p} t^{-1}\right]^{-(p-1)}\right\} \\
&= 2^{p+1} c_p^{1+(1/p)} \frac{p}{p-1} t^{-(p+1)}.
\end{aligned}$$

We have derived the desired bound for all $t \ge 2c_p^{1/p}$. But also, for all $0 < t < 2c_p^{1/p}$, we have

$$2^{p+1} c_p^{1+(1/p)} \frac{p}{p-1} t^{-(p+1)} \ge \frac{p}{p-1} \ge 1 \ge |\phi(t)|,$$

so the estimate holds for all $t > 0$. □

Lemma 2.6 completes the proof of finiteness of every c_p in (2.1) (by induction), and of the estimate (2.3). The bound for $c_{3/2}$ obtained above now shows (using `Maple`) that $c_{5/2} < 103215$, which then gives $c_{7/2} < 197102280$.

We can rewrite (2.3) as

$$\begin{aligned}
c_{p+1}^{1/(p+1)} &\le 2c_p^{1/p}\left(1 + \frac{1}{p-1}\right)^{1/(p+1)} \le 2c_p^{1/p} \exp\left(\frac{1}{(p-1)(p+1)}\right) \\
&= 2c_p^{1/p} \exp\left(\frac{1}{2(p-1)} - \frac{1}{2(p+1)}\right).
\end{aligned}$$

Hence, by induction, if $p = n + \frac{5}{2}$ for a nonnegative integer n, then

$$c_p^{1/p} \le 2^n c_{5/2}^{2/5} e^{(1/3)+(1/5)} = C2^p,$$

where $C := 2^{-5/2} e^{8/15} c_{5/2}^{2/5} < 30.6 < 2^5$, using the above estimate of $c_{5/2}$. Consequently, $c_p^{1/p} < 2^{p+5}$ when $p = n + \frac{5}{2}$. For general $p > 3/2$ we now use (2.2) with $p_1 = p$ and $p_2 = \lceil p - \frac{5}{2} \rceil + \frac{5}{2}$, obtaining $c_p^{1/p} < 2^{p_2+5} < 2^{p+6}$; the case $p \le 3/2$ follows from (2.2) and the estimate $c_{3/2}^{2/3} < 33 < 2^6$. This completes the proof of (2.4) and hence of Theorem 2.1. □

Remark 2.7. We used (1.1) in two different ways. In the first step we conditioned on the values of Y and Z, while in the inductive steps we conditioned on U.

Remark 2.8. A variety of other bounds are possible. For example, if we begin with the inequality (2.7), use (2.6), and proceed just as in the proof of Lemma 2.6, we can easily derive the following result in the case $t \geq 8\pi$:

$$|\phi(t)| \leq \frac{32\pi^2}{t^2}\left(\ln\left(\frac{t}{4\pi}\right) + 2\right) \leq \frac{32\pi^2 \ln t}{t^2} \quad \text{for all } t \geq 1.72. \tag{2.8}$$

The result is trivial for $1.72 \leq t < 8\pi$, since then the bounds exceed unity.

Since Y has finite moments of all orders [recall (1.3)], the characteristic function ϕ is infinitely differentiable. Theorem 2.1 implies a rapid decrease of all derivatives, too.

Theorem 2.9. *For each real $p \geq 0$ and integer $k \geq 0$, there is a constant $c_{p,k}$ such that*

$$|\phi^{(k)}(t)| \leq c_{p,k}|t|^{-p} \quad \text{for all } t \in \mathbf{R}.$$

Proof. The case $k = 0$ is Theorem 2.1, and the case $p = 0$ follows by $|\phi^{(k)}(t)| \leq \mathbf{E}|Y|^k$. The remaining cases follows from these cases by induction on k and the following calculus lemma. □

Lemma 2.10. *Suppose that g is a complex-valued function on $(0, \infty)$ and that $A, B, p > 0$ are such that $|g(t)| \leq At^{-p}$ and $|g''(t)| \leq B$ for all $t > 0$. Then $|g'(t)| \leq 2\sqrt{AB}t^{-p/2}$.*

Proof. Fix $t > 0$ and let $\theta = \arg(g'(t))$. For $s > t$,

$$\text{Re}(e^{-i\theta}g'(s)) \geq \text{Re}(e^{-i\theta}g'(t)) - |g'(s) - g'(t)| \geq |g'(t)| - B(s-t)$$

and thus, integrating from t to $t_1 := t + (|g'(t)|/B)$,

$$\begin{aligned}\text{Re}\big(e^{-i\theta}(g(t_1) - g(t))\big) &\geq \int_t^{t_1} \big(|g'(t)| - B(s-t)\big)\,ds \\ &= (t_1 - t)|g'(t)| - \tfrac{1}{2}B(t_1 - t)^2 = |g'(t)|^2/(2B).\end{aligned}$$

Consequently,

$$|g'(t)|^2/(2B) \leq |g(t)| + |g(t_1)| \leq 2At^{-p},$$

and the result follows. □

In other words, the characteristic function ϕ belongs to the class S of infinitely differentiable functions that, together with all derivatives, decrease more rapidly than any power. (This is the important class of test functions for tempered distributions, introduced by Schwartz [14]; it is often called the class of *rapidly decreasing C^∞* functions.)

3 The limiting Quicksort density f and its derivatives

We can now improve the result by Tan and Hadjicostas [15] on existence of a density f for Y. It is an immediate consequence of Theorem 2.1, with $p = 0$ and $p = 2$, say, that the characteristic function ϕ is integrable over the real line. It is well-known—see, e.g., [3, Theorem XV.3.3]—that this implies that Y has a bounded continuous density f given by the Fourier inversion formula

$$f(x) = \frac{1}{2\pi} \int_{t=-\infty}^{\infty} e^{-itx} \phi(t)\, dt, \quad x \in \mathbf{R}. \tag{3.1}$$

Moreover, using Theorem 2.1 with $p = k + 2$, we see that $t^k \phi(t)$ is also integrable for each integer $k \geq 0$, which by a standard argument (cf. [3, Section XV.4]) shows that f is infinitely smooth, with a kth derivative ($k \geq 0$) given by

$$f^{(k)}(x) = \frac{1}{2\pi} \int_{t=-\infty}^{\infty} (-it)^k e^{-itx} \phi(t)\, dt, \qquad x \in \mathbf{R}. \tag{3.2}$$

It follows further that the derivatives are bounded, with

$$\sup_x |f^{(k)}(x)| \leq \frac{1}{2\pi} \int_{t=-\infty}^{\infty} |t|^k \, |\phi(t)|\, dt \qquad (k \geq 0), \tag{3.3}$$

and these bounds in turn can be estimated using Theorem 2.1. Moreover, as is well known [14], [13, Theorem 7.4], an extension of this argument shows that the class $\mathcal{S}$ discussed at the end of Section 2 is preserved by the Fourier transform, and thus Theorem 2.9 implies that $f \in \mathcal{S}$:

Theorem 3.1. *The* `Quicksort` *limiting distribution has an infinitely differentiable density function f. For each real $p \geq 0$ and integer $k \geq 0$, there is a constant $C_{p,k}$ such that*

$$|f^{(k)}(x)| \leq C_{p,k} |x|^{-p} \quad \text{for all } x \in \mathbf{R}.$$ □

For numerical bounds on f, we can use (3.3) with $k = 0$ and Theorem 2.1 for several different p (in different intervals); for example, using $p = 0,\ 1/2,\ 1,\ 3/2,\ 5/2,\ 7/2$, and taking $t_1 = 4$, $t_2 = 4\pi^2$, $t_3 = (187/(4\pi))^2$, $t_4 = 103215/187$, $t_5 = 197102280/103215$,

$$\begin{aligned}
f(x) &\leq \frac{1}{2\pi} \int_{t=-\infty}^{\infty} |\phi(t)|\, dt = \frac{1}{\pi} \int_{t=0}^{\infty} |\phi(t)|\, dt \\
&\leq \frac{1}{\pi} \int_{t=0}^{\infty} \min(1, 2t^{-1/2}, 4\pi t^{-1}, 187 t^{-3/2}, 103215\, t^{-5/2}, 197102280\, t^{-7/2})\, dt \\
&= \frac{1}{\pi} \Bigg(\int_{t=0}^{t_1} dt + \int_{t=t_1}^{t_2} 2t^{-1/2}\, dt + \int_{t=t_2}^{t_3} 4\pi t^{-1}\, dt + \int_{t=t_3}^{t_4} 187\, t^{-3/2}\, dt \\
&\qquad + \int_{t=t_4}^{t_5} 103215\, t^{-5/2}\, dt + \int_{t=t_5}^{\infty} 197102280\, t^{-7/2}\, dt \Bigg) \\
&\leq 18.2.
\end{aligned} \tag{3.4}$$

Remark 3.2. We can do somewhat better by using the first bound in (2.8) over the interval $(103.18, 1984)$ instead of (as above) Theorem 2.1 with $p = 1, 3/2, 5/2, 7/2$ over $(103.18, t_3)$, (t_3, t_4), (t_4, t_5), $(t_5, 1984)$, respectively. This gives

$$f(x) < 15.3.$$

Similarly, (3.3) with $k = 1$ and the same estimates of $|\phi(t)|$ as in (3.4) yield

$$|f'(x)| \le \frac{1}{2\pi} \int_{t=-\infty}^{\infty} |t||\phi(t)|\, dt = \frac{1}{\pi} \int_{t=0}^{\infty} t|\phi(t)|\, dt < 3652.1,$$

which can be reduced to 2492.1 by proceeding as in Remark 3.2. The bound can be further improved to 2465.9 by using also $p = 9/2$.

Somewhat better bounds are obtained by using more values of p in the estimates of the integrals, but the improvements obtained in this way seem to be slight. We summarize the bounds we have obtained.

Theorem 3.3. *The limiting* `Quicksort` *density function* f *satisfies* $\max_x f(x) < 16$ *and* $\max_x |f'(x)| < 2466$. □

The numerical bounds obtained here are far from sharp; examination of Figure 4 of [15] suggests that $\max f < 1$ and $\max |f'| < 2$. Our present technique cannot hope to produce a better bound on f than $4/\pi > 1.27$, since neither Lemma 2.3 nor (2.6) can improve on the bound $|\phi(t)| \le 1$ for $|t| \le 4$. Further, *no* technique based on (3.3) can hope to do better than the actual value of $(2\pi)^{-1} \int_{t=-\infty}^{\infty} |\phi(t)|\, dt$, which from cursory examination of Figure 6 of [15] appears to be about 2.

4 An integral equation for the density f

Our estimates are readily used to justify rigorously the following functional equation.

Theorem 4.1. *The continuous limiting* `Quicksort` *density* f *satisfies (pointwise) the integral equation*

$$f(x) = \int_{u=0}^{1} \int_{y \in \mathbf{R}} f(y)\, f\left(\frac{x - g(u) - (1-u)y}{u}\right) \frac{1}{u}\, dy\, du, \qquad x \in \mathbf{R},$$

where $g(\cdot)$ *is as in* (1.2).

Proof. For each u with $0 < u < 1$, the random variable

$$uY + (1-u)Z + g(u) \tag{4.1}$$

[with notation as in (1.1)] has the density function

$$f_u(x) := \int_{z \in \mathbf{R}} f(z)\, f\left(\frac{x - g(u) - (1-u)z}{u}\right) \frac{1}{u}\, dz, \tag{4.2}$$

where the integral converges for each x since, using Theorem 3.3, the integrand is bounded by $f(z)(\max f)/u \le 16f(z)/u$; dominated convergence using the continuity of f and the same bound shows further that f_u is continuous.

This argument yields the bound $f_u(x) \le 16/u$, and since $f_u = f_{1-u}$ by symmetry in (4.1), we have $f_u(x) \le 16/\max(u, 1-u) \le 32$. This uniform bound, (1.1), and dominated convergence again imply that $\int_0^1 f_u(x)\,du$ is a continuous density for Y, and thus equals $f(x)$ for every x. □

It was shown in [15] that f is positive almost everywhere; we now can improve this by removing the qualifier "almost."

Corollary 4.2. *The continuous limiting* `Quicksort` *density function is everywhere positive.*

Proof. We again use the notation (4.2) from the proof of Theorem 4.1. Fix $x \in \mathbf{R}$ and $u \in (0,1)$. Since f is almost everywhere positive [15], the integrand in (4.2) is positive almost everywhere. Therefore $f_u(x) > 0$. Now we integrate over $u \in (0,1)$ to conclude that $f(x) > 0$. □

Alternatively, Corollary 4.2 can be derived directly from Theorem 4.1, without recourse to [15]. Indeed, if $f(y_0) > 0$ and $u_0 \in (0,1)$, set $x = y_0 + g(y_0)$; then the integrand in the double integral for $f(x)$ in Theorem 4.1 is postive for (u, y) equal to (u_0, y_0), and therefore, by continuity, also in some small neighborhood thereof. It follows that $f(y_0 + g(u_0)) > 0$. Since u_0 is arbitrary and the image of $(0,1)$ under g is $(-(2\ln 2 - 1), 1)$, an open interval containing the origin, Corollary 4.2 follows readily.

Remark 4.3. In future work, we will use arguments similar to those of this paper, together with other arguments, to show that when one applies the method of successive substitutions to the integral equation in Theorem 4.1, the iterates enjoy exponential-rate uniform convergence to f. This will settle an issue raised in the third paragraph of Section 3 in [15].

References

[1] van der Corput, J. G. Zahlentheoretische Abschätzungen. Math. Ann. **84** (1921), 53–79.

[2] Eddy, W. F. and Schervish, M. J. How many comparisons does Quicksort use? *J. Algorithms* **19** (1995), 402–431.

[3] Feller, W. *An Introduction to Probability Theory and its Applications. Vol. II.* Second edition. Wiley, New York, 1971.

[4] Flajolet, Ph. See `http://algo.inria.fr/AofA/Research/05-99.html` .

[5] Hennequin, P. Combinatorial analysis of quicksort algorithm. *RAIRO Inform. Théor. Appl.* **23** (1989), 317–333.

[6] Hennequin, P. Analyse en moyenne d'algorithmes, tri rapide et arbres de recherche. Ph.D. dissertation, L'École Polytechnique Palaiseau.

[7] Hoare, C. A. R. Quicksort. *Comput. J.* **5** (1962), 10–15.

[8] Knessl, C. and Szpankowski, W. Quicksort algorithm again revisited. *Discrete Math. Theor. Comput. Sci.* **3** (1999), 43–64.

[9] Knuth, D. E. *The Art of Computer Programming. Volume 3. Sorting and searching.* Second edition. Addison–Wesley, Reading, Mass., 1998.

[10] Montgomery, H. L. *Ten Lectures on the Interface Between Analytic Number Theory and Harmonic Analysis.* CBMS Reg. Conf. Ser. Math. 84, AMS, Providence, R.I., 1994.

[11] Régnier, M. A limiting distribution for quicksort. *RAIRO Inform. Théor. Appl.* **23** (1989), 335–343.

[12] Rösler, U. A limit theorem for 'Quicksort'. *RAIRO Inform. Théor. Appl.* **25** (1991), 85–100.

[13] Rudin, W. *Functional Analysis.* Second edition. McGraw–Hill, New York, 1991.

[14] Schwartz, L. *Théorie des Distributions.* Second edition. Hermann, Paris, 1966.

[15] Tan, K. H. and Hadjicostas, P. Some properties of a limiting distribution in Quicksort. *Statist. Probab. Lett.* **25** (1995), 87–94.

Trends in Mathematics, © 2000 Birkhäuser Verlag Basel/Switzerland

The Number of Descendants in Simply Generated Random Trees

BERNHARD GITTENBERGER[1] *Department of Geometry, TU Wien, Wiedner Hauptstrasse 8-10/113, A-1040 Wien, Austria.*
Bernhard.Gittenberger@tuwien.ac.at

Abstract. *We derive asymptotic results on the distribution of the number of descendants in simply generated trees. Our method is based on a generating function approach and complex contour integration.*

1 Introduction

The aim of this note is to generalize some recent results for binary trees by Panholzer and Prodinger [15] to a larger class of rooted trees. The number of descendants of a node j is the number of nodes in the subtree rooted at j, and the number of ascendants is the number of nodes between j and the root. Recently, Panholzer and Prodinger [15] studied the behavior of these parameters in binary trees during various traversal algorithms. The case of binary search trees was treated by Martínez, Panholzer and Prodinger [14]. In this paper we will study the number of descendants in simply generated trees (defined below). The number of ascendants is already treated in [1] and [10].

Let us start with a description of the traversal algorithms we will investigate. In the binary case there are basically three traversal algorithms. All of them are recursive algorithms treating the left subtree before the right subtree. They differ with respect to the visit of the root: first (*preorder*), middle (*inorder*), and last (*postorder*). We will study the number of descendants in simply generated trees during preorder and postorder traversal. Since the outdegree of any node in a simply generated tree need not be equal to zero or two, inorder traversal cannot be well defined for that class of trees.

Let us recall the definition of simply generated trees. Let $\mathcal{A}$ be a class of plane rooted trees and define for $T \in \mathcal{A}$ the size $|T|$ by the number of nodes of T. Furthermore there is assigned a weight $\omega(T)$ to each $T \in \mathcal{A}$. Let a_n denote the quantity

$$a_n = \sum_{|T|=n} \omega(T)$$

Besides, let us define the generating function (GF) corresponding to $\mathcal{A}$ by $a(z) = \sum_{n\geq 0} a_n z^n$. According to Meir and Moon [13] we call a family of trees *simply generated* if its GF satisfies a functional equation of the form $a(z) = z\varphi(a(z))$, where $\varphi(t) = \sum_{i\geq 0} \varphi_i t^i$ with $\varphi_i \geq 0, \varphi_0 > 0$.

Let $n_k(T)$ denote the number of nodes $v \in T$ with outdegree k (the outdegree of v is the number of edges incident with v that lead away from the root). Then

[1]This research has been supported by the Austrian Science Foundation FWF, grant P10187-MAT, and by the Stiftung Aktion Österreich–Ungarn, grant 34oeu24.

we can equivalently define simply generated trees as trees with weight

$$w(T) = \prod_{k \geq 0} \varphi_k^{n_k(T)}. \tag{1}$$

Another correspondence which was pointed out by Aldous [1] is considering simply generated trees as representations of Galton-Watson branching processes conditioned on the total progeny. Under this point of view the offspring distribution induces the weights (1) (for more details see [1] or also [4]).

In order to prove our results we will employ a generating function approach and singularity analysis in a similar fashion as used in [7]. For an introduction to the combinatorial techniques see e.g. [8, 11]. For an extensive presentation of marking techniques in combinatorial constructions with applications to random mappings see [5, 6]. Random mapping statistics similar to the tree statistics studied in this paper can be found in [2, 9].

2 Main results and Preliminaries

Choose a tree with n nodes at random (according to the distribution induced by (1)) and let $\alpha_j(n)$ and $\omega_j(T)$ denote the number of descendants of the jth node during preorder and postorder traversal, respectively, of the tree. We will study the distributions of these random variables and prove the following theorem:

Theorem 2.1 *Assume that $\varphi(t)$ has a positive radius of convergence R and that the equation $t\varphi'(t) = \varphi(t)$ has a minimal positive solution $\tau < R$. Then we have for $j \sim \rho n$:*

$$\mathbf{E}\alpha_j(n) \sim \frac{\sqrt{2}}{\sigma\sqrt{\pi}} \frac{\sqrt{1-\rho}}{\sqrt{\rho}} \sqrt{n} \quad \text{and} \quad \mathbf{E}\omega_j(n) \sim \frac{\sqrt{2}}{\sigma\sqrt{\pi}} \frac{\sqrt{\rho}}{\sqrt{1-\rho}} \sqrt{n}$$

where $\sigma^2 = \tau^2\varphi''(\tau)/\varphi(\tau)$. The variances satisfy the asymptotic relations

$$\mathbf{Var}\alpha_j(n) \sim \frac{\sqrt{2}}{\sigma\sqrt{\pi}} \left(\frac{\sqrt{1-\rho}}{\sqrt{\rho}} - \arcsin\sqrt{1-\rho} \right) n^{3/2}$$

and

$$\mathbf{Var}\omega_j(n) \sim \frac{\sqrt{2}}{\sigma\sqrt{\pi}} \left(\frac{\sqrt{\rho}}{\sqrt{1-\rho}} + \arcsin\sqrt{1-\rho} - \frac{\pi}{2} \right) n^{3/2}.$$

Furthermore a local limit theorem holds: Let the singularity of $a(z)$ on the circle of convergence be denoted by $z_0 = 1/\varphi'(\tau)$, then we have

$$\begin{aligned} \mathbf{P}\{\alpha_j(n) = m\} &= \frac{a_m z_0^m}{\tau} \left(1 + O\left(\frac{m \log^2 n}{n} \right) \right) \\ &= \frac{1}{\sigma\sqrt{2\pi m^3}} \left(1 + O\left(\frac{1}{m} \right) + O\left(\frac{m \log^2 n}{n} \right) \right) \end{aligned} \tag{2}$$

$$\begin{aligned} \mathbf{P}\{\omega_j(n)=m\} &= \frac{a_m z_0^m}{\tau}\left(1+O\left(\frac{m\log^2 n}{n}\right)\right), \quad m \le j, \\ &= \frac{1}{\sigma\sqrt{2\pi m^3}}\left(1+O\left(\frac{1}{m}\right)+O\left(\frac{m\log^2 n}{n}\right)\right), \quad m \le j, \end{aligned} \qquad (3)$$

uniformly for $m \ll n/\log^2 n$.

Remark 1 *Note that if simply generated trees are viewed as conditioned branching processes, then* σ^2 *is just the variance of the offspring distribution.*

Remark 2 *Note that here an interesting phenomenon occurs: the distributions in the local limit theorem do not depend on* j. *This is no contradiction to the formulas for expectation and variance, since on the one hand the variances are very large* ($\mathbf{Var}\alpha_j(n) \gg (\mathbf{E}\alpha_j(n))^2$) *and thus the knowledge of the expectation tells us only little about the distribution. On the other hand due to the heavy tail in (2) and (3) the local limit theorem cannot be used to derive expressions for the moments.*

Let us first set up the generating functions for the preorder case. Therefore denote the by a_{nkm} the (weighted) number of trees with n nodes such that the jth node x_j has m descendants. We are interested in the generating function

$$a_1(z,u,v) = \sum_{n,j,m\ge 0} a_{nkm} z^n u^j v^m.$$

It is easier to work with

$$a_1^{(m)}(z,u) = [v^m]a_1(z,u,v),$$

where the symbol $[x^n]f(x)$ denotes the coefficient of x^n in the formal power series $f(x)$. Thus we will build this function now: Note that there is a unique path connecting x_j with the root. To each of these nodes there are attached subtrees of the whole tree. The path itself and those subtrees which lie left from the path contains only nodes which are traversed before x_j, while the nodes in the subtrees on the right-hand side from the path are traversed after x_j. Thus a node with degree i on this path and j_1 subtrees on the left-hand side and j_2 subtrees on the right-hand side contributes $zu\varphi_i a(zu)^{j_1} a(z)^{j_2}$ to the generating function. Summing up over all possible configurations we get

$$a_1^{(m)}(z,u) = \frac{uz^m a_m}{1-\phi_1(z,u,1)},$$

where

$$\phi_1(z,u,v) = zu\sum_{i\ge 1}\varphi_i \sum_{j_1+j_2=i-1} a(zu)^{j_1} a(zv)^{j_2} = \frac{a(zu)-ua(zv)/v}{a(zu)-a(zv)}$$

The postorder case can be treated in an analogous way. In this case we get

$$\tilde{a}_1^{(m)}(z,u) = \frac{u^{m+1} z^m a_m}{1-\phi_1(z,u,1)/u}$$

3 Proof of Theorem 2.1

Since the generating functions for the preorder and the postorder case are so closely related it suffices to consider the preorder case.

3.1 The Expected value of $\alpha_j(n)$

We have

$$\begin{aligned}\mathbf{E}\alpha_j(n) &= \frac{1}{a_n}[z^n u^j]\sum_{m\geq 0} m a_1^{(m)}(z,u) = \frac{1}{a_n}[z^n u^j]\frac{zu(a(zu)-a(z))a'(z)}{a(z)(u-1)} \\ &= \frac{1}{a_n}[z^n u^j]\frac{u(a(zu)-a(z))}{(u-1)(1-z\varphi'(a(z)))} \qquad (4)\end{aligned}$$

In order to compute this coefficient we will use Cauchy's integral formula with the following integration contour. Let z run through the contour $\Gamma_0 = \Gamma_{01} \cup \Gamma_{02} \cup \Gamma_{03} \cup \Gamma_{04}$ defined by

$$\begin{aligned}\Gamma_{01} &= \left\{ z = z_0\left(1+\frac{t}{n}\right) \middle| \Re t \leq 0 \text{ und } |t| = 1 \right\} \\ \Gamma_{02} &= \left\{ z = z_0\left(1+\frac{t}{n}\right) \middle| \Im t = 1 \text{ und } 0 \leq \Re t \leq \log^2 n \right\} \\ \Gamma_{03} &= \overline{\Gamma}_{02} \\ \Gamma_{04} &= \left\{ z \middle| |z| = z_0\left|1+\frac{\log^2 n + i}{n}\right| \text{ und } \arg\left(1+\frac{\log^2 n+i}{n}\right) \leq |\arg(z)| \leq \pi \right\}.\end{aligned}$$

and since the location of the singularity changes when z varies, the appropriate contour for u is $\Gamma_1 = \Gamma_{11} \cup \Gamma_{12} \cup \Gamma_{13} \cup \Gamma_{14}$ defined by

$$\begin{aligned}\Gamma_{11} &= \left\{ u = \left(1+\frac{s}{j}\right) \middle| \Re s \leq -R(t) \text{ and } |s+R(t)+I(t)i| = 1 \right\} \\ \Gamma_{12} &= \left\{ u = \left(1+\frac{s}{j}\right) \middle| \Im s = -I(t)+1, -R(t) \leq \Re s \text{ and } |u| \leq \left|1+\frac{\log^2 j + i}{j}\right| \right\} \\ \Gamma_{13} &= \left\{ u = \left(1+\frac{s}{j}\right) \middle| \Im s = -I(t)-1, -R(t) \leq \Re s \text{ and } |u| \leq \left|1+\frac{\log^2 j + i}{j}\right| \right\} \\ \Gamma_{14} &= \left\{ u \middle| |u| = \left|1+\frac{\log^2 j+i}{j}\right| \text{ and } \arg u \in [-\pi, \arg z_{13}] \cup [\arg z_{12}, \pi] \right\},\end{aligned}$$

where

$$R(t) = \max\left(0, \frac{j}{n}\Re t\right) \quad \text{and} \quad I(s,\cdots,s_p,t) = \max\left(n^{2/3}, \frac{j}{n}\Im t\right)$$

and z_{1k} denotes the point of Γ_{1k} with maximal absolute value. For convenience, set $\gamma_0 = \Gamma_{01} \cup \Gamma_{02} \cup \Gamma_{03}$ and $\gamma_1 = \Gamma_{11} \cup \Gamma_{12} \cup \Gamma_{13}$.

Now we use well known expansions (see e.g. [13]) for the tree function $a(z)$ and related functions in order to get the local behaviour of the integrand near its singularity: we have for $z \to z_0$ inside the domain $\{z : |z| \le z_0 + \varepsilon, \arg(1 - z/z_0) \ne \pi\}$ for some $\varepsilon > 0$ the local expansions

$$a(z) = \tau - \frac{\tau\sqrt{2}}{\sigma}\sqrt{1 - \frac{z}{z_0}} + O\left(\left|1 - \frac{z}{z_0}\right|\right) \tag{5}$$

and

$$z\varphi'(a(z)) = 1 - \sigma\sqrt{2}\sqrt{1 - \frac{z}{z_0}} + O\left(\left|1 - \frac{z}{z_0}\right|\right) \tag{6}$$

Inserting this into (4) yields for $z \in \gamma_0$ and $u \in \gamma_1$

$$\begin{aligned}
&\frac{1}{a_n(2\pi i)^2}\int_{\gamma_0}\int_{\gamma_1} \frac{u(a(zu) - a(z))}{(u-1)(1 - z\varphi'(a(z)))}\frac{du}{u^{j+1}}\frac{dz}{z^{n+1}} \\
&= \frac{1}{a_n z_0^n (2\pi i)^2}\int_{\gamma_0}\int_{\gamma_1} \frac{\left(\sqrt{-\frac{t}{n}} - \sqrt{-\frac{t}{n} - \frac{s}{j}}\right)\frac{\tau\sqrt{2}}{\sigma}}{\frac{s}{j}\sigma\sqrt{2}\sqrt{-\frac{t}{n}}} e^{-t-s}\frac{dt\,ds}{nj} \\
&\quad \times \left(1 + O\left(\left|\frac{t}{n}\right| + \left|\frac{s}{j}\right|\right)\right) \\
&= \frac{\tau}{a_n z_0^n \sigma^2 \sqrt{n}(2\pi i)^2}\int_{\gamma_0}\int_{\gamma_0} \frac{\sqrt{-\frac{t}{n}} - \sqrt{-\frac{v}{j}}}{\left(v - \frac{tj}{n}\right)\sqrt{-t}} e^{-t(1-j/n)-v}\,dt\,dv \\
&\quad \times \left(1 + O\left(\frac{\log^2 n}{n} + \frac{\log^2 j}{j}\right)\right)
\end{aligned}$$

Extending the integration contour to ∞ (call the new contour γ and expanding the denominator into a series and using the fact (Hankel's representation of the Gamma function, see e.g. [16]) that for any positive constant A and integers k, l, one of which is nonnegative, we have

$$\int_\gamma\int_\gamma t^k v^l e^{-tA-v}\,dt\,dv = 0,$$

yields after some elementary calculations

$$\begin{aligned}
&\frac{\tau}{a_n z_0^n \sigma^2 \sqrt{n}(2\pi i)^2}\int_{\gamma_0}\int_{\gamma_0} \frac{\sqrt{-\frac{t}{n}} - \sqrt{-\frac{v}{j}}}{\left(v - \frac{tj}{n}\right)\sqrt{-t}} e^{-t(1-j/n)-v}\,dt\,dv \\
&= \frac{\tau}{a_n z_0^n \sigma^2 \sqrt{jn}(2\pi i)^2}\sum_{k\ge 0}\left(\frac{j}{n}\right)^k\left(1 - \frac{j}{n}\right)^{-k-1/2} \\
&\quad \times \int_\gamma\int_\gamma (-w)^{k-1/2}(-v)^{-k-1/2}e^{-w-v}\,dw\,dv\left(1 + O\left(e^{-\frac{1}{2}\log^2 n}\right)\right) \\
&= \frac{\tau}{a_n z_0^n \sigma^2 \pi\sqrt{jn}}\sum_{k\ge 0}\left(\frac{j}{n}\right)^k\left(1 - \frac{j}{n}\right)^{-k-1/2}(-1)^k\left(1 + O\left(e^{-\frac{1}{2}\log^2 n}\right)\right) \\
&= \frac{\sqrt{2}}{\sigma\sqrt{\pi}}\frac{\sqrt{1 - j/n}}{\sqrt{j/n}}\sqrt{n}\left(1 + O\left(e^{-\frac{1}{2}\log^2 n}\right)\right),
\end{aligned} \tag{7}$$

where we used again Hankel's representation

$$\frac{1}{2\pi i}\int_\gamma (-s)^{-\alpha}e^{-s}\,ds = \frac{1}{\Gamma(\alpha)}$$

as well as

$$\frac{1}{\Gamma(-k+1/2)\Gamma(k+1/2)} = \frac{\sin\pi(k+1/2)}{\pi} = \frac{(-1)^k}{\pi}$$

and $a_n = \tau/\sigma z_0^n\sqrt{2\pi n^3}(1+O(1/\sqrt{n}))$ which can be easily obtained by applying [7, Theorem 3.1] to (5). (7) is already the desired expression, thus what remains to be shown is that the integrals where $z \in \Gamma_{04}$ or $u \in \Gamma_{14}$ are negligibly small. On Γ_{04} and Γ_{14} the estimates $|u|^{-j-1} \ll e^{-\log^2 j}$ and $|z|^{-n-1} \ll z_0^{-n}e^{-\log^2 n}$, respectively, hold. Moreover, observe that we have $1/|u-1| \le 1/j \ll 1/n$ along the integration contour. Furthermore, $a(z)$ (and hence $1/(1-z\varphi'(a(z)))$ is analytic in the set surrounded by the integration contour. This in conjunction with (6) yields $1/(1-z\varphi'(a(z)) \ll n$ and therefore

$$\begin{aligned}&\frac{1}{a_n}\int\int_{(z,u)\in\Gamma_0\times\Gamma_1\setminus\gamma_0\times\gamma_1}\frac{u(a(zu)-a(z))}{(u-1)(1-z\varphi'(a(z)))}\frac{du}{u^{j+1}}\frac{dz}{z^{n+1}}\\ &\ll n^{7/2}e^{-\log^2 n-\log^2 j}\\ &= o\left(\frac{1}{a_n}\int\int_{(z,u)\in\gamma_0\times\gamma_1}\frac{u(a(zu)-a(z))}{(u-1)(1-z\varphi'(a(z)))}\frac{du}{u^{j+1}}\frac{dz}{z^{n+1}}\right)\end{aligned}$$

which completes the proof.

3.2 The Variance (sketch)

We need an expression for the second moment. We have

$$\mathbf{E}\alpha_j(n)^2 = \frac{1}{a_n}[z^n u^j]\frac{uz^2a''(z)(a(zu)-a(z))}{(u-1)a(z)}$$

By elementary calculations we get

$$a''(z) = \frac{2\varphi'(a(z))\varphi(a(z))}{(1-z\varphi'(a(z)))^2} + \frac{z\varphi''(a(z))\varphi^2(a(z))}{(1-z\varphi'(a(z)))^3}$$

and thus

$$\mathbf{E}\alpha_j(n)^2 = \frac{1}{a_n}[z^n u^j]\left(\frac{2uz\varphi'(a(z))(a(zu)-a(z))}{(u-1)(1-z\varphi'(a(z)))^2} + \frac{uza(z)\varphi''(a(z))(a(zu)-a(z))}{(u-1)(1-z\varphi'(a(z)))^3}\right)$$

Obviously, the dominant singularity in this expression comes from the second term. Proceeding as in the previous section and using $z\varphi''(a(z)) \sim \sigma^2/\tau$ for $z \to z_0$ gives

$$\frac{1}{a_n}[z^n u^j]\frac{2uz\varphi'(a(z))(a(zu)-a(z))}{(u-1)(1-z\varphi'(a(z)))^2}$$

$$\sim \frac{\sigma^2}{\tau a_n z_0^n (2\pi i)^2} \int_{\gamma_0}\int_{\gamma_1} \frac{\tau^{\frac{\tau\sqrt{2}}{\sigma}}\left(\sqrt{-\frac{t}{n}} - \sqrt{-\frac{t}{n}-\frac{s}{j}}\right)}{\frac{s}{j}\cdot 2\sqrt{2}\sigma^3\left(-\frac{t}{n}\right)^{3/2}} e^{-t-s}\frac{dt\,ds}{nj}$$

$$= \frac{\tau\sqrt{n}}{2a_n z_0^n \sigma^2 (2\pi i)^2} \int_{\gamma_0}\int_{\gamma_0} \frac{\sqrt{-\frac{t}{n}} - \sqrt{-\frac{v}{j}}}{\left(v - \frac{tj}{n}\right)(-t)^{3/2}} e^{-t(1-j/n)-v}\,dt\,dv$$

$$\sim \frac{\tau}{2a_n z_0^n \sigma^2 (2\pi i)^2} \sum_{k\geq 0}\left(\frac{j}{n}\right)^k \int_\gamma (-t)^{k-1} e^{-t(1-j/n)}\,dt \int_\gamma (-1)(-v)^{-k-1} e^{-v}\,dv$$

$$+ \frac{\tau\sqrt{n}}{2a_n z_0^n \sigma^2 (2\pi i)^2 \sqrt{j}} \sum_{k\geq 0}\left(\frac{j}{n}\right)^k \int_\gamma (-t)^{k-3/2} e^{-t(1-j/n)}\,dt$$

$$\times \int_\gamma (-1)(-v)^{-k-1/2} e^{-v}\,dv$$

$$= -\frac{\tau}{2a_n z_0^n \sigma^2} + \frac{\tau\sqrt{n}}{2a_n z_0^n \sigma^2 \pi\sqrt{j}} \sum_{k\geq 0}\left(\frac{j}{n}\right)^k \frac{(-1)^k}{(-k+1/2)}\left(1-\frac{j}{n}\right)^{-k-1/2}$$

$$= -\frac{\tau}{2a_n z_0^n \sigma^2} + \frac{\tau}{a_n z_0^n \sigma^2 \pi\sqrt{j/n}}\sqrt{1-j/n}\left(1 + \arctan\frac{\sqrt{j/n}}{\sqrt{1-j/n}}\right)$$

$$= \frac{\tau}{a_n z_0^n \sigma^2 \pi}\left(\frac{\sqrt{1-j/n}}{\sqrt{j/n}} + \arctan\frac{\sqrt{j/n}}{\sqrt{1-j/n}} - \frac{\pi}{2}\right)$$

$$= \frac{n^{3/2}\sqrt{2}}{\sigma\sqrt{\pi}}\left(\frac{\sqrt{1-j/n}}{\sqrt{j/n}} - \arcsin\sqrt{1-j/n}\right)$$

and we are done.

3.3 The distribution (sketch)

We need to evaluate

$$\mathbf{P}\{\alpha_j(n) = m\} = \frac{1}{a_n}[z^n u^j v^m] a_1(z,u,v) = \frac{1}{a_n}[z^n u^j]\frac{uz^m a_m(a(zu) - a(z))}{a(z)(u-1)}.$$

We use the same integration contour as in the previous sections and get for $m \ll n/\log^2 n$

$$\mathbf{P}\{\alpha_j(n) = m\} = \frac{a_m\sqrt{2}}{\sigma z_0^{n-m} a_n (2\pi i)^2} \int_\gamma\int_\gamma \frac{\left(\sqrt{-\frac{t}{n}} - \sqrt{-\frac{v}{n}}\right)}{v - \frac{tj}{n}} e^{-v-t(1-j/n)}\,dt\,dv$$

$$\times \left(1 + O\left(e^{-\log^2 n/2}\right) + O\left(\frac{m\log^2 n}{n}\right)\right)$$

$$= -\frac{a_m\sqrt{2}\left(1 + O\left(m\log^2 n/n\right)\right)}{\sigma z_0^{n-m} a_n n^{3/2} (2\pi i)^2}$$

$$
\times \sum_{k\ge 0} \left(\frac{j}{n}\right)^k \int_\gamma \int_\gamma (-t)^{k+1/2}(-v)^{-k-1} e^{-t(1-j/n)-v}\,dt\,dv
$$

$$
= -\frac{a_m\sqrt{2}\left(1+O\left(m\log^2 n/n\right)\right)}{\sigma z_0^{n-m} a_n n^{3/2}(1-j/n)^{3/2}} \sum_{k\ge 0}\left(\frac{j/n}{1-j/n}\right)^k \frac{1}{\Gamma(-k-1/2)\Gamma(k+1)}
$$

$$
= \frac{a_m\left(1+O\left(m\log^2 n/n\right)\right)}{\sqrt{2\pi}\sigma z_0^{n-m} a_n n^{3/2}(1-j/n)^{3/2}} \sum_{k\ge 0}\left(\frac{j/n}{1-j/n}\right)^k \binom{-3/2}{k}
$$

$$
= \frac{a_m}{\sqrt{2\pi}\sigma z_0^{n-m} a_n n^{3/2}}\left(1+O\left(m\log^2 n/n\right)\right)
$$

$$
= \frac{1}{\sigma\sqrt{2\pi m^3}}\left(1+O\left(\frac{1}{m}\right)+O\left(\frac{m\log^2 n}{n}\right)\right)
$$

as desired.

4 Concluding remarks

It would be interesting to get also expressions for the joint distributions of $(\alpha_{j_1}(n),\ldots,\alpha_{j_d}(n))$ and joint moments, as were derived in [1, 10] for the number of ascendants. But since an invariance property similar to [10, Lemma 3.3]) is not true in this case, we are not able to derive a general and simple shape for the generating functions which occur when we compute these joint distributions. The method presented here only in principle allows us to compute these joint distributions and joint moments, but the expressions we would encounter are terribly involved.

References

[1] D. J. Aldous, The continuum random tree II: an overview, *Stochastic Analysis,* M. T. Barlow and N. H. Bingham, Eds., Cambridge University Press 1991, 23–70.

[2] G. Baron, M. Drmota and L. Mutafchief, Predecessors in random mappings, *Combin. Probab. Comput. 5, No.4,* (1996), 317–335.

[3] M. Drmota, The height distribution of leaves in rooted trees, *Discr. Math. Appl. 4* (1994), 45–58 (translated from *Diskretn. Mat. 6* (1994), 67–82).

[4] M. Drmota and B. Gittenberger, On the profile of random trees, *Random Struct. Alg. 10* (1997), 421–451.

[5] M. Drmota and M. Soria, Marking in combinatorial constructions: generating functions and limiting distributions, *Theor. Comput. Sci. 144* (1995), 67–99.

[6] M. Drmota and M. Soria, Images and preimages in random mappings, *SIAM J. Discrete Math. 10, No.2* (1997), 246–269.

[7] P. FLAJOLET AND A. M. ODLYZKO, Singularity analysis of generating functions, *SIAM J. Discrete Math. 3,* 2 (1990), 216–240.

[8] P. FLAJOLET AND J. S. VITTER, Average-Case Analysis of Algorithms and Data Structures, in *Handbook of Theoretical Computer Science,* J. van Leeuwen, Ed., vol. A: Algorithms and Complexity. North Holland, 1990, ch. 9, pp. 431–524.

[9] B. GITTENBERGER, On the number of predecessors in constrained random mappings, *Stat. Probab. Letters 36* (1997), 29–34.

[10] B. GITTENBERGER, On the contour of random trees, *SIAM J. Discrete Math. 12, No.4* (1999), 434–458.

[11] I. P. GOULDEN AND D. M. JACKSON, *Combinatorial Enumeration,* Wiley, New York, 1983.

[12] W. GUTJAHR AND G. CH. PFLUG, The asymptotic contour process of a binary tree is a Brownian excursion, *Stochastic Process. Appl. 41* (1992), 69–89.

[13] A. MEIR AND J. W. MOON, On the Altitude of Nodes in Random Trees, *Can. J. Math. 30* (1978), 997–1015.

[14] C. MARTÍNEZ, A. PANHOLZER AND H. PRODINGER, On the number of descendants and ascendants in random search trees, *Electronic Journal of Combinatorics 5* (1998), Research paper R20, 26 p.; printed version *J. Comb. 5* (1998), 267–302.

[15] A. PANHOLZER AND H. PRODINGER, Descendants and ascendants in binary trees, *Discrete Mathematics and Theoretical Computer Science 1* (1997), 247–266.

[16] WHITTAKER, E. T. AND WATSON, G. N., *A Course of Modern Analysis,* reprint of the 4th edn., Cambridge University Press, 1996.

Trends in Mathematics, © 2000 Birkhäuser Verlag Basel/Switzerland

An Universal Predictor Based on Pattern Matching : Preliminary results [1]

PHILIPPE JACQUET[2] *INRIA, Rocquencourt, 78153 Le Chesnay Cedex, France.* Philippe.Jacquet@inria.fr
WOJCIECH SZPANKOWSKI[3] *Dept. of Computer Science, Purdue University, W. Lafayette, IN 47907, U.S.A.* spa@cs.purdue.edu
IZYDOR APOSTOL *Baxter Hemoglobin Therapeutics Inc., 2545 Central Avenue, Boulder, CO 80301, U.S.A.* apostoi@baxter.com

Abstract. *We consider here an universal predictor based on pattern matching. For a given string $x_1, x_2, \ldots, x_n$, the predictor will guess the next symbol x_{n+1} in such a way that the prediction error tends to zero as $n \to \infty$ provided the string $x_1^n = x_1, x_2, \ldots, x_n$ is generated by a mixing source. We shall prove that the rate of convergence of the prediction error is $O(n^{-\varepsilon})$ for any $\varepsilon > 0$. In this preliminary version, we only prove our results for memoryless sources and a sketch for mixing sources. However, we indicate that our algorithm can predict equally successfully the next k symbols as long as $k = O(1)$.*

1 Introduction

Prediction is important in communication, control, forecasting, investment and other areas. We understand how to do optimal prediction when the data model is known, but one needs to design universal prediction algorithm that will perform well no matter what the underlying probabilistic model is. More precisely, let $X_1, X_2, \ldots$ be an infinite random data sequence, and let a predictor generate a sequence $\hat{X}_1, \hat{X}_2, \ldots$. We consider only nonanticipatory predictors so that $\hat{X}_i$ is determined by $X_1, \ldots, X_{i-1}$. We say that a predictor is *asymptotically consistent* if

$$\lim_{n\to\infty} |\Pr\{X_{n+1} = a|X_1, \ldots, X_n\} - \Pr\{\hat{X}_{n+1} = a|X_1, \ldots, X_n\}| = 0$$

for all symbols a belonging to a finite alphabet $\mathcal{A}$.

We say that an universal predictor is optimal if $\Pr\{\hat{X}_i \neq X_i\}$ is minimized for all i. When the probabilistic model is known, an optimal predictor is known to be (cf. [6])

$$\hat{X}_i := \arg\max_{a\in\mathcal{A}} \Pr\{X_i = a|X_1, \ldots, X_{i-1}\}.$$

In most cases, the probabilistic model of data is unknown. For such model we define the optimal predictor $N(X_i)$ as above, that is,

$$N(X_i) := \arg\max_{a\in\mathcal{A}} \Pr\{X_i = a|X_1, \ldots, X_{i-1}\}.$$

An *universal asymptotically optimal* predictor $\hat{X}_n$ is such that

$$\lim_{n\to\infty} |\Pr\{N(X_{n+1}) = a|X_1, \ldots, X_n\} - \Pr\{\hat{X}_{n+1} = a|X_1, \ldots, X_n\}| = 0 \qquad (1)$$

[1]This work was supported by Purdue Grant GIFG-9919.
[2]Additional support by the ESPRIT Basic Research Action No. 7141 (ALCOM II).
[3]This author was additionally supported by NSF Grants NCR-9415491 and C-CR-9804760.

for all $a \in \mathcal{A}$. It is known that in a class of stationary data models, there exists at least one universal optimal predictor. Among them, one should look for an universal predictor with the speed of convergence as fast as possible (cf. [9]).

A large body of useful research on universal prediction was done in the last fifty years (cf. [1, 4, 8, 9, 10, 11, 12, 14, 15, 16, 19]). There exist predictors based on arithmetic coding (cf. [14, 15]), Rissanen MDL (cf. [11, 12]), nonparametric universal predictors (cf. [4]), context-weighting, and so forth. In this paper, we consider a modified prediction algorithm based on pattern matching that was described in Ehrenfeucht and Mycielski [3]. This predictor seems to be performing well in practice, however, there is not yet a theoretical justification available (cf. [10]). The algorithm described in [3] is as follows: Let the sequence $x_1, \ldots, x_n$ be given (i.e., it is a realization of a random sequence $X_1, \ldots, X_n$), and we are asked to predict x_{n+1}. Let $D_n := n - \ell$ be maximal such that $x_\ell, \ldots, x_n = x_{\ell-i}, \ldots, x_{n-i}$ for some $1 \leq i \leq n$. In other words, we find the maximal suffix of $x_\ell, x_{\ell+1}, \ldots, x_n$ that occurs earlier in the sequence $x_1, \ldots, x_n$. Then, we take the smallest i (the most recent occurrence), say I, for which we found the longest match, and set $x_{n+1} = x_{n-I+1}$ (cf. [3]). It was conjectured in [3, 6] that this is an optimal predictor. However, Jacquet [5] proved that the above algorithm is a good density estimator but not an optimal predictor. More precisely, Jacquet proved that for memoryless sources $\Pr\{X_{n+1} = a\} = \Pr\{X_{n-I+1} = a\}$ for all $a \in \mathcal{A}$.

In this paper, we modified the above algorithm to make it asymptotically optimal predictor. Briefly, we consider a fractional maximal suffix, say of length αD_n for $1/2 < \alpha < 1$. We shall show that such a shorter suffix occurs $O(n^{1-\alpha})$ times in the strings $X_1, \ldots, X_n$. We find all occurrence of such shorter suffixes, called further *markers*, in $X_1, \ldots, X_n$ and then apply majority rule to all symbols that occur just after the markers (i.e., we select the most likely symbol). We shall prove that such a predictor is asymptotically optimal for mixing sources and the rate of convergence is $O(n^{-\varepsilon})$ for any $\varepsilon > 0$.

2 Main Results

We start with a precise description of our prediction algorithm. We assume that a sequence $x_1^n = x_1, \ldots, x_n$ is given. Our goal is to predict the next symbol x_{n+1} such that the error of the prediction is as small as possible. To formalize this criterion, we asume that x_1^n is a realization of a random sequence $X_1^n = X_1, \ldots, X_n$ generated by a source, and the prediction sequence $\hat{X}_1^n$ is also random. The prediction error is then the probability $\Pr\{\hat{X}_{n+1} \neq X_{n+1}\}$ that should be minimized.

Let us fix $\alpha < 1$. The prediction algorithm discussed in this paper works in four steps:

1. Find the largest suffix of x_1^n that appears somewhere in the string x_1^n. We call this the *maximal suffix* and we denote its length by D_n.

2. Take an α fraction of the maximal suffix. Its length is $k_n = \lceil \alpha D_n \rceil$. Such a suffix occurs L_n times in the string x_1^n and we call these substrings *markers*. The *marked position* is a position that occurs just after the end of a marker.

3. Create a subsequence consisting of all symbols that occur at the marked positions. We shall call such a subsequence the *sampled sequence.*

4. Set x_{n+1} to the symbol that occurs most frequently in the sampled sequence. If there is a tie, break it arbitrary.

The algorithm just described will be further called the *Sampled Pattern Matching* predictor, or in short SPM predictor.

Before we proceed, let us consider an example. Below the text and the maximal suffix defined in Step 1 is shown

SLJZGGDL[YGSJSLJZ]KGSSLJZIDSLJZJGZ[YGSJSLJZ]

where the maximal suffix and its copy are framed. Observe that $D_{40} = 8$. We set $\alpha = 0.5$ to get the suffix SLJZ that is used to find all markers. They are shown below:

[SLJZ]GGDLYGSJ[SLJZ]KGS[SLJZ] KLJZJGZYGSJ[SLJZ]

The sampled sequence is GKK, thus the SPM predicts $x_{41} = K$.

The prime goal of this paper is to prove asymptotic optimality of the SPM algorithm. To formulate it precisely, we need some additional notation. We assume that x_1^n is a realization of a random sequence X_1^n generated by a probabilistic source. We assume that the source is a mixing source that can be defined as follows:

(MX) (STRONGLY) ψ-MIXING SOURCE

Let $\mathcal{R}_m^n$ be a σ-field generated by $X_{k=m}^n$ for $m \leq n$. The source is called *mixing*, if there exists a bounded function $\psi(g)$ such that for all $m, g \geq 1$ and any two events $A \in \mathcal{R}_1^m$ and $B \in \mathcal{R}_{m+g}^\infty$ the following holds

$$(1 - \psi(g))\Pr\{A\}\Pr\{B\} \leq \Pr\{AB\} \leq (1 + \psi(g))\Pr\{A\}\Pr\{B\}. \quad (2)$$

If, in addition, $\lim_{g\to\infty} \psi(g) = 0$, then the source is called *strongly* mixing. Hereafter, we consider only strongly ψ mixing sources and we shall call them *mixing sources.*

To simplify the presentation of our results we introduce δ-discriminant distribution.

Definition 1 *Let $\delta > 0$. A distribution over a finite alphabet of size V with vector probability $(p_i)_{i\leq V}$ is said to be δ-discriminant if:*

- *There is only one integer $i_{\max}$ such that $p_{i_{\max}} = \max_j\{p_j\}$;*
- *For all $j \neq i_{\max}$ we have $p_j < p_{i_{\max}} - \delta$.*

We apply this definition to text sources.

We say that a string X_1^n is δ-discriminant if the distribution of X_{n+1} conditioned on X_1^n is δ-discriminant. Furthermore, we say that the source is *asymptotically* δ-discriminant if the probability that X_1^n is not δ-discriminant tends to zero when $n \to \infty$.

For example, it is easy to see that memoryless and Markov sources are δ-discriminant for some $\delta > 0$ depending on the model. We denote $\mu_n = P(X_1^n$ is not δ-discriminant), the rate at which μ_n tends to zero depends on the probabilistic model. For δ-discriminant memoryless sources $\mu_n = \psi(n) = 0$ for all $n \geq 1$. For Markov sources we have $\mu_n = 0$ when $n > 1$. More generally, Markov sources of memory K have $\mu_n = 0$ as soon as $n\,K$ which is lower than $\psi(n)$ which exponentially decays but is non-zero.

Our main results is summarized next. It asserts that the SPM predictor is asymptotically optimal for δ-discriminant mixing sources. Extension to non-discriminant sources is possible and will be discussed in the final version of the paper.

Theorem 1 *For all $\alpha > 1/2$ and $\delta > 0$ the pattern matching predictor is asymptotically optimal for δ-discriminant mixing sources. More precisely, for any $\varepsilon > 0$ and large n*

$$\Pr\{N(X_{n+1}) = a|X_1,\ldots,X_n\} - \Pr\{\hat{X}_{n+1} = a|X_1,\ldots,X_n\} = \mu_n + O(n^{-\varepsilon}) \quad (3)$$

for all $a \in \mathcal{A}$.

Remark. Our proof, presented in the next section, shows that the optimality of the SPM predictor can be extended to a larger class of predictors. Namely, instead of predicting only symbol x_{n+1} we can predict the next m symbols x_{n+1}^{n+m} as long as $m = O(1)$ (and even this can be relaxed). This modified algorithm works in a similar manner except that we consider m marked positions after each marker, and apply the majority rule to the sampled sequence built over the modified alphabet of cardinality V^m.

We did perform some experiments on DNA and protein sequences using the SPM predictor. We provide a detailed discussion in the final version of the paper while here we only present some of our conclusions.

Proteins are sequences of amino acids over an alphabet of size twenty, while DNA sequences are built over four bases. The following two experiments summarize our findings:

- We analyzed the protein *human adenovirus* of length 807 with relative frequencies of amino acids vary from 0.7% to 11%. Over 100 predictions were done with the relative prediction success of 22%. This is a very good score. When we reduced the alphabet to ten (by lumping similar amino acids), the relative frequencies vary from 0.7% to 17%. The prediction success was 23%. Finally, we reduced the alphabet to a binary alphabet (i.e., polar/nonpolar alphabet) with relative frequencies 43% and 57%. This time we predicted three symbols at a time with the prediction success of 54%.

- We considered a DNA sequence of length 10603 with relative frequencies vary from 24% to 26%. The prediction success was 29%.

3 Proof of the Main Result

We prove here are main results. We start with a memoryless source and later extend it to mixing sources.

3.1 Memoryless Source

We start with some technical lemmas.

Lemma 1 *The length D_n of the maximal suffix for a memoryless source is $\frac{\log n}{h}$ with high probability for $n \to \infty$, where h is the entropy of the alphabet ($h = -\sum_i p_i \log p_i$). More precisely,*

$$\lim_{n\to\infty} \Pr\left\{(1-\varepsilon)\frac{\log n}{h} < D_n < (1+\varepsilon)\frac{\log n}{h}\right\} = 1 - O\left(\frac{\log n}{n^{\varepsilon}}\right)$$

for any $\varepsilon > 0$.

Proof. Follows from Szpankowski [17]. This is also true for more general sources (e.g., mixing sources). ■

The next lemma is at the heart of our proof.

Lemma 2 *For all ε' there is $\varepsilon > 0$ such that the probability that a string X_1^n contains two markers of length greater than $\alpha\frac{\log n}{h}$ that are separated by less than $\varepsilon \log n$ symbols is $O(n^{1-2\alpha+\varepsilon'})$ for $\alpha > \frac{1}{2}$.*

Proof. We shall use the Asymptotic Equipartition Property (AEP) (cf. [2, 18]). It asserts that for any given ε, the set of all strings of length n can be partition into the set of good states $\mathcal{G}_n$ and the set of bad states $\mathcal{B}_n$ such that the probability of being in the bad states tends to zero when $n \to \infty$, while for strings $x_1^n \in \mathcal{G}_n$ the probability $P(x_1^n) = \Pr\{X_1^n = x_1^n\}$ for fixed n is between $2^{-h(1+\varepsilon)n}$ and $2^{-h(1-\varepsilon)n}$. For sources satisfying the so called *Blowing-up Property* Marton and Shields [7] proved that the convergence rate in the AEP is exponential, that is, $P(\mathcal{B}_n)$ converges exponentially fast to zero for such processes. In passing, we mention that Marton and Shields also shown in [7] that aperiodic Markov sources, finite-state sources, and m-dependent processes satisfy the blowing-up property.

We also define for a given $\varepsilon > 0$, the ε-overlap set $\mathcal{O}_n$ that consists of strings of length n that overlap with themselves on more than εn. We will use the fact that for all $\varepsilon > 0$ the probability $P(\mathcal{O}_n)$ tends to zero exponentially fast.

Let now $k = \lceil \alpha D_n \rceil$ such that $D_n \geq (1-\varepsilon)\frac{1}{h}\log n$. We also set $d = \lceil \varepsilon \log n \rceil$. We investigate the probabilities of the following two events:

- $\mathcal{E}_1$: A marker is at distance smaller than d from the suffix X_{n-k}^n of length k;
- $\mathcal{E}_2$: Two markers are within distance smaller than d.

Let $w_k \in \mathcal{A}^k$ be a word of length k. The probability $P(\mathcal{E}_1)$ of $\mathcal{E}_1$ is equal to

$$P(\mathcal{E}_1) = \sum_{w_k \in \mathcal{A}^k} \Pr\{\exists_{0<j\leq k+d} : X_{n-k-i}^{n-i} = X_{n-k}^{n} = w_k\}.$$

Then, using the overlap condition, we arrive at

$$P(\mathcal{E}_1) \leq P(\mathcal{O}_k) + \sum_{w_k \notin \mathcal{O}_k(\varepsilon)} \Pr\{\exists_{i\leq 2d} : X_{n-2k+2d-i}^{n-k+d-i} = X_{n-k+d}^{n}, X_{n-k}^{n} = w_k\}. \quad (4)$$

By the AEP property we have

$$\begin{aligned} P(\mathcal{E}_1) &\leq P(\mathcal{O}_k) + P(\mathcal{B}_{k-d}) \\ &\quad + \sum_{w_{k-d} \in \mathcal{G}_{k-d}} \Pr\{\exists_{i\leq 2d} : X_{n-2k+2d-i}^{n-k+d-i} = X_{n-k+d}^{n} = w_{k-d}\} \\ &\leq P(\mathcal{O}_k) + P(\mathcal{B}_{k-d}) \\ &\quad + \sum_{w_{k-d} \in \mathcal{G}_{k-d}} \sum_{i\leq 2d} \Pr\{X_{n-2k+2d-i}^{n-k+d-i} = X_{n-k+d}^{n} = w_{k-d}\}. \end{aligned}$$

For a word $w_{k-d} \in \mathcal{G}_{k-d}$ the probability $P(w_k) \leq 2^{-(1-\varepsilon)(k-d)h}$, and therefore

$$\begin{aligned} P(\mathcal{E}_1) &\leq P(\mathcal{O}_k) + P(\mathcal{B}_{k-d}) + \sum_{w_{k-d} \in \mathcal{G}_{k-d}} dP(w_{k-d})^2 \\ &\leq P(\mathcal{O}_k) + P(\mathcal{B}_{k-d}) + 2\varepsilon \log n 2^{-h(1-\varepsilon)(k-d)} \\ &\leq P(\mathcal{O}_k) + P(\mathcal{B}_{k-d}) + n^{-\alpha+O(\varepsilon)} \end{aligned}$$

where the last line follows after setting $k = \alpha\frac{1}{h}\log n$ and $d = \varepsilon \log n$.

The probability $P(\mathcal{E}_2)$ of $\mathcal{E}_2$, formally satisfies the following identity

$$P(\mathcal{E}_2) = \sum_{w_k \in \mathcal{A}^k} \Pr\{\exists_{m<n}\exists_{0<j\leq k+d} : \ X_{m-k-j}^{m-j} = X_{m-k}^{m} = X_{n-k}^{n} = w_k\}. \quad (5)$$

Using the same arguments as above we conclude that

$$\begin{aligned} P(\mathcal{E}_2) &\leq P(\mathcal{O}_k) + P(\mathcal{B}_{k-d}) + \sum_{w_{k-d} \in \mathcal{G}_{k-d}} ndP(w_{k-d})^3 \\ &\leq P(\mathcal{O}_k) + P(\mathcal{B}_{k-d}) + 2n\varepsilon \log n 2^{-2h(1-\varepsilon)(k-d)} \\ &\leq P(\mathcal{O}_k) + P(\mathcal{B}_{k-d}) + n^{1-2\alpha+O(\varepsilon)} \end{aligned}$$

where the last line is a consequence of the values of k and d. The proof follows since for memoryless sources $P(\mathcal{B}_k)$ decays exponentially fast with k. ∎

In passing we observe that the main arguments of the above proof remain valid for mixing sources.

Now, we show that the number L_n of markers is $O(n^{1-\alpha})$ with high probability.

Lemma 3 *The number L_n of markers is such that there exists $\varepsilon > 0$ and $\varepsilon' > 0$ so that*

$$\Pr\{L_n < n^{1-\alpha-\varepsilon'}\} = O(n^{\alpha-1+\varepsilon})$$

for large n.

Proof. Let $\widetilde{L}_n$ be the number of non-overlapping markers that may occur only at positions that are an integer multiply of k. Let also $w_k \in \mathcal{A}^k$ be the suffix of X_1^n of length k. Observe that $\widetilde{L}_n \leq L_n$ and the number $\widetilde{L}_n$ of nonoverlapping markers is binomially distributed $B(n/k, P(w_k))$ with mean $P(w_k)n/k$ and variance $(1 - P(w_k))P(w_k)n/k$. Given that $w_k \in \mathcal{G}_k$ such that $P(w_k) \geq n^{-(1+\varepsilon)\alpha}$, we use Chebyshev's inequality to yield (for properly chosen ε')

$$\begin{aligned}
\Pr\{L_n \leq n^{1-(1-\varepsilon')\alpha}(1-\varepsilon')\} &\leq \Pr\{\widetilde{L}_n \leq \frac{n^{1-(1-\varepsilon)\alpha}}{k}(1-\varepsilon)\} \\
&\leq \Pr\{\widetilde{L}_n \leq \mathbf{E}[\widetilde{L}_n] - \sqrt{n\varepsilon \mathbf{Var}(\widetilde{L}_n)}\} \\
&\leq \frac{k}{\varepsilon n^{1-(1+\varepsilon)\alpha}}.
\end{aligned}$$

Since $P(w_k \notin \mathcal{G}_k) = O(n^{-\varepsilon})$ for memoryless sources, the proof is complete. ∎

To proceed we need to introduce the important notion of *stable strings* and then *paired strings.*

Definition 2 *A string X_1^n is stable if a modification of any sampled symbol does not change the positions of all markers in the new string.*

We observe that a string is stable with high probability as stated below.

Lemma 4 *A string is stable with probability $1 - O(n^{-\varepsilon})$.*

Proof : By changing a symbol on a marked position we either create new markers that overlap with the previous marker or delete existing markers that were overlapping with the previous marker. From Lemma 2 we know that a string of length n contains overlapping markers with probability $O(n^{-\varepsilon})$, thus a string is stable with probability $1 - O(n^{-\varepsilon})$. ∎

Using stable strings, we define now an important notion of paired strings that are used to define an orbit of strings.

Definition 3 *A string X_1^n is paired to a string $\widetilde{X}_1^n$ if*

- *X_1^n and $\widetilde{X}_1^n$ are both stable strings;*
- *X_1^n and $\widetilde{X}_1^n$ have their markers at the same positions;*
- *X_1^n and $\widetilde{X}_1^n$ match on every positions except the sampled symbols.*

Definition 4 *An orbit for X_1^n is the set of the strings paired to X_1^n.*

Let $L_n(\mathcal{F})$ be the number of markers in a string $X_1^n \in \mathcal{F}$. Clearly, the cardinality of the orbit $\mathcal{F}$ is $V^{L_n(\mathcal{F})}$ where $L_n(\mathcal{F})$ is the number of markers in $X_1^n \in \mathcal{F}$.

The next lemma summarizes our knowledge about the sampled sequence.

Lemma 5 *Let $\mathcal{F}$ be an orbit of size $V^{L_n(\mathcal{F})}$. Under the condition that the string X_1^n belongs to $\mathcal{F}$, the sampled sequence is i.i.d. with probability p_i for symbol $i \in \mathcal{A}$.*

Proof. Since X_1^n is generated by a memoryless source, the distribution of the sampled sequence is just equivalent to the joint distribution of $L_n(\mathcal{F})$ symbols at fixed marked positions in X_1^n. More precisely, let $i_1, i_2, \ldots, i_\ell$ be the marked positions, ℓ being a short-hand notation for $L_n(\mathcal{F})$. The sampled sequence is $X_{i_1}X_{i_2}\ldots X_{i_\ell}$. Observe that all the other values X_j for $j \notin \{i_1, \ldots, i_\ell\}$ are fixed when X_1^n belongs to the orbit $\mathcal{F}$. We denote by $X(\mathcal{F})_1^{i_1-1}$ the fixed substring $X_1^{i_1-1}$, $X(\mathcal{F})_{i_k+1}^{i_{k+1}-1}$ the fixed substring $X_{i_k+1}^{i_{k+1}-1}$, and $X(\mathcal{F})_{i_\ell+1}^n$ the fixed substring $X_{i_\ell+1}^n$ when $X_1^n \in \mathcal{F}$. We have

$$\Pr\{X_{i_1}\ldots X_{i_\ell} = x_1\ldots x_\ell | X_1^n \in \mathcal{F}\} = \frac{P(X(\mathcal{F})_1^{i_1}x_1\ldots X(\mathcal{F})_{i_{\ell-1}+1}^{i_\ell-1}x_\ell X_{i_\ell+1}^n)}{P(X_1^n \in \mathcal{F})} . \quad (6)$$

Since X_1^n is generated by a memoryless source source, we have

$$\Pr\{X(\mathcal{F})_1^{i_1}x_1\ldots x_\ell X(\mathcal{F})_{i_\ell+1}^n\} = P(X(\mathcal{F})_1^{i_1})P(x_1)\cdots P(x_\ell)P(X(\mathcal{F})_{i_\ell+1}^n`w) \quad (7)$$

Furthermore, $P(X_1^n \in \mathcal{F}) = P(X(\mathcal{F})_1^{i_1}\cdots P(X(\mathcal{F})_1^{i_1})$ thus

$$P(X_{i_1}\ldots X_{i_\ell} = x_1\ldots x_\ell | X_1^n \in \mathcal{F}) = \frac{P(x_1)\cdots P(x_\ell)P(X_1^n \in \mathcal{F})}{P(X_1^n \in \mathcal{F})} \quad (8)$$

that yields the desired result. ■

To complete the proof we need a simple fact that will allow us to conclude that the sampled sequence contains $i_{\max}$ with high probability.

Lemma 6 *The probability that $\hat{X}_n = i_{\max}$ under the condition that X_1^n belongs to an $\mathcal{F}$ is $1 - O(\beta^{L_n(\mathcal{F})})$ for some $\beta > 1$.*

Proof. It suffices to show that $i_{\max}$ occurs in the sampled sequence with probability $1-\beta^\ell$ for $L_n(\mathcal{F}) = \ell$. But the sampled sequence is i.i.d. of length $O(n^{1-\alpha})$, thus any symbol $i \in \mathcal{A}$ occurs with probability p_i and the convergence is exponential. In particular, $i_{\max}$ occurs with probability $p_{\max} > p_i$ for all $i \neq i_{\max}$. By a standard large deviation estimate (e.g., Chernoff's bound), we prove the lemma. ■

Now we are ready to **prove Theorem 1**. The unconditional probability that the predictor does not predict $i_{\max}$ satisfies the inequality:

$$\begin{aligned}
\Pr\{\hat{X}_n \neq i_{\max}\} &\leq \Pr\{X_1^n \text{ is not stable}\} \\
&\quad +\Pr\{X_1^n \text{ is paired and } \hat{X}_n \neq i_{\max}\} \\
&\leq O(n^{-\varepsilon}) + \sum_{\mathcal{F}} P(\mathcal{F})O(\beta^{L_n(\mathcal{F})}) \\
&\leq O(n^{-\varepsilon}) + O(\mathbf{E}[\beta^{L_n}])
\end{aligned}$$

where the second line follows from Lemma 4. But by Lemma 3

$$\mathbf{E}[\beta^{L_n}] \leq \Pr\{L_n < n^{1-\alpha-\varepsilon}\} + \beta^{n^{1-\alpha-\varepsilon}} = O(n^{\alpha-1-\varepsilon'}).$$

This completes the proof of Theorem 1 for memoryless sources.

3.2 Mixing Sources

In this conference version of the paper, we provide only a sketch of the proof for the mixing model.

In the sequel we set $k = \lceil \gamma \log n \rceil$ for some $\gamma > 0$. We consider the distribution of X_{n+1} under the condition X_{n-k}^n. If w_k is a string of length k, we denote $i_n(w_k) = \arg\max_i\{P(X_{n+1} = i | X_{n-k}^n = w_k)\}$.

We will use the following easy and technical lemma.

Lemma 7 *We consider a δ-discriminant distribution over an alphabet of size V and a string of length ℓ from a memoryless source based on this distribution. $i_{\max}$ is the most likely symbol. For all $\delta > 0$ there exists a non-negative $\omega < 1$ such that the probability that the most frequent symbol in the string is equal to $i_{\max}$ is $O(\omega^\ell)$.*

In order to prove Theorem 1 for mixing sources, we use similar definitions and lemmas as for the memoryless sources. However, we need some important modifications. We start with a generalization of stable and paired strings.

Definition 5 *A string X_1^n is k-stable if a modification of the k marked symbols after a marker does not change the positions of markers in the new string.*

Definition 6 *A string X_1^n is k-paired to a string $\widetilde{X}_1^n$ if*

- *X and $\widetilde{X}$ are both k-stable strings;*
- *X and $\widetilde{X}$ have their markers at the same positions;*
- *X and $\widetilde{X}$ match on every positions except the marked symbols.*

Lemma 1 to Lemma 4 of previous section are easy to extend to mixing model. In Lemma 3 we set $\gamma < \frac{\alpha}{h}$. We now rephrase Lemma 5. We consider an orbit $\mathcal{F}$ such that all markers are of lengths greater than $k = \lceil \gamma \log n \rceil$, therefore each marked symbol is proceeded by a full copy of a suffix of X_1^n of length k. This will allow us to reduce the model to the memoryless one. This is stated below in a rather vague form (details will be provided in the final version).

Lemma 8 *Under the condition that $X_1^n \in \mathcal{F}$, the probability distribution of the marked sequence is within the factor $(1 \pm O(\psi(k))^{L_n(\mathcal{F})}$ form the memoryless model (i.e., the probability of any event in the mixing model is equal the probability of the same event under the memoryless model modulo the multiplicative factor $(1 \pm O(\psi(k))^{L_n(\mathcal{F})})$.*

Using above and Lemma 7 we have

$$\begin{aligned}
\Pr\{\hat{X}_n \neq i_{\max}\} &\leq \Pr\{X_n \text{ is not } k\text{-stable or is not } \delta\text{-discriminant}\} \\
&+ \Pr\{X_n \text{ is } k\text{-paired and } \delta\text{-discriminant and } \hat{X}_n \neq i_{\max}(X_n)\} \\
&\leq O(n^{-\varepsilon}) + \mu_n + \sum_{\mathcal{F}} P(\mathcal{F}) O((1+\psi(L_n(\mathcal{F})))\omega)^{L_n(\mathcal{F})}) \\
&\leq O(n^{-\varepsilon}) + \mu_n + O(\mathbf{E}[((1+\psi(k))\omega)^{L_n}]) = O(n^{-\varepsilon})) + \mu_n.
\end{aligned}$$

This gives a sketch of the proof of Theorem 1 for mixing sources.

References

[1] P. Algoet, Universal Schemes for Prediction, Gambling and Portfolio Selection, *Ann. Prob.*, 20, 901–941, 1992.

[2] T.M. Cover and J.A. Thomas, *Elements of Information Theory*, John Wiley&Sons, New York (1991).

[3] A. Ehrenfeucht and J. Mycielski, A Pseudorandom Sequence — How Random Is It? *Amer. Math. Monthly*, 99, 373-375, 1992

[4] M. Feder, N. Merhav, and M. Gutman, Universal Prediction of Individual Sequences, *IEEE Trans. Information Theory*, 38, 1258–1270, 1992.

[5] P. Jacquet, Average Case Analysis of Pattern Matching Predictor, INRIA TR, 1999.

[6] J. Kieffer, Prediction and Information Theory, preprint, 1998.

[7] K. Marton and P. Shields, The Positive-Divergence and Blowing-up Properties, *Israel J. Math*, 80, 331-348 (1994).

[8] N. Merhav, M. Feder, and M. Gutman, Some Properties of Sequential Predictors for Binary Markov Sources, *IEEE Trans. Information Theory*, 39, 887–892, 1993.

[9] N. Merhav, M. Feder, Universal Prediction, *IEEE Trans. Information Theory*, 44, 2124-2147, 1998.

[10] G. Moravi, S. Yakowitz, and P. Algoet, Weak Convergent Nonparametric Forecasting of Stationary Time Series, *IEEE Trans. Information Theory*,

[11] J. Rissanen, Complexity of Strings in the Class of Markov Sources, *IEEE Trans. Information Theory*, 30, 526–532, 1984.

[12] J. Rissanen, Universal Coding, Information, Prediction, and Estimation, *IEEE Trans. Information Theory*, 30, 629–636, 1984.

[13] B. Ryabko, Twice-Universal Coding, *Problems of Information Transmission*, 173–177, 1984.

[14] B. Ryabko, Prediction of Random Sequences and Universal Coding, *Problems of Information Transmission*, 24, 3–14, 1988.

[15] B. Ryabko, The Complexity and Effectiveness of Prediction Algorithms, *J. Complexity*, 10, 281–295, 1994.

[16] C. Shannon, Prediction and Entropy of Printed English, *Bell System Tech. J.*, 30, 50–64, 1951.

[17] W. Szpankowski, A Generalized Suffix Tree and Its (Un)Expected Asymptotic Behaviors, *SIAM J. Compt.*, 22, 1176–1198, 1993.

[18] W. Szpankowski, *Average Case Analysis of Algorithms on Sequences*, John Wiley & Sons, New York 2000 (to appear).

[19] J. Vitter and P. Krishnan, Optimal Prefetching via Data Compression, *J. ACM*, 43, 771–793, 1996.

Part II

Combinatorics and Random Generation

Trends in Mathematics, © 2000 Birkhäuser Verlag Basel/Switzerland

A bijective proof for the arborescent form of the multivariable Lagrange inversion formula

MICHEL BOUSQUET *LaCIM, Université du Québec à Montréal, 201 Avenue Président-Kennedy, Montréal (Québec), Canada.* bousq2@math.uqam.ca
CEDRIC CHAUVE *LaBRI, Université Bordeaux I, 351 Cours de la Libération, 33405 Talence, France.* chauve@labri.u-bordeaux.fr
GILBERT LABELLE *LaCIM, Université du Québec à Montréal, 201 Avenue Président-Kennedy, Montréal (Québec), Canada.* gilbert@math.uqam.ca
PIERRE LEROUX *LaCIM, Université du Québec à Montréal, 201 Avenue Président-Kennedy, Montréal (Québec), Canada.* leroux@math.uqam.ca

Abstract. *In [12], Goulden and Kulkarni propose a bijective proof of a form of the Lagrange-Good's multivariable inversion formula [9]. In this paper, we propose a new bijective proof of this formula, simpler than Goulden and Kulkarni's proof and we illustrate the interest of this proof in the enumeration of multisort rooted trees such that the edges partition is given.*

1 Introduction

Lagrange's inversion formula for formal power series is a classical tool in enumerative combinatorics (see for example [11, 2]). In 1960, Good [9] proposed an extension that could handle the case of multivariable formal power series, called the Lagrange-Good's formula, and is well adapted for the enumeration of multisort (or multicolored) structures. It has been extensively used by Goulden and Jackson [13, 10, 11]. Let $d \geq 1$ be a fixed integer, $\mathbf{x} = (x_1, \ldots, x_d)$ be a vector of formal variables, $\mathbf{n} = (n_1, \ldots, n_d)$ a vector of integers and let $\mathbf{x}^\mathbf{n} = x_1^{n_1} \cdots x_d^{n_d}$. For a given multivariable formal power series $h(\mathbf{x})$, the coefficient of $\mathbf{x}^\mathbf{n}$ in $h(\mathbf{x})$ is denoted by $[\mathbf{x}^\mathbf{n}]\mathbf{h}(\mathbf{x})$. There are several forms of Lagrange-Good's formula, and we give the most classical in the following theorem.

Theorem 1 *Let $F(\mathbf{x})$, $R_1(\mathbf{x}), \ldots, R_d(\mathbf{x})$ be $d+1$ multivariable formal power series such that $R_i(\mathbf{0}) \neq 0$ for $i \in [d]$, $\mathbf{0} = (0, \ldots, 0)$ and $A_i(\mathbf{t}) = t_i R_i(\mathbf{A}(\mathbf{t}))$ for $i \in [d]$. Then we have the implicit form of Lagrange-Good's formula*

$$[\mathbf{t}^\mathbf{n}] \frac{F(\mathbf{A}(\mathbf{t}))}{\det\left(\delta_{i,j} - t_i \dfrac{\partial R_i(\mathbf{t})}{\partial t_j}\right)_{d \times d}} = [\mathbf{x}^\mathbf{n}] F(\mathbf{x}) \mathbf{R}^\mathbf{n}(\mathbf{x}), \qquad (1)$$

where $\delta_{i,j} = 1$ if $i = j$ and 0 otherwise, and the explicit form

$$[\mathbf{t}^\mathbf{n}] F(\mathbf{A}(\mathbf{t})) = [\mathbf{x}^\mathbf{n}] F(\mathbf{x}) \mathbf{R}^\mathbf{n}(\mathbf{x}) \det\left(\delta_{i,j} - \frac{x_i}{R_i(\mathbf{x})} \frac{\partial R_i(\mathbf{x})}{\partial x_j}\right)_{d \times d}. \qquad (2)$$

In [8] Gessel gives a bijective proof of formula (1). This proof is also given in the terms of the theory of combinatorial species in [2, Section 3.2], where it appears as a natural generalization of the proof of the Lagrange's inversion formula in one variable due to Labelle [15]. In [7], Ehrenborg and Méndez give a combinatorial proof of (2) in the case of formal power series in an infinity of variables, in which the determinant in the right-hand-side of (2) is computed using an involution. The first direct bijective proof is due to Goulden and Kulkarni [12]. In fact, Goulden and Kulkarni show that formula (2) is equivalent to a form of Lagrange-Good's inversion which can be called *arborescent*, based on the notion of derivative according to a rooted tree (which can also be called *arborescence*).

Definition 1 Let $\mathcal{G}$ be a directed graph having $S = \{1, \ldots, d\}$ as set of vertices and A as set of (directed) edges, $\mathbf{x}$ and $\mathbf{f}(\mathbf{x})$ two vectors of formal variables and formal power series indexed by S. We define the derivative of $\mathbf{f}(\mathbf{x})$ according to $\mathcal{G}$ by

$$\frac{\partial \mathbf{f}(\mathbf{x})}{\partial \mathcal{G}} = \prod_{j \in S} \left\{ \left(\prod_{(i,j) \in A} \frac{\partial}{\partial x_i} \right) f_j(\mathbf{x}) \right\},$$

where (i, j) stands for an edge directed from the vertex i to the vertex j.

Theorem 2 *Let $F(\mathbf{x})$, $R_1(\mathbf{x}), \ldots, R_d(\mathbf{x})$ be $d+1$ formal power series such that $R_i(\mathbf{0}) \neq 0$ for $i \in [d]$ and $A_i(\mathbf{t}) = t_i R_i(\mathbf{A}(\mathbf{t}))$ for $i \in [d]$. Then we have*

$$[\mathbf{t}^{\mathbf{n}}]F(\mathbf{A}(\mathbf{t})) = \left(\prod_{i=1}^{d} \frac{1}{n_i} \right) [\mathbf{x}^{\mathbf{n}-\mathbf{1}}] \sum_{\mathcal{T}} \frac{\partial(F, R_1^{n_1}, \ldots, R_d^{n_d})}{\partial \mathcal{T}}, \tag{3}$$

where $\mathbf{n} - \mathbf{1} = (n_1 - 1, \ldots, n_d - 1)$, the sum being taken over all rooted trees $\mathcal{T}$ having $\{0, 1, \ldots, d\}$ as set of vertices, rooted in 0, and in which all edges are directed towards 0.

Goulden and Kulkarni prove this formula by establishing a bijection between rooted trees on d sorts (or d colors) and some endofunctions on d sorts, which is an extension of the classical bijection between "vertebrates"and endofunctions [14, 15]. This arborescent form of Lagrange-Good's inversion formula has been independently discovered by Bender and Richmond [1], who notice that the determinant appearing in the right-hand-side of (2) can cause problems if one wants to obtain asymptotical informations about the coefficients in the A_i's. On the other hand, the arborescent formulation (3) contains only positive terms and therefore is better suited for an asymptotical analysis.

The interest of bijective proofs for Lagrange-Good's inversion formula is at least twofold. First, they give a better understanding of the combinatorial signification of the enumerative formulas that one obtains when applying them. Second, a bijection usually leads to an algorithm for random or exhaustive generation. This then justifies the research for simple and efficient proofs of Lagrange-Good's formula (at least from an algorithmic point of view).

In this paper, we first propose a new bijective proof of formula (3). The bijection that we describe here is based on the same principle as Goulden and Kulkarni's bijection [12], namely the manipulation of multisort rooted trees and endofunctions, and has similar properties from an algorithmic point of view. However, it has two advantages: it is easier to put into application and the proof of its validity is immediate. In a second step, we show that this bijection, combined with the Matrix-Tree Theorem, leads to a combinatorial interpretation of various enumeration formulas for rooted trees with a given edges partition, and also, a version of Lagrange-Good's formula due to Goulden and Jackson [13] (see also [11, Section1.2.13]).

2 A bijective proof of formula (3)

In order to prove (3), we work in the context of the theory of species, and more precisely, of multisort species (see [2, 14, 15]). In a first step, we slightly modify formula (3) in order to take into account the fact that we are manipulating labelled structures and exponential generating functions. If we denote $\mathbf{n}! = \prod_{i=1}^{d} n_i!$, then we obtain the following formula, which is equivalent to (3),

$$\mathbf{n}![\mathbf{t}^{\mathbf{n}}]F(\mathbf{A}(\mathbf{t})) = (\mathbf{n}-\mathbf{1})![\mathbf{x}^{\mathbf{n}-\mathbf{1}}]\sum_{\mathcal{T}} \frac{\partial(F, R_1^{n_1}, \dots, R_d^{n_d})}{\partial \mathcal{T}}. \tag{4}$$

2.1 Multisort rooted trees and endofunctions.

We now give a combinatorial interpretation of both sides of this identity. First, we introduce for $i \in [d]$, the species X_i of singletons of sort i, a vector $\mathbf{R} = (R_1, \dots, R_d)$ of species on d sorts having generating exponential functions $(R_1(\mathbf{x}), \dots, R_d(\mathbf{x}))$, and a species F on d sorts having generating exponential function $F(\mathbf{x})$. Using F and $\mathbf{R}$ we construct two new species $F(\mathbf{A_R})$ and $End_{F,\mathbf{R}}^{\mathcal{R}}$.

Definition 2 We denote by $A_{\mathbf{R},i}$ the species of $\mathbf{R}$-enriched rooted trees (on d sorts): for $j \in [d]$, there is an R_j-structure on the fiber (the set of sons) of a vertex of sort j, and the root is of sort i. These species hence verify the functional equations $A_{\mathbf{R},i} = X_i R_i(\mathbf{A_R})$, which is the translation, in terms of species, of the equations $A_i(\mathbf{t}) = t_i R_i(\mathbf{A}(\mathbf{t}))$ of the Lagrange-Good's formula.

The species $F(\mathbf{A_R})$ of $(F, \mathbf{R})$-enriched rooted trees is the species of F-assemblies of $A_{\mathbf{R},i}$-structures. The set of $F(\mathbf{A_R})$-structures having n_i elements of sort i, for $i \in [d]$, is denoted by $F(\mathbf{A_R})[\mathbf{n}]$.

In the rest of this paper, a $(F, \mathbf{R})$-enriched rooted tree over $[\mathbf{n}]$ will be represented as a rooted tree on $\{0\} \cup [\mathbf{n}]$, rooted at 0, such that there is a F-structure on the fiber of 0. We denote by k_i the kth element of sort i. Figure 1 gives an example of a $(F, \mathbf{R})$-enriched rooted tree over $[\mathbf{n}] = [4, 3, 1]$. By convention, all edges in a rooted tree are directed towards the root and the enrichment on the fiber of an element (which is determinated by its sort) is represented by an arc of a circle. For example the fiber of 2_1 is the set $\{1_1, 1_2\}$ and is embedded into a R_1-structure, while the fiber of 0 is embedded into a F-structure.

By recalling that, for a species G on d sorts, $\mathbf{n}![\mathbf{x}^{\mathbf{n}}]G(\mathbf{x}) = |G[\mathbf{n}]|$, we can interpret the left-hand-side of (4) in the following way.

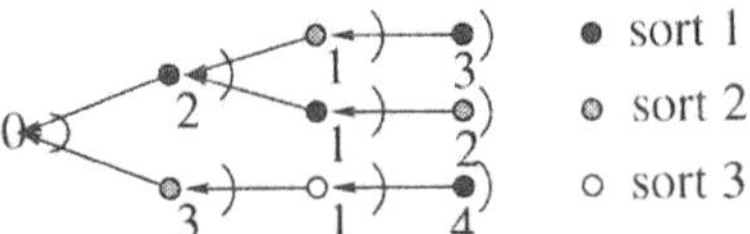

Figure 1: A $F(\mathbf{A_R})$-structure.

Lemma 1

$$\mathbf{n}![\mathbf{t^n}]F(\mathbf{A}(\mathbf{t})) = |F(\mathbf{A_R})[\mathbf{n}]|\,.$$

Definition 3 A $(F,\mathbf{R})$-*enriched partial endofunction* over $[\mathbf{n}]$ is a function f of $[\mathbf{n}]$ into $\{0\} \cup [\mathbf{n}]$ in which each fiber $f^{-1}(u)$ is embedded into a R_i-structure if u is of sort i ($i \in [d]$) and into a F-structure if $u = 0$. We denote by $End^{\mathcal{P}}_{F,\mathbf{R}}[\mathbf{n}]$ the set of $(F,\mathbf{R})$-enriched partial endofunctions over $[\mathbf{n}]$.

Definition 4 Let $f \in End^{\mathcal{P}}_{F,\mathbf{R}}[\mathbf{n}]$. We call *graph of sorts* of f, denoted by $G(f)$, the directed graph having $[d] \cup \{0\}$ as set of vertices and having, for $i \in [d]$, a directed edge from i to $j \neq 0$ (resp. from i to 0), if and only if 1_i is in the fiber of an element of sort j (resp. in the fiber of 0).

One usually represents a partial endofunction f by a directed graph: there is a directed edge from u to v if $f(u) = v$. For example, if $d = 3$ and $[\mathbf{n}] = [4,4,3]$, the following figure represents a partial endofunction $f \in End^{\mathcal{P}}_{F,\mathbf{R}}[\mathbf{n}]$ and its related graph of sorts $G(f)$.

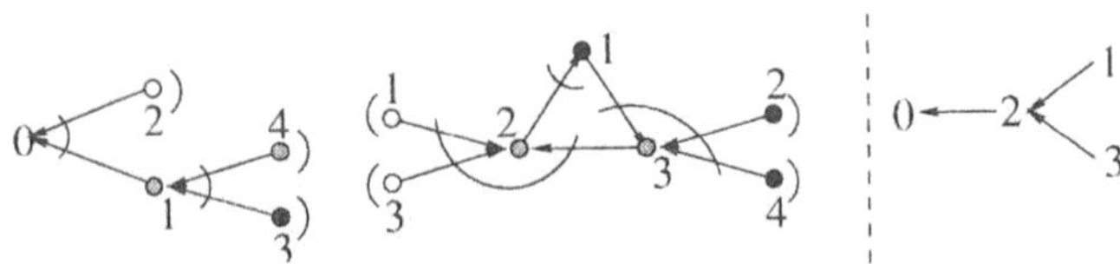

Figure 2: A $End^{\mathcal{P}}_{F,\mathbf{R}}$-structure and its graph of sorts.

Definition 5 A partial endofunction $f \in End^{\mathcal{P}}_{F,\mathbf{R}}[\mathbf{n}]$ is said to be *restricted* if $G(f)$ is a rooted tree, rooted at 0, and in which all edges are directed towards the root. We denote by $End^{\mathcal{R}}_{F,\mathbf{R}}[\mathbf{n}]$ the set of $(F,\mathbf{R})$-enriched restricted partial endofunctions over $[\mathbf{n}]$.

The partial endofunction given in Figure 2 is therefore a restricted partial endofunction. From now on, unless stated otherwise, we use the terms F-rooted tree for $(F,\mathbf{R})$-enriched rooted tree and endofunction for $(F,\mathbf{R})$-enriched restricted partial endofunction.

The following lemma gives a combinatorial interpretation of the right-hand-side of (4) in terms of endofunctions (see [12, 5] for a proof of this lemma).

Lemma 2

$$(\mathbf{n}-\mathbf{1})![\mathbf{x}^{\mathbf{n}-\mathbf{1}}]\sum_{\mathcal{T}} \frac{\partial(F,R_1^{n_1},\ldots,R_d^{n_d})}{\partial \mathcal{T}} = \left|End_{F,\mathbf{R}}^{\mathcal{R}}[\mathbf{n}]\right|,$$

the sum being taken over all rooted trees $\mathcal{T}$ having $\{0,1,\ldots,d\}$ as set of vertices, rooted in 0, and in which all edges are directed towards 0.

We deduce from Lemmas 1 and 2 that in order to prove formula (4) it is sufficient to show that

$$F(\mathbf{A_R})[\mathbf{n}] \simeq End_{F,\mathbf{R}}^{\mathcal{R}}[\mathbf{n}],$$

which can be considered as an expression of the arborescent form of Lagrange-Good's inversion formula in terms of species. In the rest of this section, devoted to a bijective proof of this combinatorial identity, which will be illustrated using an example, we call minimal elements of $\{0\}\cup[\mathbf{n}]$ the set $Min([\mathbf{n}]) = \{1_1, 1_2, \ldots, 1_d\}$.

2.2 From F-rooted trees to endofunctions.

Let $\mathcal{A} \in F(\mathbf{A_R})[\mathbf{n}]$ be a F-rooted tree. We call *skeleton* of $\mathcal{A}$, denoted by $S(\mathcal{A})$, the set of paths starting from an element of $Min([\mathbf{n}])$ and ending at the root 0. $S(\mathcal{A})$ is then a rooted tree having 0 as root and in which all the leaves are in the set $\{1_1, 1_2, \ldots, 1_d\}$.

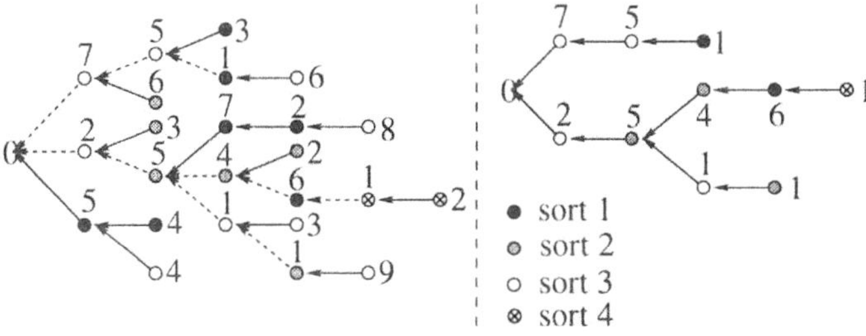

Figure 3: A F-rooted tree and its skeleton.

The first step in the transformation of $\mathcal{A}$ into an endofunction $f_{\mathcal{A}}$ belonging to $End_{F,\mathbf{R}}^{\mathcal{R}}[\mathbf{n}]$ consists in isolating $S(\mathcal{A})$. In a second step, for each leaf 1_i in $S(\mathcal{A})$ (covered according to the increasing order of their sort), we denote by C_i the path going from 1_i to the closest ancestor of 1_i belonging to another path C_j with $j < i$ (or to 0 if $i = 1$). In our example,

$$C_1 = 0 \leftarrow 7_3 \leftarrow 5_3 \leftarrow 1_1,$$

$$C_2 = 0 \leftarrow 2_3 \leftarrow 5_2 \leftarrow 1_3 \leftarrow 1_2,$$

$$C_4 = 5_2 \leftarrow 4_2 \leftarrow 6_1 \leftarrow 1_4.$$

In a third step, each path C_i of length l_i is transformed into a biword B_i of length $l_i - 1$ in the following way: for each element u in C_i different from 1_i, if its predecessor (on the path starting from 1_i) is the element k_j, then we perform the transformation

$$u \leftarrow k_j \longrightarrow \begin{array}{c} *_j \\ \downarrow \\ u \end{array},$$

which indicates that in the fiber of u, k_j will be replaced by an element of sort j, possibly different from k_j. We concatenate the B_i's, which results in a single biword, denoted by $B(\mathcal{A})$:

$$B(\mathcal{A}) = \begin{pmatrix} *_3 & *_3 & *_1 & *_3 & *_2 & *_3 & *_2 & *_2 & *_1 & *_4 \\ \downarrow & \downarrow & \downarrow & \downarrow & \downarrow & \downarrow & \downarrow & \downarrow & \downarrow & \downarrow \\ 0 & 7_3 & 5_3 & 0 & 2_3 & 5_2 & 1_3 & 5_2 & 4_2 & 6_1 \end{pmatrix}.$$

Finally, for $i \in [d]$, we replace the $*_i$'s in $B(\mathcal{A})$ by the elements of $S(\mathcal{A})$ of sort i taken by increasing order. We then obtain the biword $B'(\mathcal{A})$:

$$B'(\mathcal{A}) = \begin{pmatrix} 1_3 & 2_3 & 1_1 & 5_3 & 1_2 & 7_3 & 4_2 & 5_2 & 6_1 & 1_4 \\ \downarrow & \downarrow & \downarrow & \downarrow & \downarrow & \downarrow & \downarrow & \downarrow & \downarrow & \downarrow \\ 0 & 7_3 & 5_3 & 0 & 2_3 & 5_2 & 1_3 & 5_2 & 4_2 & 6_1 \end{pmatrix}.$$

From this biword, one can define an endofunction $f_{\mathcal{A}}$ over $[\mathbf{n}]$ in the following way: for each $u \in [\mathbf{n}]$, if $u \in S(\mathcal{A})$, then $f_{\mathcal{A}}(u)$ is the image of u in $B'(\mathcal{A})$, otherwise $f_{\mathcal{A}}(u)$ is the father of u in $\mathcal{A}$. In our example, we obtain the following endofunction $f_{\mathcal{A}}$.

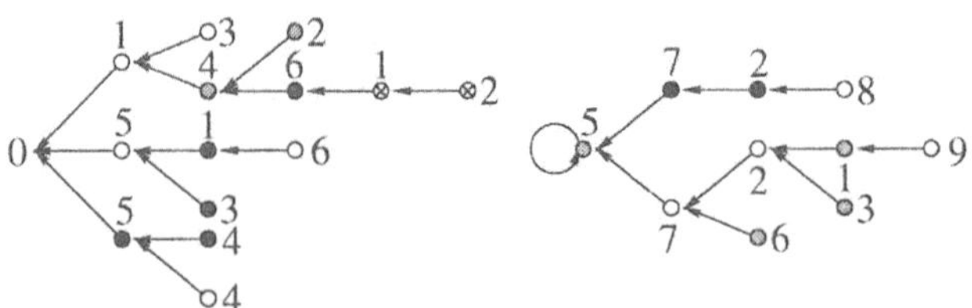

Figure 4: The endofunction $f_{\mathcal{A}}$.

It remains to show that the partial endofunction $f_{\mathcal{A}}$ is indeed restricted. We call *biletter* (u, v) an element

$$\begin{array}{c} u \\ \downarrow \\ v \end{array}$$

of $B(\mathcal{A})$ or $B'(\mathcal{A})$ (u can then be of the form $*_j$ or k_j). By construction, we can say that:

- in $B(\mathcal{A})$, at the left of a biletter $(*_i, k_j)$, there always exists a biletter $(*_j, u)$;
- in $B'(\mathcal{A})$, for a biletter $(1_i, u)$, either $u = 0$, or $u = k_j$ and in the latter case, there exists, at the left of $(1_i, u)$, a biletter $(1_j, v)$ (this can be deduced

from the previous remark and from the fact that in order to go from $B(\mathcal{A})$ to $B'(\mathcal{A})$, one replaces the $*_i$'s by the elements of $S(\mathcal{A})$ of sort i *taken in increasing order*);

- the first biletter in $B(\mathcal{A})$ is of the form $(1_i, 0)$.

We then deduce that $G(f_\mathcal{A})$ is a rooted tree, having root 0 and in which all edges are directed towards 0, which proves that $f_\mathcal{A}$ belongs to $End^{\mathcal{R}}_{\mathbf{F},\mathbf{R}}[\mathbf{n}]$.

We define the *skeleton* of $f_\mathcal{A}$, denoted by $S(f_\mathcal{A})$, in the following way: $S(f_\mathcal{A})$ includes the elements of all cycles in $f_\mathcal{A}$ and all paths starting from elements belonging to $Min([\mathbf{n}])$ and ending either at 0, or in a cycle in $f_\mathcal{A}$ ($u \in S(f_\mathcal{A})$ if there is $k \geq 1$ such that $f^k_\mathcal{A}(u) = u$ or if there are $k \geq 0$ and $i \in [d]$ such that $u = f^k_\mathcal{A}(1_i)$).

Property 1 An element $u \notin Min([\mathbf{n}])$ appears k times in $B(\mathcal{A})$ (in the bottom line of $B(\mathcal{A})$) if and only if there exists a maximal subset $\{i_1, \dots, i_k\}$ of $[d]$ such that for $j = 1, \dots, k$, $u \in C_{i_j}$.

Property 2 An element u belongs to $S(f_\mathcal{A})$ if and only if it belongs to $S(\mathcal{A})$, and its degree $d_{S(f_\mathcal{A})}(u)$ in $S(f_\mathcal{A})$ is equal to its degree $d_{S(\mathcal{A})}(u)$ in $S(\mathcal{A})$ (in particular, the leaves of $S(f_\mathcal{A})$ are the leaves of $S(\mathcal{A})$).

2.3 From endofunctions to F-rooted trees.

Let $f \in End^{\mathcal{R}}_{\mathbf{F},\mathbf{R}}[\mathbf{n}]$ be an endofunction. The first step of the transformation of f into a F-rooted tree $\mathcal{A}_f$ in $F(\mathbf{A_R})[\mathbf{n}]$ consists in isolating its skeleton $S(f)$. Let f be the endofunction $f_\mathcal{A}$ in Figure 4.

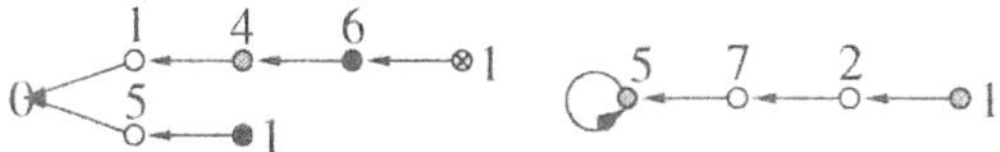

Figure 5: The skeleton of f.

We recall that for each element u of $S(f)$, we denote by $d_{S(f)}(u)$ the degree of u in $S(f)$. We then create a set $\mathcal{P}(f)$ and a multiset $\mathcal{M}(f)$:

- $\mathcal{P}(f)$ contains all the elements of $S(f)$ different from 0,
- for each $u \in S(f)$, $\mathcal{M}(f)$ contains k occurrences of u if and only if there are k elements v in $S(f)$ such that $f(v) = u$ (i.e. $d_{S(f)}(u) = k$).

In our example we have $\mathcal{M}(f) = \{0, 0, 6_1, 4_2, 5_2, 5_2, 1_3, 2_3, 5_3, 7_3\}$ and $\mathcal{P}(f) = \{1_1, 6_1, 1_2, 4_2, 5_2, 1_3, 2_3, 5_3, 7_3, 1_4\}$. Now let l be the number of leaves in $S(f)$ and $1_{i_1}, \dots, 1_{i_l}$ these leaves, with $i_1 < \dots < i_l$ (here, $l = 3$, $i_1 = 1$, $i_2 = 2$ and $i_3 = 4$). We perform the following transformations on the skeleton of f (the other edges of f are not modified):

- for j from 1 to l, let $u = 1_{i_j}$ and repeat

1. let v be the greatest element in $\mathcal{P}(f)$ of the same sort as u,
2. suppress in f the edge $f(v) \leftarrow v$ and add the edge $f(v) \leftarrow u$,
3. suppress v of $\mathcal{P}(f)$, one occurrence of $f(v)$ in $\mathcal{M}(f)$, and let $u = f(v)$,

as long as $u \neq 0$ and $u \notin \mathcal{M}(f)$.

We then have to show that the structure $\mathcal{A}_f$ obtained in this way is a F-rooted tree. By definition of the skeleton of an endofunction, and from the fact that the only modifications made on f to obtain $\mathcal{A}_f$ concern its skeleton, it suffices to show that $S(f)$ is transformed into a directed rooted tree with root 0 and in which all edges are directed towards 0. This comes from the fact that for each step $j = 1, \ldots, l$ the previous algorithm creates a path starting at 1_{i_j} and ending either at 0 (at least one of the paths created in this way ends at 0), or at an element which also belongs to a path created during a subsequent step (in our example, we create successively the paths $0 \leftarrow 7_3 \leftarrow 5_3 \leftarrow 1_1$, $0 \leftarrow 2_3 \leftarrow 5_2 \leftarrow 1_3 \leftarrow 1_2$ and $5_2 \leftarrow 4_2 \leftarrow 6_1 \leftarrow 1_4$). Finally, it remains to verify that $\mathcal{A}_{f_\mathcal{A}} = \mathcal{A}$ and that $f_{\mathcal{A}_f} = f$. It is a consequence of Properties 1 and 2 and of the following facts:

- during the transformation of $\mathcal{A}$ into $f_\mathcal{A}$, we replace all elements $*_j$ of $B(\mathcal{A})$ by the elements of $S(\mathcal{A})$ of sort j *taken in increasing order*;
- during the transformation of f into $\mathcal{A}_f$, we remove from $\mathcal{P}(f)$ the elements of sort j *in decreasing order.*

By construction, this bijection has the property (also verified in the bijection of Goulden and Kulkarni [12]) that the structure of the fibers of the elements is preserved. We will make use of this property of *fiber structure preservation* in the next section.

Property 3 Let f be an endofunction in $End^{\mathcal{R}}_{\mathbf{F},\mathbf{R}}[\mathbf{n}]$ and u any element of $\{0\} \cup [\mathbf{n}]$. For all $i \in [d]$, the restriction of $f^{-1}(u)$ to the elements of sort i and the restriction of the fiber of u in $\mathcal{A}_f$ to the elements of sort i have the same cardinality.

3 Rooted trees with a given edges partition

In this last section, we apply the previous bijection to the enumeration of rooted trees having a given edges partition. The results in this section are already present, implicitly or explicitly, in [13, 10]. However, the simple combinatorial proofs we give here illustrate well the fact that *restricted partial endofunctions are structures that are easier to describe and enumerate than rooted trees*, whence one of the interests of a bijective proof of Lagrange-Good's formula.

Let M be a square matrix of size $(d+1) \times (d+1)$, $M = (m_{i,j})_{0 \leq i,j \leq d}$, and $\mathcal{A}$ a F-rooted tree over $[\mathbf{n}]$. We say that M is the *edges partition* of $\mathcal{A}$ if there is, in $\mathcal{A}$, $m_{i,j}$ (resp. $m_{i,0}$) edges (recall that all edges are directed towards the root) going from an element of sort i to an element of sort j (resp. to 0). We say that a matrix $M = (m_{i,j})_{0 \leq i,j \leq d}$ is a *valid edges partition for* $[\mathbf{n}]$ if it verifies that for all $j \in [d]$, $m_{0,j} = 0$, and for all $i \in [d]$, $\sum_{j=0}^{d} m_{i,j} = n_i$. We denote by $\mathbf{m_i}$ the vector corresponding to the ith column $(m_{0,i}, m_{1,i}, \ldots, m_{d,i})$ in the matrix M and $\mathbf{M}! = \prod_{i,j} m_{i,j}!$. Moreover, we denote by $F(\mathbf{A_R})[\mathbf{n}]_M$ the set of F-rooted trees over $[\mathbf{n}]$ having M as edges partition. We naturally extend the notion of edges partition to endofunctions and we denote by $End^{\mathcal{R}}_{F,\mathbf{R}}[\mathbf{n}]_M$ the set of endofunctions having M as edges partition.

Proposition 1 *Let M be a a valid edges partition for $[\mathbf{n}]$ and $\delta(M)$ the matrix $(\delta_{i,j}n_i - m_{i,j})_{1\le i,j\le d}$:*

$$|F(\mathbf{A_R})[\mathbf{n}]_M| = \det(\delta(M)) \frac{(\mathbf{n}-\mathbf{1})!}{\mathbf{M}!} |F[\mathbf{m_0}]| \prod_{i=1}^{d} |R_i^{n_i}[\mathbf{m_i}]| . \tag{5}$$

Proof. By virtue of the bijection between F-rooted trees and endofunctions described in the previous section, and by Property 3 (fiber structure preservation), we have $F(\mathbf{A_R})[\mathbf{n}]_M \simeq End^{\mathcal{R}}_{F,\mathbf{R}}[\mathbf{n}]_M$. In order to give an interpretation to the determinant appearing in the right-hand-side of (5), we use the Matrix-Tree Theorem for directed graphs (see [11, Section 3.3.24] for example):

- let $\lambda = (\lambda_{i,j})_{1\le i,j\le m}$ the adjacency matrix of a directed labelled graph $\mathcal{G}$ having m vertices: the number of directed spanning trees of $\mathcal{G}$ having root $c \in [m]$ in which all edges are directed towards c is given by

$$\det\left(\left\{\delta_{i,j}\sum_{k=1}^{m}\lambda_{i,k}\right\} - \lambda_{i,j}\right)_{1\le i,j\le m, i\ne c, j\ne c} .$$

Let M be a valid edges partition for $[\mathbf{n}]$, and $\mathcal{T}$ a rooted tree over $\{0\} \cup [d]$ compatible with M (there cannot be an edge from i to j if $m_{i,j} = 0$). To compute the number of endofunctions f in $End^{\mathcal{R}}_{F,\mathbf{R}}[\mathbf{n}]$ having $\mathcal{T}$ as graph of sorts, we note that for $i \in [d]$, if $\mathcal{T}$ has an edge from i to j ($j \in \{0\} \cup [d]$), we have

$$\binom{n_i - 1}{m_{i,0}, \ldots, m_{i,j-1}, m_{i,j} - 1, m_{i,j+1}, \ldots, m_{i,d}} = \frac{m_{i,j}}{n_i}\binom{n_i}{m_{i,0}, \ldots, m_{i,j}, \ldots, m_{i,d}}$$

choices for the sort of the images of the elements of sort i different from 1_i, while for 1_i, the sort of its image $f(1_i)$ is fixed by $\mathcal{T}$ (its image is of sort j). We deduce that the number of endofunctions of $End^{\mathcal{R}}_{F,\mathbf{R}}[\mathbf{n}]_M$ over $[\mathbf{n}]$ having $\mathcal{T}$ as graph of sorts is

$$\left(\prod_{(i,j)\in\mathcal{T}} m_{i,j}\right)\left(\prod_{i=1}^{d}\frac{1}{n_i}\binom{n_i}{m_{i,0}, \ldots, m_{i,d}} |R_i^{n_i}[\mathbf{m_i}]|\right)|F[\mathbf{m_0}]| . \tag{6}$$

Now, let T_M be the set of rooted trees over $\{0\} \cup [d]$ rooted at 0 that are compatible with M. By the Matrix-Tree Theorem we can say that $\det(\delta(M)) = \sum_{\mathcal{T}\in T_M}\left(\prod_{(i,j)\in\mathcal{T}} m_{i,j}\right)$, which, combined with (6), let us deduce that

$$\left|End^{\mathcal{R}}_{F,\mathbf{R}}[\mathbf{n}]_M\right| = \det(\delta(M))\left(\prod_{i=1}^{d}\left\{\frac{1}{n_i}\binom{n_i}{m_{i,0}, \ldots, m_{i,d}} |R_i^{n_i}[\mathbf{m_i}]|\right\}\right)|F[\mathbf{m_0}]| ,$$

which implies (5). □

We then immediately deduce a combinatorial proof of a version of Lagrange-Good's formula due to Goulden and Jackson [13] (see also [11, Section1.2.13]) and used, among other, in [13, 10].

Corollary 1 *Let $F(\mathbf{x}), R_1(\mathbf{x}), \ldots, R_d(\mathbf{x})$ be $d+1$ multivariable formal power series such that $R_i(\mathbf{0}) \neq 0$ and $A_i(\mathbf{t}) = t_i R_i(\mathbf{A}(\mathbf{t}))$ for $i \in [d]$. Then we have*

$$[\mathbf{t}^{\mathbf{n}}]F(\mathbf{A}(\mathbf{t})) = \left(\prod_{i=1}^{d} \frac{1}{n_i}\right) \sum_{M} \det(\delta(M)) \left(\prod_{i=1}^{d} [\mathbf{x}^{\mathbf{m_i}}] R_i(\mathbf{x})^{n_i}\right) [\mathbf{x}^{\mathbf{m_0}}]F(\mathbf{x}),$$

the sum being taken over all valid edges partitions M for $[\mathbf{n}]$.

A F-rooted tree over $[\mathbf{n}]$ is an ordered (resp. unordered) rooted tree, with root of sort k, if the fiber of 0 only contains one element, of sort k, and if the fiber of every element different from 0 is embedded into a structure of totally ordered set L (resp. unordered set E). A matrix M is a valid edges partition for rooted trees over $[\mathbf{n}]$ having root of sort k if and only if M is a valid edges partition for $[\mathbf{n}]$, $m_{k,0} = 1$ and for all $i \in [d]$ different from k, $m_{i,0} = 0$. For $i \in [d]$, let us denote $q_j = \sum_{i=1}^{d} m_{i,j}$, and, for a matrix P, $\mathrm{cof}_{k,k}(P)$ the determinant of P where the kth line and the kth column have been deleted.

Corollary 2 *Let M be a valid edges partition for rooted trees over $[\mathbf{n}]$. The number of multisort ordered rooted trees over $[\mathbf{n}]$ having root of sort k and M as edges partition is*

$$\mathrm{cof}_{k,k}(\delta(M)) \frac{(\mathbf{n}+\mathbf{q}-\mathbf{1})!}{\mathbf{M}!}. \tag{7}$$

The number of multisort unordered rooted trees over $[\mathbf{n}]$ having root of sort k and M as edges partition is

$$\mathrm{cof}_{k,k}(\delta(M)) \frac{\mathbf{n}^{\mathbf{q}}(\mathbf{n}-\mathbf{1})!}{\mathbf{M}!}. \tag{8}$$

Proof. These results are direct consequences of Proposition 1 and of the following remark: by definition, if $\mathcal{A}$ is a rooted tree having root of sort k, in the graph of sorts of $f_{\mathcal{A}}$, the root 0 has exactly one son, and this son is the element 1_k. So by the Matrix-Tree Theorem, we can say that $\det(\delta M) = \mathrm{cof}_{k,k}(\delta(M))$.

Then it suffices to remark that the number of L^{n_i}-structures (we recall that L is the species of totally ordered sets) over $[\mathbf{m_i}]$ is

$$q_i! \binom{n_i + q_i - 1}{q_i},$$

to prove (7), and to remark that the number of E^{n_i}-structures (we recall that E is the species of unordered sets) over $[\mathbf{m_i}]$ is $n_i^{q_i}$ to prove (8). □

A matrix $P = (p_{i,j})_{1 \le i,j}$ is called the *degrees partition* of a rooted tree $\mathcal{A}$ on $[\mathbf{n}]$ if the number of elements of $\mathcal{A}$ of sort i having degree j (the degree of an element is the number of elements in its fiber) is $p_{i,j}$ (this notion can naturally be extended to the endofunctions). A matrix P is a *valid degrees partitions for rooted trees over* $[\mathbf{n}]$ if it verifies that for $i \in [d]$, $\sum_{j \ge 0} p_{i,j} = n_i$ and $q_i = \sum_{j \ge 0} j p_{i,j}$.

Corollary 3 *Let M be a valid edges partition for rooted trees over* $[\mathbf{n}]$ *having root of sort k, and P a valid degrees partition for rooted trees over* $[\mathbf{n}]$. *The number of multisort ordered rooted trees over* $[\mathbf{n}]$ *having root of sort k, M as edges partition and P as degrees partition is*

$$\mathrm{cof}_{k,k}(\delta(M))\frac{\mathbf{q}!\mathbf{n}!(\mathbf{n}-\mathbf{1})!}{\mathbf{M}!\mathbf{P}!}. \tag{9}$$

The number of multisort unordered rooted trees over $[\mathbf{n}]$ *having root of sort k, M as edges partition and P as degrees partition is*

$$\mathrm{cof}_{k,k}(\delta(M))\frac{\mathbf{q}!\mathbf{n}!(\mathbf{n}-\mathbf{1})!}{\mathbf{M}!\mathbf{P}!}\prod_{i,j\geq 1}\frac{1}{j!^{p_{i,j}}}. \tag{10}$$

Proof. The proof of these two formulas is similar to the proof of Corollary 2. The only difference is the fact that the number of L^{n_i}-structures on $[\mathbf{m_i}]$ such that $p_{i,j}$ elements have degree j is

$$q_i!\binom{n_i}{p_{i,0},\ p_{i,1},\ p_{i,2},\ldots}.$$

□

By noticing that ordered rooted trees are asymmetric structures, we obtain, from the two previous corollary, the formulas for the corresponding problem for unlabelled ordered rooted trees, by dividing (7) and (9) by $\mathbf{n}!$.

4 Conclusion

In this paper, we presented a simple bijective proof of the multivariable Lagrange-Good's inversion formula and we illustrated its usefulness in the problem of the enumeration of multisort rooted trees. Using the same kind of reasoning (but with some variations), we are able to give a combinatorial explanation on two formulas enumerating m-ary cacti (a family of planar maps involved in the classification of complex polynomials, see [3] for example) and to deduce from this explanation a uniform random generation algorithm for these structures [4, 5].

However, there is a problem we were not able to solve: design a proof of the arborescent form of the Lagrange-Good's formula involving unlabelled structures and ordinary generating functions. Among all the proofs of the multivariable Lagrange-Good's formula, as far as we know, the only "unlabelled proof" is due to Chottin and is limited to the case of two variables series ($d = 2$) [6]. Such a proof would be of great interest, especially in the design of exhaustive generation algorithms of multisort unlabelled combinatorial structures (like cacti for example). We can think to solve this problem by using the theory of linear species [14]. Indeed, in [14], Joyal gives, in the case of one variable series, an unlabelled analog to the proof of Labelle using linear species. It would be interesting to extend this proof to the multivariable case.

Acknowledgments. We thank Gilles Schaeffer for many fruitful discussions.

References

[1] E. A. Bender and L. B. Richmond. A multivariate Lagrange inversion formula for asymptotic calculations. *Electron. J. Combin.*, 5(1): Paper 33, 1998.

[2] F. Bergeron, G. Labelle, and P. Leroux. *Combinatorial species and tree-like structures.* Cambridge University Press, Cambridge, 1998.

[3] M. Bóna, M. Bousquet, G. Labelle, and P. Leroux. Enumeration of m-ary cacti. *Adv. in Appl. Math.*, 24:22–56, 2000.

[4] M. Bousquet, C. Chauve, and G. Schaeffer. Enumération et génération aléatoire de cactus m-aires. In *Colloque LaCIM 2000*, volume 27 of *Publications du LaCIM.* Université du Québec à Montréal, Montréal (Québec), Canada, 2000. (to appear).

[5] C. Chauve. *Structures arborescentes : problèmes algorithmiques et combinatoires.* PhD thesis, LaBRI, Université Bordeaux I, 2000. (in preparation).

[6] L. Chottin. Énumération d'arbres et formules d'inversion de séries formelles. *J. Combin. Theory Ser. B*, 31(1):23–45, 1981.

[7] R. Ehrenborg and M. Méndez. A bijective proof of infinite variated Good's inversion. *Adv. in Math.*, 103(2):221–259, 1994.

[8] I. M. Gessel. A combinatorial proof of the multivariable Lagrange inversion formula. *J. Combin. Theory Ser. A*, 45(2):178–195, 1987.

[9] I. J. Good. Generalizations to several variables of Lagrange's expansion, with applications to stochastic processes. *Proc. Cambridge Philos. Soc.*, 56:367–380, 1960.

[10] I. P. Goulden and D. M. Jackson. The application of Lagrangian methods to the enumeration of labelled trees with respect to edges partition. *Canad. J. Math.*, 34(3):513–518, 1982.

[11] I. P. Goulden and D. M. Jackson. *Combinatorial enumeration.* John Wiley & Sons Inc., New York, 1983.

[12] I. P. Goulden and D. M. Kulkarni. Multivariable Lagrange inversion, Gessel-Viennot cancellation, and the matrix tree theorem. *J. Combin. Theory Ser. A*, 80(2):295–308, 1997.

[13] D. M. Jackson and I. P. Goulden. The generalisation of Tutte's result for chromatic trees, by Lagrangian methods. *Canad. J. Math.*, 33(1):12–19, 1981.

[14] A. Joyal. Une théorie combinatoire des séries formelles. *Adv. in Math.*, 42(1):1–82, 1981.

[15] G. Labelle. Une nouvelle démonstration combinatoire des formules d'inversion de Lagrange. *Adv. in Math.*, 42(3):217–247, 1981.

Trends in Mathematics, © 2000 Birkhäuser Verlag Basel/Switzerland

Counting paths on the slit plane (extended abstract)

MIREILLE BOUSQUET-MÉLOU *CNRS, LaBRI, Université Bordeaux 1, 351 cours de la Libération, 33405 Talence Cedex, France.* bousquet@labri.u-bordeaux.fr
GILLES SCHAEFFER *CNRS,* Loria, Campus Scientifique, 615 rue du Jardin Botanique – B.P. 101, 54602 Villers–lès–Nancy Cedex, France. Gilles.Schaeffer@loria.fr

Abstract. *We present a method, based on functional equations, to enumerate paths on the square lattice that avoid a horizontal half-line. The corresponding generating functions are algebraic, and sometimes remarkably simple: for instance, the number of paths of length* $2n+1$ *going from* $(0,0)$ *to* $(1,0)$ *and avoiding the nonpositive horizontal axis (except at their starting point) is* C_{2n+1}*, the* $(2n+1)$*th Catalan number.*

More generally, we enumerate exactly all paths of length n *starting from* $(0,0)$ *and avoiding the nonpositive horizontal axis. We then obtain limit laws for the coordinates of their endpoint: in particular, the average abscissa of their endpoint grows like* $\sqrt{n}$ *(up to an explicit multiplicative constant), which shows that these paths are strongly repelled from the origin.*

We derive from our results the distribution of the position where a random walk, starting from a given point, hits for the first time the horizontal half-line.

1 Introduction

In January 1999, Rick Kenyon posted on the "domino" mailing-list the following e-mail:

> "Take a simple random walk on $\mathbb{Z}^2$ starting on the y-axis at $(0,1)$, and stopping when you hit the nonpositive x-axis. Then the probability that you end at the origin is $1/2$.
> Since this result was obtained from a long calculation involving irrational numbers, I wonder if there is an easy proof? By way of comparison, if you start at $(1,0)$ the probability of stopping at the origin in $2-\sqrt{2}$."

This mail led Olivier Roques, a graduate student at LaBRI, to investigate the *number* of such walks of fixed length: he soon conjectured that exactly $4^n C_n$ walks of length $2n+1$ go from $(0,1)$ to $(0,0)$ without hitting the nonpositive x-axis before they reach their endpoint, where $C_n = \binom{2n}{n}/(n+1)$ is the nth Catalan number. Similarly, he conjectured that, if the starting point is chosen to be $(1,0)$, then the number of walks is even more remarkable, being C_{2n+1}. Let us mention that these conjectures directly imply Rick Kenyon's results.

We attack them by studying the problem in its full generality: denoting by $a_{i,j}(n)$ the number of n-step walks starting from $(0,0)$, ending at (i,j), and avoiding the forbidden half-axis (*walks on the slit plane*), we give a closed form expres-

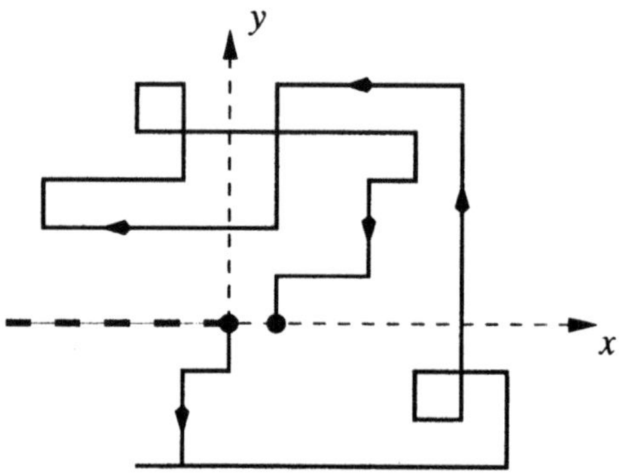

Figure 1: A path on the slit plane joining (0,0) to (1,0).

sion for the generating function

$$S(x,y;t) = \sum_{i \in \mathbb{Z}} \sum_{j \in \mathbb{Z}} \sum_{n \geq 0} a_{i,j}(n) x^i y^j t^n,$$

which turns out to be algebraic. Our approach is based on a functional equation, which is trivial to establish, but tricky to solve (Section 4).

From this three-variate generating function, we can compute the generating function for walks ending at a prescribed position (i,j): this proves O. Roques's conjectures (Section 2), and allows us to study R. Kenyon's question for any starting point (Section 3). This gives the (bounded) solution of a discrete Dirichlet problem on the square lattice, with boundary conditions on the negative x-axis (see [12, Section 1.4] for definitions). We also study the asymptotic properties of n-step walks on the slit plane: for instance, we prove and quantify their *transience* (the point $(1,0)$ is only visited with probability $2-\sqrt{2}$). We show that their endpoint lies, on average, at distance $\sqrt{n}$ from the origin. More precisely, this endpoint, normalized by $\sqrt{n}$, converges in law towards an explicit distribution (Section 5).

Our method can be applied to similar problems: in particular, we can change the "forbidden" axis into the half-line $x = y$, $x \leq 0$, as explained in the full version of the paper [3].

Let us mention that two simple proofs of the C_{2n+1} result have been given recently [2, 4], the former being bijective. Moreover, the series $S(1,1;t)$ has already been studied in the literature [12, Chap.2] but, to our knowledge, only asymptotic results had been obtained so far. See also [6] for numerical simulations.

2 Main results

2.1 The complete generating function

We consider paths (or *walks*) on the square lattice made of four kinds of steps: north, east, south and west. Let $n \geq 0$, and $(i,j) \in \mathbb{Z} \times \mathbb{Z} \setminus \{(k,0) : k < 0\}$. We denote by $a_{i,j}(n)$ the number of walks of length n on the square lattice that start from $(0,0)$, end at (i,j), and never touch the horizontal half-axis $\{(k,0) : k \leq 0\}$ once they have left their starting point: we call them *paths on the slit plane*. Fig. 1 shows such a walk, with $(i,j) = (1,0)$ and $n = 59$. We denote by $a(n)$ the number of paths of length n on the slit plane, starting from $(0,0)$, regardless of their endpoint.

Let $S(x,y;t)$ be the generating function for paths on the slit plane, counted by their length and the position of their endpoint:

$$\begin{aligned} S(x,y;t) &\equiv S(x,y) = \sum_{i\in\mathbb{Z}} \sum_{j\in\mathbb{Z}} \sum_{n\geq 0} a_{i,j}(n) x^i y^j t^n, \\ &= 1 + t(x+y+\bar{y}) + t^2(x^2 + 2xy + 2x\bar{y} + \bar{x}y + \bar{x}\bar{y} + y^2 + \bar{y}^2) \\ &+ (5x + x^3 + 4y + 4\bar{y} + y^3 + \bar{y}^3 + 3xy^2 + 3x\bar{y}^2 + 3x^2y + 3x^2\bar{y} \\ &+ \bar{x}^2y + \bar{x}^2\bar{y} + 2\bar{x}y^2 + 2\bar{x}\bar{y}^2)t^3 + O(t^4) \end{aligned}$$

with $\bar{x} = x^{-1}, \bar{y} = y^{-1}$. We shall prove that this series is algebraic of degree 8 over the field of rational functions in x, y and t.

Theorem 1 *The generating function $S(x,y;t)$ for paths on the slit plane is*

$$S(x,y;t) = \frac{\left(1 - 2t(1+\bar{x}) + \sqrt{1-4t}\right)^{1/2} \left(1 + 2t(1-\bar{x}) + \sqrt{1+4t}\right)^{1/2}}{2[1 - t(x + \bar{x} + y + \bar{y})]}.$$

This series is algebraic of degree 8. *When $x = y = 1$, it specializes to*

$$S(1,1;t) = \sum_{n\geq 0} a(n)t^n = \frac{(1+\sqrt{1-4t})^{1/2}(1+\sqrt{1+4t})^{1/2}}{2(1-4t)^{3/4}},$$

so that the asymptotic growth of the number of paths of length n on the slit plane is

$$a(n) \sim \frac{\sqrt{1+\sqrt{2}}}{2\Gamma(3/4)} 4^n n^{-1/4}.$$

In other words, the probability that a random walk on the square lattice, starting from $(0,0)$, has not met the forbidden half-line after n steps is asymptotic to $cn^{-1/4}$ with $c = \sqrt{1+\sqrt{2}}/2/\Gamma(3/4)$. The decay in $n^{-1/4}$ was known [10, 12, p.71], but the details of the asymptotic behaviour seem to be new.

2.2 Paths ending at a prescribed ordinate

We would like to enumerate paths on the slit plane ending at a prescribed point (i,j). Let

$$S_{i,j}(t) \equiv S_{i,j} = \sum_{n\geq 0} a_{i,j}(n)t^n$$

be the corresponding generating function. This series is obtained by extracting the coefficient of $x^i y^j$ from $S(x,y;t)$. As i and j belong to $\mathbb{Z}$, rather than $\mathbb{N}$, this is not an obvious task. In particular, let us underline that the algebraicity of $S(x,y;t)$ does *not* imply that $S_{i,j}(t)$ is algebraic. This is clearly shown by the enumeration of paths starting from $(0,0)$ in the (unslit) plane. The complete generating function is $1/[1-t(x+\bar{x}+y+\bar{y})]$. It is rational, hence algebraic. However, for $i,j \geq 0$, the coefficient of $x^i y^j$ in this series is

$$\sum_{m\geq 0} \binom{2m+i+j}{m}\binom{2m+i+j}{m+i} t^{2m+i+j},$$

which is transcendental: the coefficient of t^n grows like $4^n/n$, up to a multiplicative constant, revealing a logarithmic singularity in the generating function that implies its transcendence (see [9] for a discussion on the possible singularities of an algebraic series).

In constrast, we shall prove that for any i and j, the series $S_{i,j}(t)$ is algebraic.

Our first step will be to extract from $S(x,y;t)$ the coefficient of y^j. Before we state our result, let us introduce a few notations.

Notations. Given a ring $\mathbb{L}$ and n indeterminates $x_1,\ldots,x_n$, we denote by

- $\mathbb{L}[x_1,\ldots,x_n]$ the ring of polynomials in $x_1,\ldots,x_n$ with coefficients in $\mathbb{L}$,
- $\mathbb{L}[[x_1,\ldots,x_n]]$ the ring of formal power series in $x_1,\ldots,x_n$ with coefficients in $\mathbb{L}$,

and if $\mathbb{L}$ is a field, we denote by

- $\mathbb{L}(x_1,\ldots,x_n)$ the field of rational functions in $x_1,\ldots,x_n$ with coefficients in $\mathbb{L}$.

It will be convenient to express the series $S_{i,j}(t)$ in terms of the following power series in t:

$$u = \frac{\sqrt{1+4t}-1}{\sqrt{1-4t}+1} = \sum_{n\geq 0}(2.4^n C_n - C_{2n+1})t^{2n+1}.$$

Note that u is quartic over $\mathbb{Q}(t)$:

$$t = \frac{u(1-u^2)}{(1+u^2)^2}.$$

This equation allows us to write any rational function in t as a rational function in u. Moreover, the generating function of Catalan numbers,

$$C(t) = \frac{1-\sqrt{1-4t}}{2t} = \sum_{n\geq 0}\frac{1}{n+1}\binom{2n}{n}t^n,$$

satisfies

$$C(t) = \frac{1+u^2}{1-u}, \quad C(-t) = \frac{1+u^2}{1+u} \quad \text{and} \quad tC(t)C(-t) = u,$$

so that $\mathbb{Q}(u) = \mathbb{Q}(t, \sqrt{1-4t}, \sqrt{1+4t})$. Also, let

$$\Delta(x) = [1 - x(C(t)-1)]\,[1 + x(C(-t)-1)] = \left(1 - xu\,\frac{1+u}{1-u}\right)\left(1 - xu\,\frac{1-u}{1+u}\right).$$

Corollary 2 *Let $j \in \mathbb{Z}$. The generating function for paths on the slit plane ending at ordinate j is*

$$S_j(x,t) = \sum_{i\in\mathbb{Z},n\geq 0} a_{i,j}(n)x^i t^n = \frac{M(x)^j}{\sqrt{\Delta(x)}}$$

where

$$M(x) \equiv M(x;t) = \frac{1 - t(x+\bar{x}) - \sqrt{[1-t(x+\bar{x}+2)]\,[1-t(x+\bar{x}-2)]}}{2t}.$$

The generating function $S_j^+(x;t)$ for paths on the slit plane ending at ordinate j and at a positive abscissa is of the form

$$S_j^+(x;t) = \sum_{i>0,n\geq 0} a_{i,j}(n)x^i t^n = f(x,u) + \frac{g(x,u)}{\sqrt{\Delta(x)}},$$

where $f(x,u)$ and $g(x,u)$ are Laurent polynomials in x with coefficients in $\mathbb{Q}(u)$. Similarly, the generating function $S_j^-(x;t)$ for paths on the slit plane ending at ordinate j and at a non-positive abscissa is of the form

$$S_j^-(x;t) = \sum_{i\leq 0,n\geq 0} a_{i,j}(n)x^i t^n = S_j^-(x;t) = f(x,u) + g(x,u)\sqrt{\Delta(\bar{x})},$$

with the same conditions on $f(x,u)$ and $g(x,u)$. The series S_j^+ and S_j^- can be computed explicitely. For instance,

$$\begin{aligned} S_0^+(x;t) &= \frac{1}{\sqrt{\Delta(x)}} - 1, \\ S_1^+(x;t) &= \frac{1}{2}\left[\frac{1-t(x+\bar{x})}{t\sqrt{\Delta(x)}} - \frac{1}{t(1+u^2)} + \bar{x}\right], \\ S_1^-(x;t) &= \frac{1}{2}\left[\frac{1}{t(1+u^2)} - \bar{x} - \frac{\sqrt{\Delta(\bar{x})}}{u}\right]. \end{aligned}$$

Sketch of the proof. We start from the expression of $S(x,y;t)$ given in Theorem 1:

$$S(x,y;t) = \frac{t\sqrt{\Delta(\bar{x})}}{u[1 - t(x+\bar{x}+y+\bar{y})]},$$

and convert the denominator $1/[1 - t(x+\bar{x}+y+\bar{y})]$ into partial fractions of y.

■

2.3 Paths ending at a prescribed position

The series $S_j^+(x;t)$ and $S_j^-(x;t)$ can now be expanded in x (resp. $\bar{x}$), and their form implies the following result.

Corollary 3 *For all i and j, the generating function $S_{i,j}(t) = \sum_n a_{i,j}(n)t^n$ for paths on the slit plane ending at (i,j) belongs to $\mathbb{Q}(u) = \mathbb{Q}(t, \sqrt{1-4t}, \sqrt{1+4t})$, and can be computed explicitely. It is either quadratic, or quartic over $\mathbb{Q}(t)$. In particular,*

$$S_{0,1}(t) = \frac{u}{1-u^2} = \frac{1-\sqrt{1-16t^2}}{8t} = \sum_{n\geq 0} 4^n C_n t^{2n+1},$$

and

$$S_{1,0}(t) = \frac{u(1+u^2)}{1-u^2} = \frac{2-\sqrt{1-4t}-\sqrt{1+4t}}{4t} = \sum_{n\geq 0} C_{2n+1} t^{2n+1},$$

as conjectured by O. Roques. Some other values are

$$S_{-1,1}(t) = \frac{u^2}{1-u^2} = \frac{\sqrt{1+4t}-\sqrt{1-4t}-4t}{8t} = \frac{1}{2}\sum_{n\geq 1} C_{2n} t^{2n},$$

$$\begin{aligned} S_{1,1}(t) &= \frac{u^2(2-u^2)}{(1-u^2)^2} = \frac{1-24t^2+4t\sqrt{1+4t}-4t\sqrt{1-4t}-\sqrt{1-16t^2}}{32t^2} \\ &= \sum_{n\geq 1} (4^{n-1}C_n + C_{2n}/2)t^{2n}. \end{aligned}$$

Proof. Expanding in x (or $\bar{x}$) the series $S_0^+(x;t)$, $S_1^+(x;t)$ and $S_1^-(x;t)$ given in the previous corollary provides the expressions of $S_{0,1}$, $S_{1,0}$, $S_{-1,1}$ and $S_{1,1}$. In theory, the series $S_{i,j}$ could be rational, but this can be ruled out by a simple asymptotic argument on the numbers $a_{i,j}(n)$.

∎

Using our results, we can actually enumerate walks starting from *any* point of the nonpositive x-axis. This will be useful in the following section.

Corollary 4 *Let $a_{i,j}^{[k]}(n)$ be the number of paths of length n that go from $(-k,0)$ to (i,j) and do not meet the nonpositive horizontal axis once they have left their starting point. Let*

$$D_{i,j}(x;t) = \sum_{k,n\geq 0} a_{i,j}^{[k]}(n) x^k t^n$$

be the corresponding generating function. Then the series $D_{i,j}(x;t)$ can be computed from the partial sections

$$S_{i,j}^+(x;t) = \sum_{k\geq 0} x^k S_{i+k,j}(t)$$

through the following relation, valid for any i, j:

$$S_{i,j}^+(x;t) = D_{i,j}(x;t) + xS_{i,j}^+(x;t) D_{1,0}(x;t). \tag{1}$$

Proof. Let P be a path joining $(0,0)$ to $(i+k,j)$ in the slit plane. Such paths are counted by $S^+_{i,j}(x;t)$. If P does not meet the segment $[1,k] = \{(\ell,0) : 1 \leq \ell \leq k\}$, then P is isomorphic to a path joining $(-k,0)$ to (i,j) in the slit plane. This gives the term $D_{i,j}(x;t)$ in the right-hand side of (1). Otherwise, let $(\ell,0)$ be the leftmost point of $[1,k]$ that belongs to P. We can split P into two factors: from $(0,0)$ to the last time $(\ell,0)$ is reached, we get a path that is isomorphic to a path going from $(-\ell+1,0)$ to $(1,0)$ in the slit plane; this corresponds to the coefficient of $x^{\ell-1}$ in $D_{1,0}(x;t)$. The second factor goes from $(0,\ell)$ to $(i+k,j)$ and is isomorphic to a path going from $(0,0)$ to $(i+k-\ell,j)$ in the slit plane; it corresponds to the coefficient of $x^{k-\ell}$ in $S^+_{i,j}(x;t)$. This completes the proof of Relation (1).

■

2.4 Walks on the slit plane are transient

It is well-known that the random walk on the square lattice is *recurrent*: any given point (i,j) of the lattice is visited by the walk with probability 1. In more enumerative terms, the proportion of walks of length n visiting (i,j) tends to 1 as n goes to infinity. This is no longer the case for walks on the slit plane.

Corollary 5 *Let $b(n)$ denote the number of walks of length n on the slit plane that visit the point $(1,0)$. Then, as n goes to infinity,*

$$\frac{b(n)}{a(n)} \to 2-\sqrt{2} < 1.$$

Proof. A walk visiting $(1,0)$ can be seen in a unique way as the concatenation w_1w_2 of a walk w_1 going from $(0,0)$ to $(1,0)$, and a walk w_2 starting from $(1,0)$, that not only avoids the horizontal half-axis but also the point $(1,0)$ itself. This implies that

$$\sum_n b(n)t^n = S_{1,0}(t)S(1,1;t),$$

and the result follows, as $S_{1,0}(1/4) = 2-\sqrt{2}$.

■

3 The hitting distribution of a half-line

The above results allow us to solve a number of probabilistic questions "à la Kenyon". Let (i,j) be a point of $\mathbb{Z}^2$ not belonging to the forbidden half-axis. A random walk starting from (i,j) hits this half-axis with probability 1. The probability that the *first* hitting point is $(0,0)$ is

$$p_{i,j} = \sum_{n\geq 0} \frac{a_{i,j}(n)}{4^n} = S_{i,j}(1/4).$$

More precisely, $a_{i,j}(n)/4^n$ is the probability that this event occurs after n steps. Corollary 3 implies that $p_{i,j} \in \mathbb{Q}[\sqrt{2}]$. In particular, we find $p_{0,1} = 1/2$ and $p_{1,0} = 2-\sqrt{2}$, as stated in R. Kenyon's e-mail.

More generally, given $k \geq 0$, one can ask about the probability $p_{i,j}^{[k]}$ that the first hitting point is $(-k, 0)$. Using the notations of Corollary 4,

$$\sum_{k\geq 0} p_{i,j}^{[k]} x^k = D_{i,j}(x; 1/4),$$

and this series can be computed explicitely.

Let us, for instance, derive the distribution of the hitting abscissa for a walk starting from $(1,0)$. Using Relation (1) with $(i,j) = (1,0)$, we express $D_{1,0}(x;t)$ in terms of $S_{1,0}^+(x;t) = S_0^+(x;t)/x$:

$$D_{1,0}(x;t) = \frac{1}{x}\frac{S_0^+(x;t)}{1 + S_0^+(x;t)}.$$

The value of $S_0^+(x;t)$ is given in Corollary 2. Hence the probability distribution of the hitting abscissa for random walks starting from $(1,0)$ is

$$\begin{aligned}\sum_{k\geq 0} p_{1,0}^{[k]} x^k &= D_{1,0}(x;1/4) = \frac{1}{x}\frac{S_0^+(x;1/4)}{1 + S_0^+(x;1/4)} \\ &= \frac{1}{x}\left[1 - \sqrt{(1-x)\left(1 - x\left(\sqrt{2}-1\right)^2\right)}\right].\end{aligned}$$

Observe that the smallest singularity of this series is at $x = 1$, and that $p_{1,0}^{[k]}$ decays like $k^{-3/2}$ as $k \to \infty$. This asymptotic behaviour actually holds for all starting points:

Proposition 6 *For any starting point (i,j), the probabilities $p_{i,j}^{[k]}$ decay like $k^{-3/2}$ as k goes to infinity.*

The proof is based on Corollaries 2 and 4. Equivalently, the study of the dominant singularity of $D_{i,j}(x;1/4)/(1-x)$ shows that the probability that the hitting abscissa is smaller than $-k$ decays like $k^{-1/2}$. This is related to the (already known) fact that the probability that a random walk starting from the origin reaches a point at distance k of the origin before it hits the horizontal half-line also decays like $k^{-1/2}$ [10, 12, Chap.2].

4 Derivation of the complete generating function

4.1 Functional equations

We obtain a functional equation for the series $S(x,y;t)$ defined in Section 2.1 by saying that a path of length n is obtained by adding a step to another path of length $n-1$. However, when generating paths via this procedure, one must be careful not to produce paths ending on the forbidden half-line. This gives:

$$S(x,y) = 1 + t(x + \bar{x} + y + \bar{y})S(x,y) - t\bar{x} - tS_{1,0} - 2tS_1^-(x),$$

where, as above,

$$S_{1,0} = \sum_{n\geq 0} a_{1,0}(n)t^n \quad \text{and} \quad S_1^-(x) = \sum_{i\leq 0, n\geq 0} a_{i,1}(n)x^i t^n.$$

That is,

$$[1 - t(x + \bar{x} + y + \bar{y})]\, S(x,y) = 1 - t(\bar{x} + S_{1,0} + 2S_1^-(x)). \tag{2}$$

Observe that this equation is equivalent to a recurrence relation defining the numbers $a_{i,j}(n)$ by induction on n: hence, this equation completely determines the series $S(x,y)$, and in particular its sections $S_{1,0}$ and $S_1^-(x)$.

However, we found some difficulties working with Eq. (2), because $S(x,y)$ involves simultaneously positive and negative exponents of y. Instead, we are going to work with the series

$$T(x,y;t) \equiv T(x,y) = \sum_{i,j,n} a_{i,j}(n)x^i y^{|j|} t^n,$$

which, by symmetry of the model, contains as much information as $S(x,y)$ itself. Again, one writes easily a functional equation for the series $T(x,y)$:

$$[1 - t(x + \bar{x} + y + \bar{y})]\, T(x,y) = 1 - t[\bar{x} + S_{1,0} + 2S_1^-(x) + (\bar{y} - y)S_0(x)], \tag{3}$$

with

$$S_0(x) = \sum_{i\geq 0, n\geq 0} a_{i,0}(n)x^i t^n.$$

The combinatorial part of the proof is now achieved. What we shall do from now on to solve Eq. (3) resorts to algebra.

4.2 First application of the kernel method

Let us call $K(x,y) := [1 - t(x + \bar{x} + y + \bar{y})]$ the *kernel* of the functional equation (3). The principle of the kernel method is to cancel the kernel so as to obtain certain relations between the sections that occur on the right-hand side of the equation. See [7, 8, 11] for early uses of this method, and [1, 5] for more recent developments.

Here, the kernel $K(x,y)$, as a rational function of y, it has two roots. One of them is a formal power series in t with coefficients in $\mathbb{Z}[x,\bar{x}]$, and will be denoted $M(x)$:

$$M(x) = \frac{1 - t(x+\bar{x}) - \sqrt{(1 - t(x+\bar{x}))^2 - 4t^2}}{2t} = t + (x+\bar{x})t^2 + (x^2+3+\bar{x}^2)t^3 + O(t^4).$$

Observe that $T(x, M(x))$ is a well-defined series belonging to $\mathbb{Z}[x, \bar{x}][[t]]$. Let us replace y by $M(x)$ in Eq. (3): the kernel vanishes, and we obtain

$$0 = 1 - t\left[\bar{x} + S_{1,0} + 2S_1^-(x) + \left(\frac{1}{M(x)} - M(x)\right) S_0(x)\right]. \tag{4}$$

4.3 From linear to algebraic equations

The series $M(x)$ is defined as one of the roots of the kernel $K(x,y) = 1 - t(x + \bar{x} + y + \bar{y})$. We eliminate $M(x)$ between (4) and the equation $K(x, M(x)) = 0$ and obtain

$$[1-t(x+\bar{x}+2)][1-t(x+\bar{x}-2)]S_0(x)^2 - 4t^2 S_1^-(x)^2 + 4t(1 - t\bar{x} - tS_{1,0})S_1^-(x) - (1 - t\bar{x} - tS_{1,0})^2 = 0.$$

Now, remember that $S_0(x)$ is a formal power series in t and x, while $S_1^-(x)$ is a formal power series in t and $\bar{x}$. Thanks to the absence of terms involving simultaneously a series in x and a series in $\bar{x}$, we can easily extract the nonnegative part of the above equation, that is, the terms in x^i with $i \geq 0$. This forces us, however, to introduce the generating function $S_{0,1}$ for paths on the slit plane ending at $(0,1)$. We obtain:

$$[1-t(x+\bar{x}+2)][1-t(x+\bar{x}-2)]S_0(x)^2 - 4t^2 S_{1,0}S_{0,1} - (1-2tS_{0,1})^2 - (1-t\bar{x}-tS_{1,0})^2 + 1 = 0. \tag{5}$$

The extraction of the negative part yields, after dividing by $4t$:

$$-tS_1^-(x)^2 + (1 - t\bar{x} - tS_{1,0})S_1^-(x) + S_{0,1}[tS_{0,1} + tS_{1,0} - 1] = 0. \tag{6}$$

4.4 Second application of the kernel method

Let us focus on Eq. (5). Its kernel $[1 - t(x + \bar{x} + 2)][1 - t(x + \bar{x} - 2)]$, as a rational function of x, has four roots. Two of them are formal power series in t. With the notations used in Section 2.2, these two roots are

$$C(t) - 1 = u\,\frac{1+u}{1-u} \quad \text{and} \quad 1 - C(-t) = u\,\frac{1-u}{1+u}.$$

Replacing x by these roots in Eq. (5) provides two relations between the unknown functions $S_{1,0}$ and $S_{0,1}$, from which we compute:

$$S_{0,1} = \frac{u}{1-u^2} \quad \text{and} \quad S_{1,0} = \frac{u(1+u^2)}{1-u^2}.$$

4.5 The complete solution

We now replace $S_{1,0}$ and $S_{0,1}$ by their values in Eqs. (5) and (6). This yields, with the notations of Section 2.2,

$$S_0(x) = \frac{1}{\sqrt{\Delta(x)}}$$

and

$$S_1^-(x) = \frac{1}{2}\left[\frac{1}{t(1+u^2)} - \bar{x} - \frac{\sqrt{\Delta(\bar{x})}}{u}\right].$$

Having computed $S_0(x)$, $S_1^-(x)$ and $S_{1,0}$, we can now express the series $S(x,y)$ and $T(x,y)$ using Eqs. (2) and (3). In particular, we obtain

$$S(x,y;t) = \frac{t\sqrt{\Delta(\bar{x})}}{u[1-t(x+\bar{x}+y+\bar{y})]}$$

which completes the proof of Theorem 1.

■

5 The limit distribution of the endpoint

We wish to describe the distribution of the endpoint of a random path in the slit plane. More precisely, when all paths of length n are taken equally likely, the endpoint becomes a random variable (X_n, Y_n). With the notations used in the previous sections,

$$\Pr((X_n, Y_n) = (i,j)) = \frac{[x^i y^j t^n]S(x,y;t)}{[t^n]S(1,1;t)} = \frac{a_{i,j}(n)}{a(n)}.$$

By expanding the series $S(x,y;t)$, we can plot the histograms of X_n and Y_n. This suggests to normalize by $\sqrt{n}$ (Fig. 2) and that the normalized random variable $(X_n/\sqrt{n}, Y_n/\sqrt{n})$ converges in distribution.

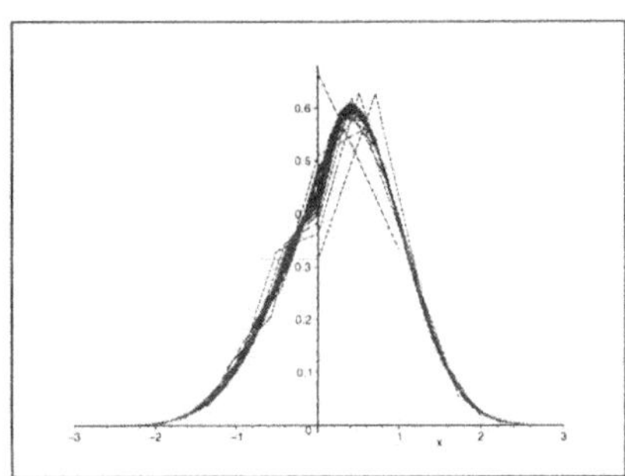

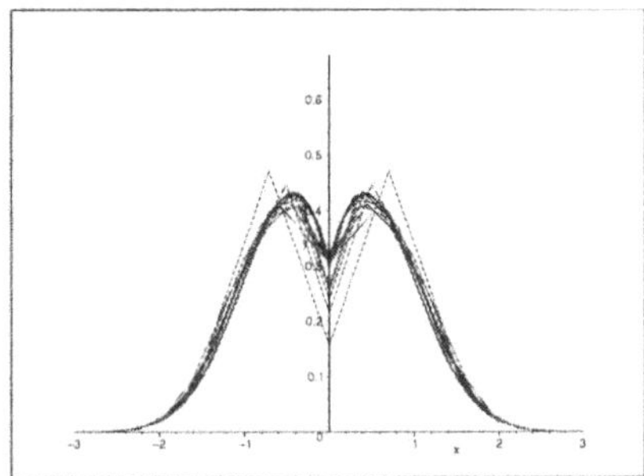

Figure 2: The convergence of $\sqrt{n}\Pr(X_n = i)$ and $\sqrt{n}\Pr(Y_n = i)$ against $i/\sqrt{n}$, for $n = 2, 3, \ldots, 10, 20, 30, \ldots, 100$.

We have, indeed, proved the existence of a limit law, and obtained a (big) expression for its density. However, comparing with the corresponding continuous problem (a Brownian motion conditioned so that it avoids a half-line) suggests that this expression can be simplified. Let us simply state a result on the average abscissa of the endpoint:

$$< X_n > \sim \frac{\Gamma(3/4)}{\Gamma(1/4)}\sqrt{n}.$$

The behaviour in $\sqrt{n}$ had been observed numerically [6], and is also known to hold in the corresponding continuous model.

Acknowledgements. We are indebted to Olivier Roques who discovered the remarkable conjectures that were the starting point of this work. We also thank Philippe Duchon for bringing them to our attention, and Tony Guttmann amd Greg Lawler for providing extremely useful references to previous work.

References

[1] C. Banderier, M. Bousquet-Mélou, A. Denise, P. Flajolet, D. Gardy and D. Gouyou-Beauchamps, On generating functions of generating trees, Proceedings of the 11th conference "Formal power series and algebraic combinatorics", Barcelona, June 1999, pp. 40–52.

[2] E. Barcucci, E. Pergola, R. Pinzani and S. Rinaldi, A bijection for some paths on the slit plane, in preparation.

[3] M. Bousquet-Mélou and G. Schaeffer, Counting paths of the slit plane, in preparation.

[4] M. Bousquet-Mélou, Counting paths of the slit plane via the cycle lemma, in preparation.

[5] M. Bousquet-Mélou and M. Petkovšek, Linear recurrences with constant coefficients: the multivariate case. *Discrete Mathematics*, to appear.

[6] D. Considine and S. Redner, Repulsion of random and self-avoiding walks from excluded points and lines, *J. Phys. A: Math. Gen* **22** (1989) 1621–1638.

[7] R. Cori and J. Richard, Énumération des graphes planaires à l'aide des séries formelles en variables non commutatives, *Discrete Math.* **2** No.2 (1972) 115–162.

[8] G. Fayolle and R. Iasnogorodski, Solutions of functional equations arising in the analysis of two-server queueing models, In *Performance of computer systems*, pages 289–303. North-Holland, 1979.

[9] P. Flajolet, Analytic models and ambiguity of context–free languages, *Theoret. Comput. Sci.* **49** (1987) 283–309.

[10] H. Kesten, Hitting probabilities of random walks on $\mathbb{Z}^d$, *Stoch. Proc. and Appl.* **25** (1987) 165–184.

[11] D. E. Knuth, *The Art of Computer Programming, Vol. 1: Fundamental Algorithms.* Addison-Wesley, Reading Mass., 1968. Exercises 4 and 11, Section 2.2.1.

[12] G. F. Lawler, *Intersections of random walks*, Probabilities and its applications, Birkhäuser Boston, 1991.

Trends in Mathematics, © 2000 Birkhäuser Verlag Basel/Switzerland

Random generation of words of context-free languages according to the frequencies of letters

ALAIN DENISE *LRI, UMR CNRS 8623, Université Paris-Sud XI.*
Alain.Denise@lri.fr
OLIVIER ROQUES *LaBRI, UMR CNRS 5800, Université Bordeaux I.*
roques@labri.u-bordeaux.fr
MICHEL TERMIER *IGM, UMR CNRS 8621, Université Paris-Sud XI.*
termier@igmors.u-psud.fr

Abstract. *Let L be a context-free language on an alphabet $X = \{x_1, x_2, \ldots, x_k\}$ and n a positive integer. We consider the problem of generating at random words of L with respect to a given distribution of the number of occurrences of the letters. We consider two alternatives of the problem. In the first one, a vector of natural numbers $(n_1, n_2, \ldots, n_k)$ such that $n_1 + n_2 + \cdots + n_k = n$ is given, and the words must be generated uniformly among the set of words of L which contain exactly n_i letters x_i $(1 \leq i \leq k)$. The second alternative consists, given $\mathbf{v} = (v_1, \ldots, v_k)$ a vector of positive real numbers such that $v_1 + \cdots + v_k = 1$, to generate at random words among the whole set of words of L of length n, in such a way that the* expected *number of occurrences of any letter x_i equals nv_i $(1 \leq i \leq k)$, and two words having the same distribution of letters have the same probability to be generated. For this purpose, we design and study two alternatives of the* recursive method *which is classically employed for the uniform generation of combinatorial structures. This type of "controlled" non-uniform generation can be applied in the field of statistical analysis of genomic sequences.*

1 Introduction

The problem of *uniform* random generation of combinatorial structures has been extensively studied in the past few years. To our knowledge, random generation according to a given distribution was much less treated (except for random numbers, for which one finds an abundant literature –see [3] for example). We are interested here in a problem of this type. Let L be a language on an alphabet $X = \{x_1, x_2, \ldots, x_k\}$, and n an integer. Let us denote L_n the set of words of L of length n. The problem consists in generating words of L_n while respecting a distribution of the letters given by a vector of k positive numbers. We consider two alternatives:

1. *Generation according to exact frequencies.* The distribution of the number of letters of any word must respect exactly a given vector of integers $(n_1, \ldots, n_k)$. In other words, we generate words uniformly at random in a subset of L_n constituted of all the words $w \in L_n$ such that $|w|_{x_i} = n_i$ for all $i \in \{1, 2, \ldots, k\}$.

2. *Generation according to expected frequencies.* The words must respect *on average* a distribution given by a vector $\mathbf{v} = (v_1, \ldots, v_k)$ such that $v_1 + \ldots + v_k = 1$. More precisely, we generate words at random in such a way that

(a) any word of L_n has a positive probability to be generated;

(b) for all $i \in \{1, 2, \ldots, k\}$, the expected number of occurrences of x_i in the words is equal (or asymptotically equivalent) to nv_i: if $p(w)$ is the probability of the word w to be generated by the algorithm, we must have $\sum_{w \in L_n} |w|_{x_i} p(w) \sim nv_i$;

(c) two words having the same distribution of letters have the same probability of being generated.

These generation schemes lead to applications in genomics. An important problem in the field of analysis of genomes consists in determining whether some properties observed in genomic sequences are biologically significant or not. The main idea is as follows: if a property observed in natural sequences is really relevant from a biological point of view, one should not observe it significantly in random sequences. Thus for example, in order to evaluate the biological significance of similarities between proteinic sequences of different organisms, one compares the scores of their alignments with scores obtained on random sequences [15, 14]. Similarly, the comparison of the frequencies of certain motifs in natural and random sequences can contribute to determine if these motifs are biologically relevant [7, 22, 23].

Traditionally, sequences are generated according to purely statistical considerations. The fundations of these models and the first algorithms of generation were described in [8]. The parameters which are taken into account are the frequencies of nucleotides (letters in DNA) or oligonucleotides (factors) of fixed length l observed in a natural sequence which is taken as reference. Thus for example, for $l = 2$ and the natural sequence *aatgtaacgt*, the frequencies are $aa = 2$, $at = 1$, $tg = 1$, $gt = 2$, $ta = 1$, $ac = 1$ and $cg = 1$. Random sequences are generated with respect to these frequencies, either exactly (generation according to exact frequencies) or in average (generation according to expected frequencies).

In this last case, the generation is carried out according to a Markov chain of order $l-1$: the sequence is generated letter by letter and, at any step, the probability of generating a given letter depends on the $l-1$ previous letters. The process is clearly linear according to the length of the sequence. In fact, in this model, recent works [17, 20] allow to compute analytically some parameters (expected number of occurrences, variance, etc) concerning the frequencies of appearance of given motifs in random sequences. This allows to avoid the generation of a large number of sequences, and to obtain exact values of the required parameters. However, random generation remains useful when the studied properties do not relate to relatively simple motifs.

Random generation according to exact frequencies is a more difficult problem if one wishes the generation to be uniform among all the allowed sequences. This problem is solved in linear expected time in [13]. One of the main ideas is the fact, stated in [8], that it can be reduced to the generation of a random Eulerian trail in a directed graph. An implementation of the algorithm is presented in [1].

These methods do not handle syntactic constraints: the words are generated in X^*. But it is of interest to generate words in particular languages because genomic sequences can be syntactically constrained. In this context, the aim of our work is to generate more "realistic" random sequences by taking into account syntactic criteria as well as statistical ones.

Our approach is based on the so-named *recursive* method, which was initiated by Nijenhuis and Wilf [18] and then generalized and formalized by Flajolet, Zimmermann and Van Cutsem [9]. Section 2 is devoted to a short presentation of this methodology within the framework of the context-free languages. We present in section 3 a simple adaptation which allows to generate words in exact frequencies of the letters. In section 4, we focus on generation according to expected frequencies.

2 Uniform generation

The general methodology of uniform random generation of decomposable structures (which include context-free languages) is presented in detail in [9]. Some variations which deal with the special case of context-free languages are studied in [16] and [10]. We present here a simple version of the method. The reader will find more powerful alternatives in the referenced papers.

The starting point is a non-ambiguous context-free grammar in Chomsky Normal Form: any right member of a rule is either the empty word ε, or a letter of X, or a non-terminal symbol, or a product (concatenation) of two non-terminal symbols. Moreover, any non-terminal symbol can be left member of at most two rules; and in this case each right member is a single non-terminal symbol.

The first stage consists in counting words: for any non-terminal symbol T and for all $0 \leq j \leq n$, the number $T(j)$ of words of length j which derive from T is computed. This can be done by using recurrence relations that result directly from the rules of the grammar:

$$T \to \varepsilon \quad \Rightarrow \quad T(0) = 1\,; \tag{1}$$

$$T \to x_i \quad \Rightarrow \quad T(1) = 1\,; \tag{2}$$

$$T \to T' \mid T'' \quad \Rightarrow \quad T(n) = T'(n) + T''(n)\,; \tag{3}$$

$$T \to T'T'' \quad \Rightarrow \quad T(n) = \sum_{n'+n''=n} T'(n')T''(n''). \tag{4}$$

Note that, since the generating series of any context-free language is holonomic, these same coefficients can be calculated by using linear recurrences. This leads to faster computations (see [10] for example).

This preprocessing stage is done only once, whatever the number of words of size n (or less) than one wishes to generate. A random word is generated by carrying out a succession of derivations starting from the axiom of the grammar. At each step, a derivation is chosen with the appropriate probability. Suppose for instance that, at a given step of generation of a word of length j, one has to choose a rewriting rule for the symbol T. If $T \to T' \mid T''$, one chooses to generate a word of length j either deriving from T' with probability $T'(j)/T(j)$, or deriving from T'' with probability $T''(j)/T(j)$. If $T \to T'T''$, one chooses an integer h with probability $T'(h)T''(j-h)/T(h)$, and then one generates a word of length h deriving from of T', concatenated to a word of length $j-h$ deriving from T''. Details of the process are given in [9].

The best algorithms derived from this method carry out the generation of a word of length n in $O(n)$ arithmetic operations after a preprocessing stage in $O(n^2)$ [16], or the generation of a word in $O(n \log n)$ arithmetic operations after a preprocessing stage in $O(n)$ [10]. Note however that the coefficients which are computed are extremely large, so the *bit complexity* is a more accurate measure for these algorithms, as discussed in [2].

The above method applies obviously to rational languages. In this particular case, a slightly different approach makes it possible to carry out a generation in $O(n)$ arithmetic operations with a preprocessing stage in $O(n)$ too [12]. We consider a regular grammar of the language, in which any rule has the shape $T \to T_1 \mid \ldots \mid T_m$, where either $T_i = xT_j$ or $T_i = x$ ($x \in X$), or $T_i = \varepsilon$. The $T(.)$'s are computed using the following recurrences:

$$T \to T_1 \mid \ldots \mid T_m \quad \Rightarrow \quad T(n) = T_1(n) + \ldots + T_m(n) \tag{5}$$

where

$$T_i = xT_j \quad \Rightarrow \quad T_i(n) = T_j(n-1) \tag{6}$$
$$T_i = x \quad \Rightarrow \quad T_i(n) = 1 \text{ if } n = 1 \text{ ; } T_i(n) = 0 \text{ otherwise.} \tag{7}$$
$$T_i = \varepsilon \quad \Rightarrow \quad T_i(n) = 1 \text{ if } n = 0 \text{ ; } T_i(n) = 0 \text{ otherwise.} \tag{8}$$

The generation stage is processed like in the general case. Here, each step simply consists in choosing a letter with an appropriate probability. This is equivalent to build a path of length n in a deterministic finite-state automaton of the language.

3 Generation according to exact frequencies

Let $(n_1, n_2, \ldots, n_k)$ be a vector of integers such that $n_1 + n_2 + \cdots + n_k = n$. Our goal is to generate uniformly at random a word of L_n which contains exactly n_i letters x_i, for all $1 \leq i \leq k$.

The principle of the method that we describe here is a natural extension of the general outline given in the previous section. The method is implicitely used in works like [5], were the problem of randomly generating structures while fixing more than one parameter is adressed. Our goal is simply to formalize the method within the framework of the generation of words with fixed numbers of letters in context-free languages.

Like above, the grammar is supposed to be in Chomsky Normal Form. For any non-terminal symbol T, we note $T(j_1, j_2, \ldots, j_k)$ the number of words which derive from T and which contain j_1 letters x_1, j_2 letters x_2, $\ldots$, j_k letters x_k.

The preprocessing stage consists in computing a table of the $T(j_1, j_2, \ldots, j_k)$ for $0 \le j_1 \le n_1, \ldots, 0 \le j_k \le n_k$. We use the following recurrences:

$$T \to \varepsilon \quad \Rightarrow \quad T(0, \ldots, 0) = 1 \; ; \tag{9}$$

$$T \to x_i \quad \Rightarrow \quad T(0, \ldots, 0, 1, 0, \ldots, 0) = 1 \text{ if } n_i = 1. \; ; \tag{10}$$

$$T \to T' \mid T'' \quad \Rightarrow \quad T(n_1, \ldots, n_k) = T'(n_1, \ldots, n_k) + T''(n_1, \ldots, n_k) \; ; \tag{11}$$

$$T \to T'T'' \quad \Rightarrow \quad T(n_1, \ldots, n_k) = \sum_{\substack{n'_1 + n''_1 = n_1 \\ n'_2 + n''_2 = n_2 \\ \cdots \\ n'_k + n''_k = n_k}} T'(n'_1, \ldots, n'_k) T''(n''_1, \ldots, n''_k). \tag{12}$$

Like above, each step of the generation stage consists in choosing a rewriting rule of the current symbol T. Suppose that, at a given step of generation of a word of distribution $\mathbf{j} = (j_1, \ldots, j_k)$, one has to choose a rewriting rule for the symbol T. If $T \to T' \mid T''$, one generates a word of distribution $\mathbf{j}$ deriving from T' with probability $T'(\mathbf{j})/T(\mathbf{j})$, or deriving from T'' with probability $T''(\mathbf{j})/T(\mathbf{j})$. If $T \to T'T''$, one chooses a vector $\mathbf{h} = (h_1, \ldots, h_k)$ with probability $T'(\mathbf{h})T''(\mathbf{j} - \mathbf{h})/T(\mathbf{h})$, then one generates a word deriving from T' having distribution $\mathbf{h}$, concatenated to a word deriving from T'' having distribution $\mathbf{j} - \mathbf{h}$.

The rational case is similar:

$$T \to T_1 \mid \ldots \mid T_m \quad \Rightarrow \quad T(n_1, \ldots, n_k) = T_1(n_1, \ldots, n_k) + \ldots + T_m(n_1, \ldots, n_k)$$

where

$$T_i = x_m T_j \quad \Rightarrow \quad T_i(n_1, \ldots, n_m, \ldots, , n_k) = T_j(n_1, \ldots, n_m - 1, \ldots, n_k) \; ;$$

$$T_i = x_m \quad \Rightarrow \quad T_i(0, \ldots, 0, 1, 0, \ldots, 0) = 1 \text{ if } n_m = 1 \; ;$$

$$T_i = \varepsilon \quad \Rightarrow \quad T_i(0, \ldots, 0) = 1.$$

The bottleneck of this method lies in its strong complexity in time and memory: it requires the computation of a table of $\Theta(n_1 n_2 \ldots n_k)$ values. Moreover, in the nonrational case, the multiple convolutions of the formula (12) imply a very slow generation stage too. On the other hand, if the language is rational, the generation is carried out in linear arithmetic complexity (but the problem of computing the table remains). Thus the method is feasible for an alphabet of limited size. It can also be extended to an alphabet of any size, provided that the number of constraints is limited. For example, instead of fixing $(n_1, n_2, \ldots, n_k)$, one may fix only $(n_1, \ldots, n_j)$ with $j < k$.

4 Generation according to expected frequencies

In this section, we consider the problem of generating words of L_n at random in such a way that each word is generated with positive probability $p(w)$, and the vector of expected distributions of letters equals the given vector $\mathbf{v}$. More precisely:

$$p(w) > 0 \quad \forall w \in L_n \tag{13}$$

and

$$\frac{1}{n}\sum_{w\in L_n}|w|_{x_i}p(w) \sim v_i \ \ \forall i\in\{1,2,\ldots,k\}. \tag{14}$$

Moreover, two words having the same letters distribution must be equally generated:

$$(|w|_{x_i}=|w'|_{x_i}\ \forall i\in\{1,2,\ldots,k\}) \Rightarrow p(w)=p(w'). \tag{15}$$

Our method consists in assigning a *weight* to each letter of the alphabet. For this purpose, we define a *weight function* $\pi : X \to I\!R_+^*$. It naturally acts on X^*:

$$\pi(w)=\prod_{1\le i\le k}\pi(x_i)^{|w|_{x_i}},$$

and on any finite language, in particular on L_n:

$$\pi(L_n)=\sum_{w\in L_n}\pi(w).$$

If the algorithm is such that

$$p(w)=\frac{\pi(w)}{\pi(L_n)}, \quad \forall w\in L_n, \tag{16}$$

then the larger the weight of any given letter is (with regard to the weights of the other letters), the more this letter will occur in a random sample. On the other hand, formula (16) implies conditions (13) and (15).

Now we have to solve two problems:

1. find a function π satisfying (14), providing that(16) is respected;
2. design a generation algorithm which satisfies (16).

Let us look first at the second point. Suppose that π is given. In order to generate words with the required distribution (16), we use the methodology presented in section 2, but now the rule

$$T\to x_i \quad\Rightarrow\quad T(1)=\pi(x_i).$$

stands in for the rule (2). For the particular case of rational languages, the rules

$$\begin{aligned} T_i = xT_j \quad &\Rightarrow\quad T_i(n)=\pi(x)T_j(n-1)\\ T_i = x \quad &\Rightarrow\quad T_i(n)=\pi(x) \text{ if } n=1\ ;\ T_i(n)=0 \text{ otherwise.}\end{aligned}$$

stand in for (6) and (7).

The generation process works exactly like the uniform one of section 2. In this way, the probability that a word w occurs will be proportional to its weight $\pi(w)$.

Now we have to find a function π which satisfies (14). Let us consider the following generating function:

$$L_\pi(t, \mathbf{x}) = \sum_{w \in L} \pi(w) t^{|w|} x_1^{|w|_{x_1}} \dots x_k^{|w|_{x_k}},$$

where $\mathbf{x} = (x_1, x_2, \dots, x_k)$.

For all $i \in \{1, \dots, k\}$, the expected number of occurrences of x_i in the words generated by the algorithm is equal to $n\mu_i(\pi)$, where

$$\mu_i(\pi) = \frac{\sum_{w \in L_n} |w|_{x_i} \pi(w)}{n\pi(L_n)} = \frac{[t^n]\Gamma_{\pi,x_i}(t)}{[t^n]\Gamma_\pi(t)} \qquad (17)$$

with

$$\Gamma_{\pi,x_i}(t) = \frac{\partial L_\pi(t,\mathbf{x})}{\partial x_i}(t, \mathbf{1}) \quad \text{and} \quad \Gamma_\pi(t) = t\frac{\partial L_\pi(t,\mathbf{x})}{\partial t}(t, \mathbf{1})$$

Now our problem consists in finding a function π such that $\mu_i(\pi) = v_i$ for all $i \in \{1, \dots, k\}$. Two cases have to be considered, whether L is rational or (non rational) context-free. Let us look first at the latter.

4.1 The context-free case

¿From a more general theorem due to Drmota [4, Theorem 1, p.107], we immediatly state

Theorem 1 *Let $\mathbf{y} = \mathbf{F}(t, \mathbf{y}, \mathbf{x})$ be a set of equations coming from a weighted context-free grammar, where $\mathbf{y} = (y_1, y_2, \dots, y_N)$ is the vector of generating functions for the non terminal symbols and $\mathbf{F}(t, \mathbf{y}, \mathbf{x}) = (F_1(t, \mathbf{y}, \mathbf{x}), \dots, F_N(t, \mathbf{y}, \mathbf{x}))$. Suppose that*

- $\mathbf{F}(0, \mathbf{y}, \mathbf{x}) = 0$ *and* $\mathbf{F}(t, \mathbf{0}, \mathbf{x}) \neq 0$*;*
- *the system of equations is not linear in* $\mathbf{y}$*;*
- *the system of equations is* simple, *i.e there exists a set of N $(k+1)$-dimensional cones $C_i \subseteq \mathbb{R}^{k+1}$ such that, for any $(n, m_1, \dots, m_k) \in C_j$ with n, m_1, ..., m_k large enough, the coefficient of the term $t^n x_1^{m_1} \dots x_k^{m_k}$ is positive;*
- *the grammar is* strongly connected, *i.e each non terminal symbol can be reached from any other non terminal symbol.*

Under these assumptions, if the system of equations

$$\begin{cases} \mathbf{y} &= \mathbf{F}(t, \mathbf{y}, \mathbf{1}) \\ 0 &= \det(I - \mathbf{F}_\mathbf{y}(t, \mathbf{y}, \mathbf{1})) \end{cases}$$

where $\mathbf{F}_\mathbf{y}$ is the matrix $\left(\frac{\partial F_i}{\partial y_j}\right)_{1 \leq i,j \leq N}$, admits a positive solution $(\mathbf{y}_0, t_0)$, then

$$\mu_i(\pi) = \frac{1}{t(1)} \frac{\partial t}{\partial x_i}(\mathbf{1})$$

for any $1 \leq i \leq k$, *where* $t(\mathbf{x})$ *is a solution of*

$$\left\{\begin{array}{lll} \mathbf{y} & = & \mathbf{F}(t,\mathbf{y},\mathbf{x}) \\ 0 & = & \det(I - \mathbf{F}_{\mathbf{y}}(t,\mathbf{y},\mathbf{x})) \end{array}\right. \tag{18}$$

such that $t(\mathbf{1}) = t_0$.

Now, in order to solve our problem, we just have to take the generating function $L_\pi(t,\mathbf{x})$ as the first component of the vector $\mathbf{y}$. However, Theorem 3 does not apply to rational languages neither to context-free languages which generating series is rational, because in these cases the system of equations is linear.

Here is an example of application. We consider the problem of generating rooted plane trees (see [19] for example) with n edges, setting the expected number of leaves at μn $(0 < \mu < 1)$. A classical bijection turns this problem into the problem of generating words of length $2n$ belonging to the language defined by the grammar $S \to aSbS + aSb + cdS + cd$. The factor cd represents a leaf of the tree. We can now modify the expected number of leaves in a random tree by associating a weight $\pi_1 > 0$ to the letter $c : S \to aSbS + aSb + \pi_1 cdS + \pi_1 cd$. This weighted grammar satisfies the conditions of Theorem 1 and thus we get

$$\left\{\begin{array}{lll} F(t,y,x_1) & = & xy^2 + xy + \pi_1 z_1 xy + \pi_1 z_1 x \\ F_y(t,y,x_1) & = & 2\,xy + x + \pi_1 z_1 x. \end{array}\right.$$

(Since letters are paired, we consider only a's and c's.) Solving the system (18) gives

$$\pi_1 = \left(\frac{\mu}{1-\mu}\right)^2 \quad \text{with } 0 < \mu < 1.$$

This means that in order to generate a word (resp. rooted plane tree) with $2n$ letters (resp. n edges) having μn letters c (resp. leaves) on average, we have to set $\pi_1 = (\mu/(1-\mu))^2$.

In fact, this example is particularly simple. In general, we may be unable to solve numerically the system (18) because of the unknown function π. Thus for instance, our method fails to generate words of the language defined in [4, p.112] with a given expected number of letters a and b, since we have to solve an algebraic equation of degree five whose coefficients depend of the unknown weights. Morever, Theorem 1 only applies to strongly connected grammars.

4.2 The rational case

In this section, we show that we can solve the problem of generating word according to given frequencies for a whole class of rational languages. If L is a rational language, the functions $\Gamma_{\pi,x_i}(t)$ and $\Gamma_\pi(t)$ defined in (17) are rational fractions. Therefore the nth coefficients of their Taylor series at $t = 0$ depend on the zeros of their denominators (see for example [11, chap. 7]). If we are only interested in the expected number of letters in the words of length n as n goes to infinity, we just need to know the dominant singularity of each series, that is the real zero of smallest modulus of their denominator, say s, and its multiplicity ν. Let us recall that, from Pringsheim's theorem, at least one of the dominant singularities of the generating function of any rational language is real positive.

However, in our problem, the denominator depends on the unknown variables $\pi(x_i)$. Hence, we encounter ***a priori*** the same difficulty as in the context-free case: that of solving a possibly high degree context-free equation in the variable s with unknown coefficients, in others words without the help of numerical approximation. However, thanks to the following proposition, we will be able to find a solution.

Proposition 2 *Let L be a rational language on the alphabet $X = \{x_1, x_2, \ldots, x_k\}$. Let π a weight function on X. Assume that s is the unique real pole of smallest modulus of $\Gamma_\pi(t)$ (resp. of $\Gamma_{\pi,x_i}(t)$), with multiplicity ν. It follows that $\mu_i(s\pi) = \mu_i(\pi)$ for any $i \in \{1, \ldots, k\}$, and the unique real pole of smallest modulus of $\Gamma_{s\pi}(t)$ (resp. of $\Gamma_{s\pi,x_i}(t)$) is 1, with mutiplicity ν.*

Proof. Under the assumption of Proposition 2, the denominator $Q(t)$ of $\Gamma_\pi(t)$ can be written as $Q(t) = (t-s)^\nu P(t)$ where the modulus of each root of $P(t)$ is greater than s. Hence, the denominator of $\Gamma_{s\pi}(t)$ is $Q(ts) = (ts - s)^\nu P(ts)$ and it follows that 1 is the unique root of smallest modulus of $Q(ts)$, with multiplicity ν. □

The above Proposition means that if there exists a weight function leading to the required frequency then there exists another weight function π, leading also to the required frequencies, such that the series of interest are singular at 1.

We shall now derive sufficient conditions for the series involved in Proposition 2 to have the same dominant pole with same multiplicity. This is the purpose of the following Theorem.

Theorem 3 *Let L be a rational language defined on the alphabet $X = \{x_1, \cdots, x_k\}$ and $\mathcal{A}$ the finite minimal deterministic automaton which recognizes L. Let $C = \{C_1, \cdots, C_f\}$ the set of non trivial strongly connected components of $\mathcal{A}$. Assume that for any $i \in \{1, \cdots, f\}$, C_i is aperiodic (in the sense of Markov chains). Under this assumption, for any weighting function π, we have*

- *$L_\pi(t, 1)$ has an unique pole of smallest modulus and multiplicity ν.*
- *$\mu_i(\pi) > 0$ for any $i \in \{1, \cdots, k\}$ as n goes to infinity.*

This result can be proved with the help of the Perron-Frobenius theorem on the properties of the eigenvalues of irreducible and primitive matrix (see [19, 17].)

Corollary 4 *Under the assumption of Theorem 3, each series $\Gamma_\pi(t)$ and $\Gamma_{\pi,x_i}(t)$ for $i \in \{1, \ldots, k\}$ has an unique pole of smallest modulus and multiplicity $\nu + 1$.*

Using the above results, We design below an algorithm in order to find a weight function π which gives the required frequencies as in (14) .

1. Compute the series $L_\pi(t, \mathbf{x})$, $\Gamma_\pi(t)$ and $\Gamma_{\pi,x_i}(t)$ for $i \in 1, \ldots, k$ (in which $\pi(x_i)$ are parameters which values are unknown). Let s the unique dominant pole of these series.

2. Set $s = 1$ and take ν as in Theorem 3. Compute the asymptotic values, say $\bar{\mu}_i(\pi)$, of each $\mu_i(\pi)$ according to (17) and using ([21][Theorem 4.1]). This can be done since Corollary 4 ensures that all series have the same unique dominant pole.

3. Solve the following algebraic system:

$$\begin{cases} \bar{\mu}_1(\pi) & = & v_1 \\ & \vdots & \\ \bar{\mu}_{k-1}(\pi) & = & v_{k-1} \\ D_\pi(1, \mathbf{1}) & = & 0 \end{cases}$$

in the unknown variables $(\pi(x_1), \pi(x_2), \ldots, \pi(x_k))$, where $D_\pi(t, \mathbf{x})$ is the denominator of the series $L_\pi(t, \mathbf{x})$. (The last equation forces 1 to be a pole of the series.)

4. Look for a solution of the above system such that

   - each $\pi(x_i)$ is positive;
   - the pole of smallest modulus of $L_\pi(t, \mathbf{1})$ is 1. (In step 3, we have forced 1 to be a pole, but not the dominant one!)

The system of step 3 above can be solved with numerical techniques (using GB [6] for example) If we don't find a solution that holds for the step 4 of the algorithm, then, from Proposition 2, no suitable weight function π exists.

Let us show how the algorithm works by taking a language involved in genomics. An ORF ("*Open Reading Frame*") in a sequence of DNA is a particular word which represents a "putative proteinic gene". In some simple organisms, one can approximate the "language of ORFs" by the following rational language:

$$L = atg((X^3 \backslash \{taa, tag, tga\})^*)(taa + tag + tga),$$

where $X = \{a, c, g, t\}$. The minimal automaton of L has got an unique strongly connected component, which contains all the letters of X.

¿From the regular expression above, we find the generating function

$$L(x, a, c, g, t) = \frac{atgx(2tgax + taax)}{(1 - (a + c + g + t)^3 x + taax + 2tgax)}$$

where the coefficient of x^n is the number of words of length $3n$ in order to respect the primitivity condition. The series $L(x, a, c, g, t)$ has an unique pole of smallest modulus, $\frac{1}{61}$, which is simple. Taking for instance the trivial weight function, say π_0=1, we find the root of $L(x, a, 1, 1)$ to be $\rho(a) = \frac{1}{a^3+8a^2+25a+27}$. Expanding $\rho(a)$ at a=1, we get

$$\rho(a) = \frac{1}{61} - \frac{44}{3721}(a-1) - O(a-1)^2 \quad \text{and} \quad \bar{\mu}_a = 44/183.$$

In a similar way, the frequency of the other letters of X are $\bar{\mu}_c = 16/61$; $\bar{\mu}_g = 46/183$; $\bar{\mu}_t = 15/61$. This means for example that an uniform random word of L has got $44\,n/183$ letters a on average.

We are now looking for a weight function π such that the frequencies of nucleotides are the same as those of the chromosome III of *Saccharomyces cerevisiae* (brewers' yeast), say $\mathbf{v} = (0.307\,;0.193\,;0.193\,;0.307)$. Proceeding as above, but now with the series

$$L_\pi(x,a,c,g,t) = \frac{atgx\pi(atg)(tgax2\pi(tga) + taax\pi(taa))}{(1-(a\pi(a)+c\pi(c)+g\pi(g)+t\pi(t))^3x + taax\pi(taa) + 2tgax\pi(tga))}$$

we express (setting $s = 1$) $\bar{\mu}_a(\pi)$, $\bar{\mu}_c(\pi)$, $\bar{\mu}_g(\pi)$, $\bar{\mu}_t(\pi)$ as functions of $\pi(a)$, $\pi(c)$, $\pi(g)$, $\pi(t)$. With the help of the formal calculus software Maple, we solve the system

$$\begin{cases} \bar{\mu}_a(\pi) & = & 0.307 \\ \bar{\mu}_c(\pi) & = & 0.193 \\ \bar{\mu}_g(\pi) & = & 0.193 \\ \bar{\mu}_t(\pi) & = & 0.307 \\ 1-(\pi(a)+\pi(c)+\pi(g)+\pi(t))^3 + \pi(t)\pi(a)\pi(a) + 2\pi(t)\pi(g)\pi(a)) & = & 0 \end{cases}$$

and we find the following: $\pi(a) = 0.31632$; $\pi(c) = 0.16991$; $\pi(g) = 0.18396$; $\pi(t) = 0.32707$. One can observe that the weights of the letters are close to the frequencies. This is because the language L is not much constrained in the sense that only few patterns are forbidden. Hence, if we take $\pi = \mathbf{v}$, we find $\bar{\mu}_a(\pi) = 0.29484\ldots$; $\bar{\mu}_c(\pi) = 0.20648\ldots$; $\bar{\mu}_g(\pi) = 0.19351\ldots$; $\bar{\mu}_t(\pi) = 0.30519\ldots$. A statistical test was processed to evaluate the sigificance of differences between π and μ values. We carried out the "conformity to a law" χ^2 test on occurrences of all four nucleotides in a sequence of same length as yest chromosome III, according to values of π and μ. We got the χ^2 value 0.04425 which confirms that there is no significant difference between π and μ.

5 Conclusion

To our knowledge, we present for the first time algorithms of random generation of words wich take into account both statistical and syntactic criteria.

The generation algorithm in exact frequencies presented in section 3 has a restricted sphere of activity since the complexity strongly increases with the number of statistical parameters. Moreover, it does not allow to deal with the frequencies of patterns of length > 1. However, if only few parameters are to be considered, the algorithm works well. Thus for instance, in the field of genomics, the "GC-content", i.e the sum of frequencies of letters g and c, is of great interest in some analysis.

The method given in section 4 for generating according to expected frequencies is much more efficient. After computing once and for all the weight function, the complexity is identical to the one of uniform generation algorithms. Unfortunately, in the context-free case we are only able to deal with low degree functional equations. In the case of rational languages, numerical techniques and tools (like GB [6]) allow to find the suitable weight function in most cases. Moreover, the method can be easily generalised to pattern frequencies, by modifying the automaton or the grammar like in [17].

Aknowledgements

We thank Jean-Guy Penaud, Gilles Schaeffer, Edward Trifonov and Paul Zimmermann for fruitful discussions. We also are grateful to Marie-France Sagot for her help on bibliography.

References

[1] E. Coward. Shufflet: Shuffling sequences while conserving the k-let counts. *Bioinformatics*, 15(12):1058–1059, 1999.

[2] A. Denise and P. Zimmermann. Uniform random generation of decomposable structures using floating-point arithmetic. *Theoretical Computer Science*, 218:233–248, 1999.

[3] L. Devroye. *Non-uniform random variate generation.* Springer-Verlag, 1986.

[4] M. Drmota. Systems of functional equations. *Random Structures and Algorithms*, 10:103–124, 1997.

[5] I. Dutour and J.-M. Fédou. Object grammars and random generation. *Discrete Mathematics and Theoretical Computer Science*, 2:47–61, 1998.

[6] J-.C. Faugère. GB. http://calfor.lip6.fr/GB.html.

[7] J.W. Fickett. ORFs and genes: how strong a connection? *J Comput Biol*, 2(1):117–123, 1995.

[8] W.M. Fitch. Random sequences. *Journal of Molecular Biology*, 163:171–176, 1983.

[9] Ph. Flajolet, P. Zimmermann, and B. Van Cutsem. A calculus for the random generation of labelled combinatorial structures. *Theoretical Computer Science*, 132:1–35, 1994.

[10] M. Goldwurm. Random generation of words in an algebraic language in linear binary space. *Information Processing Letters*, 54:229–233, 1995.

[11] R.L. Graham, D.E. Knuth, and O. Patashnik. *Concrete Mathematics.* Addison Wesley, 2nd edition, 1997. French translation : Mathématiques concrètes, International Thomson Publishing, 1998.

[12] T. Hickey and J. Cohen. Uniform random generation of strings in a context-free language. *SIAM J. Comput.*, 12(4):645–655, 1983.

[13] D. Kandel, Y. Matias, R. Unger, and P. Winkler. Shuffling biological sequences. *Discrete Applied Mathematics*, 71:171–185, 1996.

[14] D.J. Lipman and W.R. Pearson. Rapid and sensitive protein similarity searches. *Science*, 227:1435–1441, 1985.

[15] D.J. Lipman, W.J. Wilbur, T.F. Smith, and M.S. Waterman. On the statistical signifiance of nucleic acid similarities. *Nucleic Acids Research*, 12:215–226, 1984.

[16] H. G. Mairson. Generating words in a context free language uniformly at random. *Information Processing Letters*, 49:95–99, 1994.

[17] P. Nicodème, B. Salvy, and Ph. Flajolet. Motif statistics. In *European Symposium on Algorithms-ESA99*, pages 194–211. Lecture Notes in Computer Science vol. 1643, 1999.

[18] A. Nijenhuis and H.S. Wilf. *Combinatorial algorithms*. Academic Press, New York, 2nd edition, 1978.

[19] Ph. Flajolet and R. Sedgewick. The average case analysis of algorithms: Multivariate asymptotics and limit distribution. *RR INRIA, Number 3162*, 1997.

[20] M. Régnier. A unified approach to word occurrence probabilities. *Discrete Applied Mathematics*, 2000. To appear in a special issue on Computational Biology; preliminary version at RECOMB'98

[21] R. Sedgewick and Ph. Flajolet. *An introduction to the analysis of algorithms*. Addison Wesley, 1996. French translation: Introduction à l'analyse des algorithmes, International Thomson Publishing, 1996.

[22] M. Termier and A. Kalogeropoulos. Discrimination between fortuitous and biologically constrained Open Reading Frames in DNA sequences of *Saccharomyces cerevisiae*. *Yeast*, 12:369–384, 1996.

[23] A. Vanet, L. Marsan, and M.-F. Sagot. Promoter sequences and algorithmical methods for identifying them. *Res. Microbiol.*, 150:779–799, 1999.

Trends in Mathematics, © 2000 Birkhäuser Verlag Basel/Switzerland

An Algebra for Proper Generating Trees

D. MERLINI, R. SPRUGNOLI, M. C. VERRI. *Dipartimento di Sistemi e Informatica, via Lombroso 6/17, 50134, Firenze, Italia.*
[merlini,sprugnoli,verri]@dsi.unifi.it

Abstract. *We find an algebraic structure for a subclass of generating trees by introducing the concept of marked generating trees. In these kind of trees, labels can be marked or non marked and the count relative to a certain label at a certain level is given by the difference between the number of non marked and marked labels. The algebraic structure corresponds to a non commutative group with respect to a product operation between two generating trees. Hence we define the identity generating tree and the inverse of a given generating tree.*

1 Introduction

The concept of a *proper Riordan Array* (pRA, for short) is very useful in Combinatorics. The infinite triangles of Pascal, Catalan, Motzkin and Schröder are important and meaningful examples of pRA's, but many applications have been proposed and developed, thus proving the effectiveness of a Riordan Array approach to many combinatorial problems (see, e.g. [3, 4, 5, 9, 10]). Without doubt, the relevance of pRA's derives from the fact that a pRA D is completely and univocally described by a couple of *formal power series* (fps, for short): $D = (d(t), h(t))$, which can be seen as generating functions of sequences related to the array. This implies that a Riordan Array approach to a problem can take advantage from the concept of generating function and from the well-understood apparatus of formal power series. For example, this may concern the extraction of coefficients and the computation of asymptotic values, when fps are considered as analytic functions.

The algebraic structure of pRA's was one of the first properties pointed out (see, e.g., [8, 9]). In fact, they constitute a group by the usual row-by-column product, and this product can be easily defined through the fps describing the pRA's involved in the operation. Besides, a pRA D and its inverse $\bar{D}$ are related by the Lagrange Inversion Formula, which therefore becomes a fundamental tool in the theory and the practice of pRA's.

Recently, Merlini and Verri [5] pointed out an important connection between pRA's and *generating trees*. Generating trees are becoming more and more important in Combinatorics (see e.g., [1, 11, 12]), as a device to represent the development of many classes of combinatorial objects, which can then be counted by counting the different labels in the various levels of the tree. The proved relation between pRA's and generating trees allows to combine the counting capabilities of both approaches and thus improve our understanding of the problem under consideration. Therefore, we feel that a further, more deep insight to both pRA's and generating trees is worth studying, and this is the aim of the present paper.

There are two problems to which we dedicate our attention:

1. The definition of a pRA does not impose conditions on the fps involved, except that they should be invertible. On the other hand, it is clear that a fps

with real coefficients (where *real* is intended as opposed to *integer*) cannot be seen as a counting generating function. We have important combinatorial examples of pRA's with rational coefficients, as the triangles related to Stirling numbers of both kinds (see e.g., [9, 10]). However, in general, only pRA's with integer coefficients can have a direct combinatorial interpretation. So, we will restrict our attention to fps $f(t) = \sum_{k=0}^{\infty} f_k t^k$ having $f_0 = 1$ and $f_k \in \mathbf{Z}$, for every $k \in \mathbf{N}$. These fps will be called *monic, integer fps*, and we will prove a number of their properties. Consequently, we define *monic, integer pRA's* as pRA's whose elements are in $\mathbf{Z}$ and those in the main diagonal are 1. These arrays are the main object of our investigations.

2. Several different combinatorial interpretations can be found for monic, integer pRA's, if we can assign a combinatorial meaning to negative values. As a matter of fact, we show that a combinatorial interpretation is possible through a suitable variant of generating trees. To this purpose, we introduce the concept of a *marked label* to be considered together with usual labels: generating trees can be extended to deal with negative values if we consider a node labelled k as opposed to a node labelled $\bar{k}$ (i.e., labelled with a *marked* k) and imagine that they annihilate each other. As far as we know, the concept of marked labels has been implicitly used for the first time in [5]; the same concept has been used in [2] in relation with the definition of the *ECO-systemes signés*. A quite different approach can be found in [6] where the authors deal with what they call *coloured rules*; these correspond to generating trees in which nodes with the same label can have various colours; what makes this approach different is the fact that these labels don't annihilate each other and the count relative to a level in the tree is exactly the number of nodes at that level.

   By introducing marked labels, the correspondence between the extended generating trees and monic, integer pRA's becomes effective. We wish to point out that, as a consequence of this correspondence, we are in a position to define the *product* of two generating trees, the *inverse* of a given generating tree and the *identity* generating tree, three concepts that can be rather surprising.

The structure of this paper is straight-forward. In Section 2. we develop the concept of a monic, integer fps and that of a monic, integer pRA. Then we show some general properties of pRA's to relate them to generating trees: these new properties mainly concern the A- and Z-sequences of a pRA, are valid in the general case, but will then be used in the monic, integer case. In Section 3. we give our main theorem, relating extended generating trees to monic, integer pRA's; finally, we show a number of examples, illustrating some applications of the main theorem and the two new concepts of the inverse and the product of generating trees. In order to safe space we omit all Theorem proofs in this paper.

2 Riordan Arrays

Let $\mathcal{F} = \mathbf{C}[\![t]\!]$ be the integral domain of formal power series over the complex field. Usually, in Combinatorics we refer to this domain because its algebraic structure is well-known. However, only a proper subset of $\mathcal{F}$ has a direct interest: when we use fps as counting generating functions, we implicitly assume that their coefficients are (positive) integer numbers or, at most, are rational numbers in the case of exponential generating functions. Obviously, during some computations, more general situations can arise; the reader can think of signed binomial coefficients, signed Stirling numbers and Bernoulli numbers. However, the set of fps with integer coefficients has so a direct combinatorial relevance that it is interesting to isolate and try to characterise them in some algebraic way. In particular, we wish to study fps with integer coefficients when the t^0 coefficient is 1.

Let $\mathcal{Z} = \mathbf{Z}[\![t]\!]$ be the set of fps $f(t) = \sum_{k=0}^{\infty} f_k t^k$ such that $f_0 = 1$ and $f_k \in \mathbf{Z}$, for every $k > 0$. As we are now going to see, this set has a rather rich algebraic structure and constitutes the basis of a number of interesting developments concerning proper Riordan Arrays and generating trees.

The following lemma describes the algebraic properties of the Cauchy product.

Lemma 2.1 *Let $\cdot$ (or the simple juxtaposition) denote the Cauchy product; then $(\mathcal{Z}, \cdot)$ is a commutative group.* ∎

Generating functions are a device to deal with sequences of numbers, in the sense that we study a fps $f(t) = \sum_{k=0}^{\infty} f_k t^k$ instead of studying the sequence $\{f_0, f_1, f_2, \ldots\}$. When we pass from one dimensional sequences to two dimensional ones $\{d_{n,k}\}_{n,k \in N}$, a relevant concept is given by proper Riordan Arrays. They are defined by a couple of fps $(d(t), h(t))$, having $d(t), h(t) \in \mathcal{F}$, and the generic element of the two dimensional sequence induced by the pRA is $d_{n,k} = [t^n]d(t)(th(t))^k$. Many combinatorial triangles are examples of pRA's, like the Pascal, Catalan, Motzkin and Schröder triangles. In all these cases the elements are integer and those in the main diagonal are 1. Therefore, it seems appropriate to study monic, integer pRA's, and we can prove that these arrays are strictly related to fps in $\mathcal{Z}$. Actually, we have:

Theorem 2.2 *Let $D = (d(t), h(t))$ be a pRA: D is a monic, integer pRA if and only if $d(t)$ and $h(t)$ are monic, integer fps.* ∎

If $\mathcal{R}$ denotes the set of pRA's, it is well-known that $(\mathcal{R}, \star)$, where $\star$ is the usual row-by-column product, is a non-commutative group, called the *Riordan group*. Actually, the operation $\star$ can also be given in terms of the fps involved in the definition of the pRA's. More specifically, let $D = (d(t), h(t))$ and $E = (f(t), g(t))$ be two pRA's; the generic element in $F = D \star E$ is:

$$F_{n,k} = \sum_j D_{n,j} E_{j,k} = \sum_j [t^n]d(t)(th(t))^j [t^j] f(t)(tg(t))^k =$$

$$= [t^n]d(t) \sum_j (th(t))^j [t^j] f(t)(tg(t))^k = [t^n]d(t) f(th(t))(th(t)g(th(t)))^k.$$

This is just the generic element of a pRA:

$$(d(t), h(t)) \star (f(t), g(t)) = (d(t)f(th(t)), h(t)g(th(t))).$$

This shows that $\mathcal{R}$ is closed under $\star$ and by setting $(d(t), h(t)) \star (\bar{d}(t), \bar{h}(t)) = (1, 1)$, the neutral element with respect to $\star$, or $d(t)\bar{d}(th(t)) = 1$ and $h(t)\bar{h}(th(t)) = 1$ we find:

$$\bar{d}(y) = \left[\frac{1}{d(t)} \middle| y = th(t) \right] \qquad \bar{h}(y) = \left[\frac{1}{h(t)} \middle| y = th(t) \right].$$

This notation means that we should first solve the functional equation $y = th(t)$ in t in order to find a function $t = t(y)$, and exactly the solution for which $t(0) = 0$, and then substitute this fps into $1/d(t)$ in order to find $\bar{d}(y) = 1/d(t(y))$: this fps is the d function in the inverse pRA, i.e. in the pRA $\bar{D}$ such that $D * \bar{D} = (1, 1) = I$. An analogous procedure should be followed to find $\bar{h}(t)$. The existence and uniqueness of $\bar{d}$ and $\bar{h}$ is garanted by the fact that the functional equation $y = th(t)$ has a unique solution $t = t(y)$ with $t(0) = 0$ if and only if $h(0) \neq 0$. An important result is:

Theorem 2.3 *If D is a monic integer pRA, then also its inverse $\bar{D}$ is such.* ■

The way to represent a pRA by means of a couple of fps: $D = (d(t), h(t))$ is not the only method to deal with this sort of infinite triangles. In 1978 Rogers [7] proved that pRA have a remarkable property: every element $d_{n+1,k+1}$ not belonging to column 0 or row 0, can be expressed as a linear combination of the elements in the previous row, starting with the element in column k:

$$d_{n+1,k+1} = a_0 d_{n,k} + a_1 d_{n,k+1} + a_2 d_{n,k+2} + \ldots = \sum_{j=0}^{\infty} a_j d_{n,k+j}.$$

The sum is actually finite and the sequence $A = \{a_0, a_1, a_2, \ldots\}$ is independent of n and k, that is, it characterizes the whole triangle, except for column 0 and row 0; A is called the $A-$sequence of the pRA. Actually, column 0 is given by the function $d(t)$, and row 0 is all composed of 0, except for the element $d_{0,0} = d_0$, which is determined by $d(t)$. In this way, the pRA D can be represented by the couple $(d(t), A(t))$, if $A(t)$ is the generating function of the A-sequence. It can be shown that $A(t)$ is univocally determined by the function $h(t)$, and we have:

$$h(t) = A(th(t)) \qquad \text{or} \qquad A(y) = [h(t) \,|y = th(t)] \tag{2.1}$$

The elements of column 0 can also be expressed in terms of all the elements of the previous row, but obviously the dependence should start with the same column 0. In this case, we can show that a new sequence exists: $Z = \{z_0, z_1, z_2, \ldots\}$, called the Z-sequence of the pRA, such that:

$$d_{n+1,0} = z_0 d_{n,0} + z_1 d_{n,1} + z_2 d_{n,2} + \ldots = \sum_{j=0}^{\infty} z_j d_{n,j}.$$

The sum is obviously finite, the Z-sequence can have no relation with the A-sequence, and, in fact, it depends on both the $d(t)$ and $h(t)$ functions: for $Z(t)$, the generating function of the Z-sequence, we have:

$$d(t) = \frac{d_{0,0}}{1 - tZ(th(t))} \qquad \text{or} \qquad Z(y) = \left[\frac{d(t) - d_0}{td(t)} \middle| \, y = th(t) \right] \tag{2.2}$$

where $d_{0,0} = d_0$. At this point, we have a third characterisation of pRA: a pRA D is uniquely determined by the triple $(d_{0,0}, A(t), Z(t))$, where $A(t)$ and $Z(t)$ are the generating functions of the A- and Z-sequences of D. By the previous lemma, it is now immediate to observe that a monic integer pRA corresponds to a monic-integer A-sequence, and vice versa if also $d(t) \in \mathbf{Z}[\![t]\!]$. Instead, the Z-sequence is integer, but it can be non-monic; in fact, from the second formula in (2.2) we see that z_0 is related to d_1 and not to d_0. It can be interesting to see how the A- and Z-sequences are changed when we perform operations on pRA, essentially when we invert one triangle or when we multiply two triangles. Let us begin with the A-sequence, whose inversion is particularly easy:

Theorem 2.4 *Let $D = (d(t), h(t))$ be a pRA and $\bar{D} = (\bar{d}(t), \bar{h}(t))$ its inverse. Then $\bar{A}(t) = 1/h(t)$.* ■

More interesting is the case of row-by-column product:

Theorem 2.5 *Let $D = (d(t), h(t))$, $E = (e(t), k(t))$ be two pRA and $F = (f(t), g(t))$ their product. If $A_D(t)$,$A_E(t)$ and $A_F(t)$ are the generating functions of the corresponding A-sequences, we have:*

$$A_F(t) = A_E(t)\,[A_D(y)|\, t = yk(y)]\,.$$

■

An example is in order here. Let D be the Pascal triangle and E the Catalan triangle, so that we have

$$D = \left(\frac{1}{1-t}, \frac{1}{1-t}\right) \qquad E = \left(\frac{1-\sqrt{1-4t}}{2t}, \frac{1-\sqrt{1-4t}}{2t}\right)$$

and $A_D(t) = 1 + t$, $A_E(t) = 1/(1-t)$. Numerically we have:

$$\begin{pmatrix} 1 & & & & \\ 1 & 1 & & & \\ 1 & 2 & 1 & & \\ 1 & 3 & 3 & 1 & \\ 1 & 4 & 6 & 4 & 1 \end{pmatrix} \star \begin{pmatrix} 1 & & & & \\ 1 & 1 & & & \\ 2 & 2 & 1 & & \\ 5 & 5 & 3 & 1 & \\ 14 & 14 & 9 & -4 & 1 \end{pmatrix} = \begin{pmatrix} 1 & & & & \\ 2 & 1 & & & \\ 5 & 4 & 1 & & \\ 15 & 14 & 6 & 1 & \\ 51 & 50 & 27 & 8 & 1 \end{pmatrix}$$

In this case, the equation $t = yk(y)$ is $t = \left(1 - \sqrt{1-4y}\right)/2$ and its solution is $y = t - t^2$, so that $A_D(y) = 1 + y = 1 + t - t^2$ and

$$A_F(t) = \frac{1 + t - t^2}{1 - t} = 1 + 2t + t^2 + t^3 + t^4 + \ldots .$$

This is immediately checked against the numerical values just obtained.

For what concerns the Z-sequence, we can prove:

Theorem 2.6 *Let $D = (d(t), h(t))$ and $\bar{D} = (\bar{d}(t), \bar{h}(t))$ its inverse pRA; then if $Z(t)$, $\bar{Z}(t)$ are the generating functions of the corresponding Z-sequences, we have*

$$\bar{Z}(t) = \frac{d_0 - d(t)}{d_0 t h(t)} = \frac{-Z(th(t))}{h(t)(1 - tz(th(t))}.$$

■

For the row-by-column product we have:

Theorem 2.7 *Let $D = (d(t), h(t))$, $E = (e(t), k(t))$ be two pRA with Z-sequences $Z_D(t)$ and $Z_E(t)$. Then if $F = D \star E$, the Z-sequence of this product is*

$$Z_F(t) = \left[Z_D(y) + d_0 \frac{\bar{d}(y)}{\bar{h}(y)} Z_E(t) \middle| t = yk(y) \right].$$

■

3 Generating trees

In the paper by Merlini and Verri [5], it is shown that a particular subset of generating trees has a correspondence with some pRA's. In fact, they define an infinite matrix $\{d_{n,k}\}_{n,k \in \mathbf{N}}$ to be *associated to a generating tree* with root (c) (AGT matrix for short) if $d_{n,k}$ is the number of nodes at level n with label $k + c$ and show that, under suitable conditions, this matrix corresponds to a pRA, and vice versa. This result allows us to use the Riordan Array approach to obtain a variety of counting results on problems which can be described by such generating trees. In the present paper we extend the correspondence between generating trees and pRA's to the whole group of monic integer pRA's. The primary difficulty in doing so is the fact that coefficients in pRA's can be negative, and this requires a particular interpretation of the labels in a generating tree. The idea is rather simple, but effective. Since, as we are going to show, the construction of the generating tree related to a given monic integer pRA requires the concept of the $A-$ and $Z-$sequences, the results obtained in the previous section will be important in applying the theory.

In order to specify a generating tree we have to specify a label for the root and a set of rules explaining how to derive from the label of a parent the labels of all of its children. For example, Figure 3.1 illustrates the upper part of the generating tree which corresponds to the following specification:

$$\begin{cases} \textit{root}: & (2) \\ \textit{rule}: & (k) \quad \rightarrow (k)(k+1) \end{cases} \tag{3.3}$$

The first triangle in Table 3.1 illustrates the AGT matrix associated to the generating tree specification (3.3). Our main result is given by Theorem 3.3, which extends Theorem 3.9 in [5]. The concept of a generating tree is extended to deal with *marked* labels: a label is any positive integer, generated according to the generating tree specification; a *marked* label is any positive integer, marked by a bar, for which appropriate rules are given in the specification:

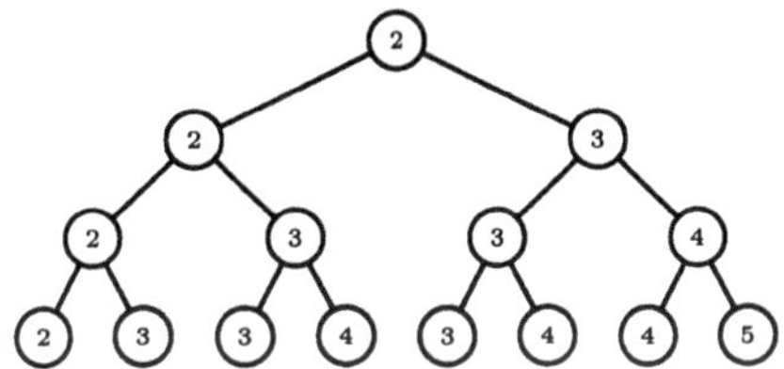

Figure 3.1: The Pascal generating tree: specification (3.3)

n/k	0	1	2	3	4
0	1				
1	1	1			
2	1	2	1		
3	1	3	3	1	
4	1	4	6	4	1

n/k	0	1	2	3	4
0	1				
1	-1	1			
2	1	-2	1		
3	-1	3	-3	1	
4	1	-4	6	-4	1

Table 3.1: The Pascal triangle and its inverse

Definition 3.1 *A marked generating tree is a rooted labelled tree (the labels can be marked or non-marked) with the property that if v_1 and v_2 are any two nodes with the same label then, for each label l, v_1 and v_2 have exactly the same number of children with label l. To specify a generating tree it therefore suffices to specify:*

1) the label of the root;

2) a set of rules explaining how to derive from the label of a parent the labels of all of its children.

From here on we will use the term "generating tree" to denote a marked generating tree.

A simple example is given by the following generating tree specification:

$$\left\{ \begin{array}{lll} root: & (2) & \\ rule: & (k) & \rightarrow (\overline{k})(k+1) \\ & (\overline{k}) & \rightarrow (k)(\overline{k+1}) \end{array} \right. \tag{3.4}$$

The first 4 levels of the corresponding generating tree are shown in Figure 3.2.

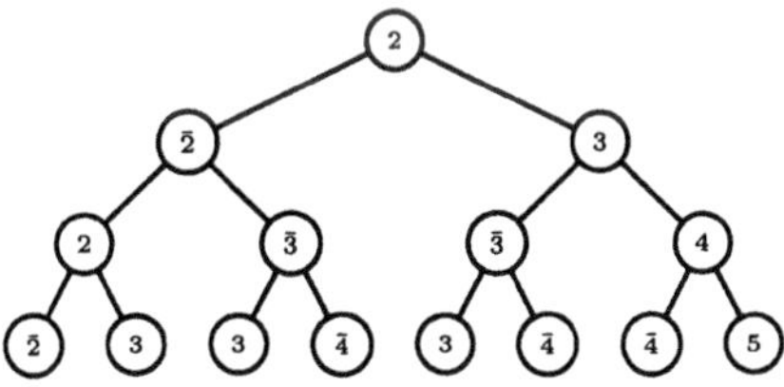

Figure 3.2: The *inverse* of Pascal generating tree: specification (3.4).

The idea is that marked labels kill or annihilate the non-marked labels with the same number, i.e. the count relative to an integer j is the difference between the number of non-marked and marked labels j at a given level. This gives a negative count if marked labels are more numerous than non-marked ones. Thus we can extend the concept of an AGT matrix in the following way:

Definition 3.2 *An infinite matrix $\{d_{n,k}\}_{n,k\in\mathbf{N}}$ is said to be "associated" to a marked generating tree with root (c) (AGT matrix for short) if $d_{n,k}$ is the difference between the number of nodes at level n with label $k+c$ and the number of nodes with label $\overline{k+c}$. By convention, the level of the root is 0.*

The second triangle in Table 3.1 corresponds to the AGT matrix associated to the specification (3.4).

Before stating our main result we introduce the following notations for generating tree specifications.

$(x) = (\bar{\bar{x}})$;
$(x)^p = \underbrace{(x)\cdots(x)}_{p},\ p \geq 0$
$(x)^p = \underbrace{(\bar{x})\cdots(\bar{x})}_{-p},\ p < 0$
$\overline{(x)^p} = (\bar{x})^p,\ p > 0$
$\overline{(x)^p} = (x)^{-p},\ p < 0$
$\prod_{j=0}^{i}(k-j)^{\alpha_j} = (k)^{\alpha_0}(k-1)^{\alpha_1}\cdots(k-i)^{\alpha_i}$

We note that $(x)^0$ is the empty sequence. We can finally state the following important theorem which relates monic integer pRA's with marked generating trees:

Theorem 3.3 *Let $c \in \mathbf{N}$, $a_j, b_k \in \mathbf{Z}$, $\forall j \geq 0$ and $k \geq c$, $a_0 = 1$, and let*

$$\left\{\begin{array}{lll} root: & (c) & \\ rule: & (k) & \rightarrow (c)^{b_k}\prod_{j=0}^{k+1-c}(k+1-j)^{a_j} \\ & (\bar{k}) & \rightarrow \overline{(c)^{b_k}\prod_{j=0}^{k+1-c}(k+1-j)^{a_j}} \end{array}\right. \tag{3.5}$$

be a marked generating tree specification. Then, the AGT matrix associated to (3.5) is a monic integer pRA D defined by the triple (d_0, A, Z), such that

$$d_0 = 1, \quad A = (a_0, a_1, a_2, \ldots), \quad Z = (b_c + a_1, b_{c+1} + a_2, b_{c+2} + a_3, \ldots).$$

Vice versa, if D is a monic integer pRA defined by the triple $(1, A, Z)$ with $a_j, z_j \in \mathbf{Z}$, $\forall j \geq 0$ and $a_0 = 1$, then D is the AGT matrix associated to the generating tree specification (3.5) with $b_{c+j} = z_j - a_{j+1}$, $\forall j \geq 0$. ■

A generating tree corresponding to the specification (3.5) will be called a *proper generating tree.*

Since monic integer pRA's constitute a non-commutative group with respect to the usual row by colum product, as we have seen in Section 2, an important consequence of Theorem 3.3 is that we can define the *product* of two proper generating trees and the *inverse* of a proper generating tree:

Definition 3.4 *Given two generating tree specifications t_1 and t_2 of type (3.5) and the corresponding AGT matrices T_1 and T_2, we define the generating tree specification product of t_1 and t_2 as the specification t_3 having $T_3 = T_1 \star T_2$ as AGT matrix.*

Definition 3.5 *Given a generating tree specification t_1 of type (3.5) and the corresponding AGT matrix T_1, we define the generating tree specification inverse of t_1 as the specification t_2 having $T_2 = T_1^{-1}$ as AGT matrix.*

To complete the definition of the algebraic structure of generating trees we only need to define an *identity*:

Definition 3.6 *The identity generating tree specification t_I is the one having the identity matrix I as AGT matrix. The specification and the corresponding generating trees are shown in (3.6) and Figure 3.3.*

$$\begin{cases} root: & (c) \\ rule: & (k) \quad \rightarrow (k+1) \end{cases} \tag{3.6}$$

Figure 3.3: The *identity* generating tree

For example, specification (3.4) is the inverse of (3.3), as can be easily verified by using formulas in Section 2. In fact, for the Pascal triangle we have $d_0 = 1$, $A = \{1,1,0,0,\ldots\}$ and $Z = \{1,0,0,\ldots\}$ and for its inverse $\bar{d_0} = 1$, $\bar{A} = \{1,-1,0,0,\ldots\}$ and $\bar{Z} = \{-1,0,0,\ldots\}$.

In what follows, we examine some other generating tree specifications, very well known in literature, by finding for each of them the corresponding *inverse*. We leave to the reader the necessary algebraic computations, which can be easily performed by hand or by some symbolic system.

The first specification is related to Motzkin numbers $M_j = \{1,1,2,4,9,\ldots\} = [t^j](1-t-\sqrt{1-2t-3t^2})/(2t)$:

$$\begin{cases} root: & (1) \\ rule: & (k) \quad \rightarrow (1)\ldots(k-1)(k+1) \end{cases} \tag{3.7}$$

(we observe that $(1) \rightarrow (2)$) and a partial generating tree is illustrated in Figure 3.4. The specification inverse of (3.7) is the following:

$$\begin{cases} root: & (1) \\ rule: & (k) \quad \rightarrow (k+1)\prod_{j=2}^{k}(\overline{k+1-j})^{M_{j-2}} \\ & (\bar{k}) \quad \rightarrow (\overline{k+1})\prod_{j=2}^{k}(k+1-j)^{M_{j-2}} \end{cases} \tag{3.8}$$

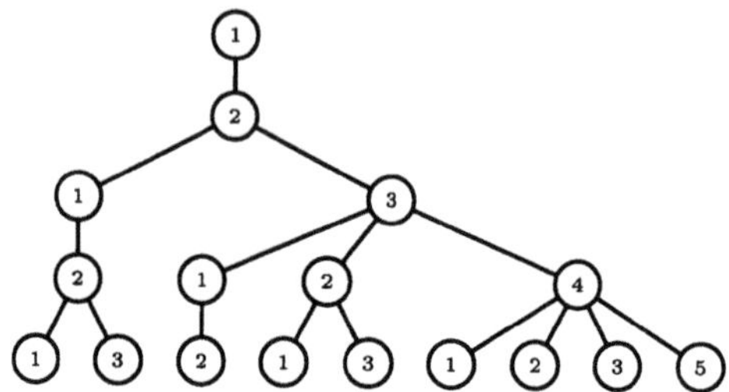

Figure 3.4: The Motzkin generating tree: specification (3.7)

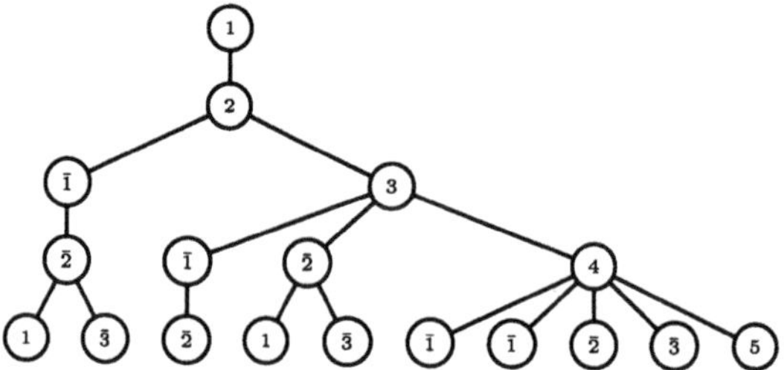

Figure 3.5: The *inverse* of Motzkin generating tree: specification (3.8)

and is illustrated in Figure 3.5. Finally, Table 3.2 shows the AGT matrices associated to the specifications (3.7) and (3.8).

The following example is related to the Catalan numbers $C_j = \{1, 1, 2, 5, 14, \ldots\} = \frac{1}{j+1}\binom{2j}{j}$. In fact, specification (3.9), illustrated in Figure 3.6, has the Catalan triangle as AGT matrix (see the first triangle in Table 3.3).

$$\begin{cases} root: & (2) \\ rule: & (k) \quad \to (2) \cdots (k)(k+1) \end{cases} \tag{3.9}$$

We point out that specification (3.9) represents the *product* between the specification (3.3) and the specification (3.7). Figure 3.7 gives the inverse of Catalan generating tree; the AGT matrix is shown in Table 3.3.

$$\begin{cases} root: & (2) \\ rule: & (k) \quad \to (k+1) \prod_{j=1}^{k-1} (\overline{k+1-j})^{C_{j-1}} \\ & (\bar{k}) \quad \to (\overline{k+1}) \prod_{j=1}^{k-1} (k+1-j)^{C_{j-1}} \end{cases} \tag{3.10}$$

n/k	0	1	2	3	4
0	1				
1	0	1			
2	1	0	1		
3	1	2	0	1	
4	3	2	3	0	1

n/k	0	1	2	3	4
0	1				
1	0	1			
2	-1	0	1		
3	-1	-2	0	1	
4	0	-2	-3	0	1

Table 3.2: The Motzkin triangle and its inverse

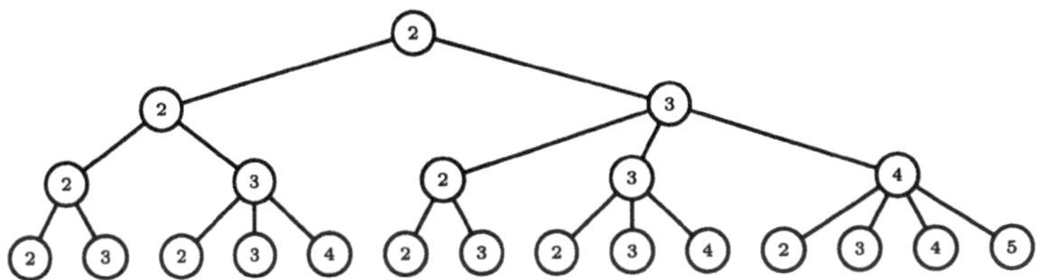

Figure 3.6: The Catalan generating tree: this is the *product* between the Pascal and the Motzkin trees: specification (3.9)

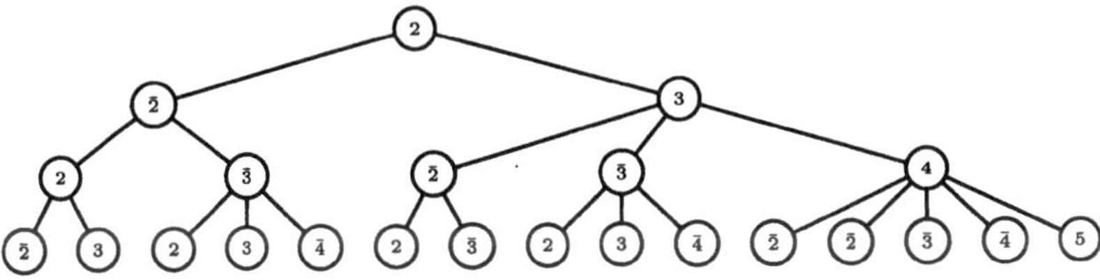

Figure 3.7: The *inverse* of Catalan generating tree: specification (3.10)

The last example is related to the Fibonacci numbers $F_j = \{1, 1, 2, 3, 5, 8, \ldots\} = [t^j]1/(1 - t - t^2)$. We give the two specifications (3.11) and (3.12), illustrated in Figures 3.8 and 3.9 respectively, and the corresponding AGT matrices in Table 3.4.

$$\begin{cases} root: & (2) \\ rule: & (k) \quad \rightarrow (2)^{k-1}(k+1) \end{cases} \tag{3.11}$$

$$\begin{cases} root: & (2) \\ rule: & (k) \quad \rightarrow (\overline{2})^{F_{2k-3}}(k+1) \\ & (\bar{k}) \quad \rightarrow (2)^{F_{2k-3}}(\overline{k+1}) \end{cases} \tag{3.12}$$

n/k	0	1	2	3	4
0	1				
1	1	1			
2	2	2	1		
3	5	5	3	1	
4	14	14	9	4	1

n/k	0	1	2	3	4
0	1				
1	-1	1			
2	0	-2	1		
3	0	1	-3	1	
4	0	0	3	-4	1

Table 3.3: The Catalan triangle and its inverse

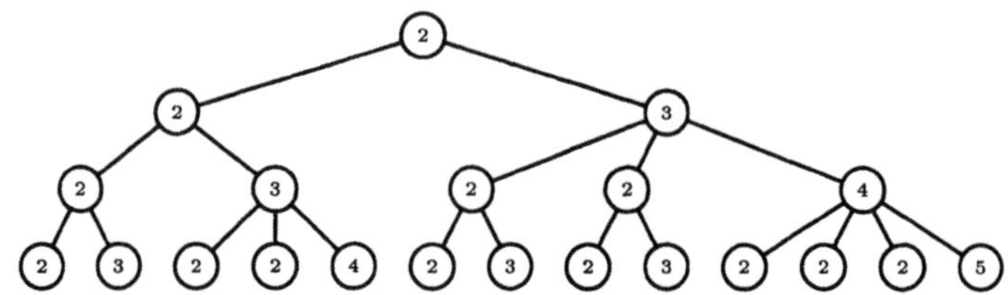

Figure 3.8: The odd Fibonacci generating tree: specification (3.11)

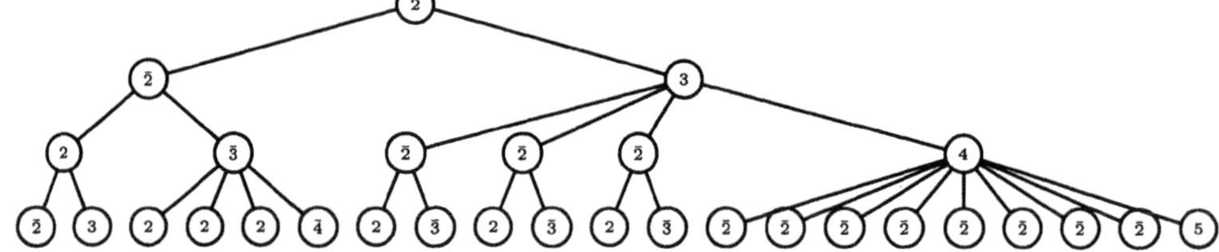

Figure 3.9: The *inverse* of odd Fibonacci generating tree: specification (3.12)

References

[1] C. Banderier, M. Bousquet-Mélou, A. Denise, P. Flajolet, D. Gardy, and D. Gouyou-Beauchamps. On generating functions of generating trees. *INRIA Research Report RR-3661. In Proceedings of Formal Power Series and Algebraic Combinatorics (FPSAC'99), Barcelona*, 1999.

[2] S. Corteel. Séries génératrices exponentielles pour les eco-systèmes signés. *Proceedings of the 12-th International Conference on Formal Power Series and Algebraic Combinatorics, Moscow*, 2000.

[3] D. Merlini, D. G. Rogers, R. Sprugnoli, and M. C. Verri. On some alternative characterizations of Riordan arrays. *Canadian Journal of Mathematics*, 49(2):301–320, 1997.

[4] D. Merlini, R. Sprugnoli, and M. C. Verri. Algebraic and combinatorial properties of simple, coloured walks. In *Proceedings of CAAP'94*, volume 787 of *Lecture Notes in Computer Science*, pages 218–233, 1994.

n/k	0	1	2	3	4
0	1				
1	1	1			
2	3	1	1		
3	8	3	1	1	
4	21	8	3	1	1

n/k	0	1	2	3	4
0	1				
1	-1	1			
2	-2	-1	1		
3	-3	-2	-1	1	
4	-4	-3	-2	-1	1

Table 3.4: The odd Fibonacci triangle and its inverse

[5] D. Merlini and M. C. Verri. Generating trees and proper Riordan Arrays. *Discrete Mathematics*, 218:167–183, 2000.

[6] E. Pergola, R. Pinzani, and S. Rinaldi. Towards an algebra of succession rules. *Proceedings of MATHINFO2000.*

[7] D. G. Rogers. Pascal triangles, Catalan numbers and renewal arrays. *Discrete Mathematics*, 22:301–310, 1978.

[8] L. W. Shapiro, S. Getu, W.-J. Woan, and L. Woodson. The Riordan group. *Discrete Applied Mathematics*, 34:229–239, 1991.

[9] R. Sprugnoli. Riordan arrays and combinatorial sums. *Discrete Mathematics*, 132:267–290, 1994.

[10] R. Sprugnoli. Riordan arrays and the Abel-Gould identity. *Discrete Mathematics*, 142:213–233, 1995.

[11] J. West. Generating trees and the Catalan and Schröder numbers. *Discrete Mathematics*, 146:247–262, 1995.

[12] J. West. Generating trees and forbidden subsequences. *Discrete Mathematics*, 157:363–374, 1996.

Trends in Mathematics, © 2000 Birkhäuser Verlag Basel/Switzerland

A set of well-defined operations on succession rules

ELISA PERGOLA, RENZO PINZANI AND SIMONE RINALDI[1] *Dipartimento di Sistemi e Informatica, Via Lombroso 6/17, Firenze.*
{elisa, pinzani, rinaldi}@dsi.unifi.it

Abstract. *In this paper we introduce a system of* well-defined operations *on the set of* succession rules. *These operations allow us to tackle combinatorial enumeration problems simply by using succession rules instead of generating functions. Finally we suggest several open problems the solution of which should lead to an algebraic characterization of the set of succession rules.*

1 Introduction

A *succession rule* Ω is a system having the form:

$$\begin{cases} (b) \\ (k) \rightsquigarrow (e_1(k))(e_2(k))\dots(e_k(k)), \end{cases}$$

where $b, k \in \mathbb{N}^+$, and $e_i : \mathbb{N}^+ \to \mathbb{N}^+$; (b) is the *axiom* and $(k) \rightsquigarrow (e_1(k))(e_2(k))\dots(e_k(k))$ is the *production*; (b), (k), $(e_i(k))$, are called *labels* of Ω. The rule Ω can be represented by means of a *generating tree*, that is a rooted tree whose vertices are the labels of Ω; (b) is the label of the root and each node labelled (k) produces k sons labelled $(e_1(k)),\dots,(e_k(k))$ respectively. We refer to [4] for further details and examples. A succession rule Ω defines such a sequence of positive integers $\{f_n\}_{n\geq 0}$, that f_n is the number of the nodes belonging to the generating tree defined by Ω and lying at level n. By convention the root is at level 0, so $f_0 = 1$. The function $f_\Omega(x) = \sum_{n\geq 0} f_n x^n$ is the *generating function* derived from Ω.

The concept of succession rules was first introduced in [6] by Chung and al. to study reduced Baxter permutations; later, West applied succession rules to the enumeration of permutations with forbidden subsequences [11]. Moreover, they are a fundamental tool used by the ECO method [4], which is a general method for the enumeration of combinatorial objects consisting essentially in the recursive construction of a class of objects. A generating tree is then associated to a certain combinatorial class, according to some enumerative parameter, so that the number of nodes appearing on level n of the tree gives the number of n-sized objects in the class. In [1] the relationships between structural properties of the rules and the rationality, algebraicity or trascendance of the corresponding generating function are studied. We wish to point out that in the present paper we deal with "pure" succession rules [4], instead of generalizations [2], or specializations [7].

Two rules Ω_1 and Ω_2 are said to be *equivalent*, $\Omega_1 \cong \Omega_2$, if they define the same number sequence, that is $f_{\Omega_1}(x) = f_{\Omega_2}(x)$. For example, the following rules are equivalent and define Schröder numbers ([5]):

[1] This work was partially supported by MURST project: *Modelli di calcolo innovativi: metodi sintattici e combinatori.*

$$\begin{cases} (2) \\ (2k) \rightsquigarrow (2)(4)^2 \dots (2k)^2(2k+2), \end{cases} \qquad \begin{cases} (2) \\ (k) \rightsquigarrow (3) \dots (k)(k+1)^2, \end{cases}$$

where the power notation is used to express repetitions: $(h)^i$ stands for (h) repeated i times.

Starting from classical succession rules we define *coloured rules* in the following way: a rule Ω is coloured when there are at least two labels (k) and $(\bar{k})$ having the same value but different productions. For example, it is easily proved, that the sequence $1, 2, 3, 5, 9, 17, 33, \dots, 2^{n-1}+1$, having $f(x) = \frac{1-x-x^2}{1-3x+2x^2}$ as generating function, can only be described by means of coloured rules, such as:

$$\Omega' : \quad \begin{cases} (2) \\ (1) \rightsquigarrow (\bar{2}) \\ (2) \rightsquigarrow (1)(2) \\ (\bar{2}) \rightsquigarrow (\bar{2})(\bar{2}). \end{cases}$$

A succession rule Ω is *finite* if it has a finite number of different labels. The number sequences $\{a_{n,k}\}_n$, defined by the recurrences:

$$\sum_{j=0}^{k} (-1)^j \binom{k}{j} a_{n-j,k} = 0 \qquad k \in \mathbb{N},$$

having $\frac{1}{(1-x)^k}$ as generating function, have finite succession rules:

$$\Omega(k) : \begin{cases} (k) \\ (1) \rightsquigarrow (1) \\ (2) \rightsquigarrow (1)(2) \\ (3) \rightsquigarrow (1)(2)(3) \\ \dots \quad \dots \\ (k) \rightsquigarrow (1)(2)(3) \dots (k-1)(k). \end{cases}$$

Moreover, let $\{a_n\}_n$ be the sequence of integers satisfying the recurrence:

$$a_n = ka_{n-1} + ha_{n-2}, \qquad k \in \mathbb{N}^+, h \in \mathbb{Z},$$

subject to the initial conditions $a_0 = 1$, $a_1 = b \in \mathbb{N}^+$; every term of the sequence is a positive number, if $k + h > 0$. In this case, the sequence $\{a_n\}_n$ is defined by the finite succession rule:

$$\Omega_{\mathcal{F}^b_{k,h}} : \begin{cases} (b) \\ (b) \rightsquigarrow (k)^{b-1}(k+h) \\ (k) \rightsquigarrow (k)^{k-1}(k+h) \\ (k+h) \rightsquigarrow (k)^{k+h-1}(k+h). \end{cases} \tag{1}$$

Finite succession rules play an important role in enumerative combinatorics, such as in the enumeration of restricted classes of combinatorial objects ([8]). Moreover, we can regard any finite succession rule Ω as a particular PDOL system

([9]), (Σ, P, w_0), where the alphabet Σ is the set of labels of Ω, P is the set of its productions and $w_0 \in \Sigma$. These remarks lead to the solution of two open problems for finite succession rules:

Equivalence. *Let Ω_1 and Ω_2 be two finite succession rules having h_1 and h_2 labels respectively, then $\Omega_1 \cong \Omega_2$, if and only if the first $h_1 + h_2$ terms of the two sequences defined by Ω_1 and Ω_2 coincide.*

For example, the number sequences defined by Ω' and by $\Omega_{\mathcal{F}^2_{1,1}}$ (the rule for Fibonacci numbers) coincide for the first 4 terms, but not for the fifth, so the rules are not equivalent.

Generating functions. *The function $f(x)$ is the generating function of a finite succession rule iff:*

1. $f(x) = \frac{P(x)}{Q(x)}$, with $P(x), Q(x) \in \mathbb{Z}[x]$, and $Q(0) = P(0) = 1$;
2. $\frac{1}{x}(f(x) - 1) - f(x)$ is $\mathbb{N}$-rational.

Roughly speaking, $\mathbb{N}$-rational functions are the generating functions of regular languages, and their analytic characterization is given by Soittola's Theorem [9] (for further details see [3]). We are so ensured that each generating function of a finite succession rule is the generating function of a regular language, while the converse does not hold. For example, let $g(x) = \frac{1}{1-10x}$ and $h(x) = \frac{1-3x+36x^2}{(1-9x)(1+2x+81x^2)}$; $h(x)$ is a rational function having all positive coefficients (see [3] for the proof) but it is not $\mathbb{N}$-rational, since the poles of minimal modulus are complex numbers. Let

$$f(x) = g(x^2) + x[g(x^2) + h(x^2)] = k_1(x^2) + xk_2(x^2); \tag{2}$$

$f(x)$ is $\mathbb{N}$-rational, since it is the merge of the two functions $k_1(x)$ and $k_2(x)$, each of them having a real positive dominating root, $x = 10$. This proves the existence of a regular language having $f(x)$ as its generating function. Moreover, it is clear that $f(x)$ defines a strictly increasing sequence of positive numbers. Neverthless $\frac{1}{x}(f(x) - 1) - f(x)$ is not $\mathbb{N}$-rational, since it is a merge of $g(x)$ and $h(x)$, and $h(x)$ is not $\mathbb{N}$-rational. Thus there are no finite succession rules having $f(x)$ as its generating function. The previous problems remain still open in the case of not finite succession rules.

2 Operations on succession rules

A n-ary operation $\circ$ on the set $\mathcal{S}$ of all succession rules is said to be *well-defined* if the equivalences $\Omega_1 \cong \Omega'_1, \ldots \Omega_n \cong \Omega'_n$ imply $\circ(\Omega_1, \ldots, \Omega_n) \cong \circ(\Omega'_1, \ldots, \Omega'_n)$. Our aim is to determine a set of well-defined operations on $\mathcal{S}$, in order to build an algebraic system on $\mathcal{S}$.

Let Ω and Ω' be two succession rules, defining the sequences $\{f_n\}_n$ and $\{g_n\}_n$, and having $f(x)$ and $g(x)$ as generating functions, respectively. In the sequel we deal with Ω and Ω' having the following general forms:

$$\Omega : \begin{cases} (a) \\ (h) \rightsquigarrow (e_1(h))(e_2(h))\dots(e_h(h)), \end{cases} \qquad \Omega' : \begin{cases} (b) \\ (k) \rightsquigarrow (c_1(k))(c_2(k))\dots(c_k(k)). \end{cases}$$

2.1 Sum of succession rules

Given two succession rules Ω and Ω', their *sum*, $\Omega \oplus \Omega'$, is the rule defining the sequence $\{h_n\}_n$ so that $h_0 = 1$ and $h_n = f_n + g_n$, if $n > 0$, and having $f(x)+g(x)-1$ as generating function. Let $(a) \rightsquigarrow (A_1)\dots(A_a)$ and $(b) \rightsquigarrow (B_1)\dots(B_b)$ be the productions for the axiom (a) in Ω and (b) in Ω', then the following succession rule:

$$\Omega \oplus \Omega' : \begin{cases} (a+b) \\ (a+b) \rightsquigarrow (A_1)\dots(A_a)(\overline{B}_1)\dots(\overline{B}_b) \\ (h) \rightsquigarrow (e_1(h))(e_2(h))\dots(e_h(h)) \\ (\overline{k}) \rightsquigarrow (\overline{c_1(k)})(\overline{c_2(k)})\dots(\overline{c_k(k)}), \end{cases}$$

gives the sequence $\{h_n\}_n$.

2.2 Bisection of succession rules

Given a succession rule Ω, its *bisection*, denoted as $\frac{\Omega}{2}$, is the rule defining the sequence $\{f_{2n}\}_n$, and having $\frac{f(\sqrt{x})+f(-\sqrt{x})}{2}$ as generating function. Let $(a) \rightsquigarrow (A_1)\dots(A_a)$ be the production for the axiom (a), and $s = A_1 + \dots + A_a$, then (s) is the axiom for $\frac{\Omega}{2}$. Let $e_i(h) \rightsquigarrow e_1^i(h)\dots e^i_{e_i(h)}(h)$ be the production for $e_i(h)$, $i = 1,\dots,h$, and $g(h) = e_1(h) + \dots + e_h(h)$, then the rule for $\frac{\Omega}{2}$ is:

$$\begin{cases} (s) \\ (g(h)) \rightsquigarrow (g(e_1^1(h)))\dots(g(e^1_{e_1(h)}(h)))\dots(g(e_1^h(h)))\dots(g(e^h_{e_h(h)}(h))). \end{cases}$$

Example 2.1

i) Bisection of the rule for the Fibonacci numbers. By applying the previous definition we obtain:

$$\Omega_{\mathcal{F}} : \begin{cases} (2) \\ (1) \rightsquigarrow (2) \\ (2) \rightsquigarrow (1)(2), \end{cases} \qquad \frac{\Omega_{\mathcal{F}}}{2} : \begin{cases} (3) \\ (2) \rightsquigarrow (2)(3) \\ (3) \rightsquigarrow (2)(3)(3). \end{cases}$$

ii) Bisection of the rule for Catalan numbers. We start from the rule Ω_C,

$$\begin{cases} (2) \\ (h) \rightsquigarrow (2)(3)\dots(h)(h+1); \end{cases}$$

the axiom for $\frac{\Omega_C}{2}$ is (5); moreover, $g(h) = 2 + 3 + \dots h + h + 1 = \frac{h^2+3h}{2}$, so the rule for Catalan numbers of even index, $\left(1, 5, 42, 429, \dots, \frac{1}{2n+2}\binom{4n+2}{2n+1}\right)$ is:

$$\frac{\Omega_C}{2} : \begin{cases} (5) \\ \left(\frac{h^2+3h}{2}\right) \rightsquigarrow (5)^h(9)^h(14)^{h-1}(20)^{h-2}\ldots\left(\frac{(h+1)^2+3(h+1)}{2}\right)^2\left(\frac{(h+2)^2+3(h+2)}{2}\right). \end{cases}$$

2.2.1 Product of succession rules

Given the rules Ω and Ω', their *product* $\Omega \otimes \Omega'$, is the succession rule defining the sequence $\{\sum_{k\leq n} f_{n-k}g_k\}_n$, and having $f(x)\cdot g(x)$ as generating function. Let $(b) \rightsquigarrow B_1 \ldots B_b$ be the production of (b), then:

$$\Omega \otimes \Omega' : \quad \begin{cases} (a+b) \\ (h+b) \rightsquigarrow (e_1(h)+b)(e_2(h)+b)\ldots(e_h(h)+b)\overline{B}_1\ldots\overline{B}_b \\ \overline{(k)} \rightsquigarrow \overline{c_1(k)}\ldots\overline{c_k(k)}. \end{cases}$$

This statement is easily proved as follows: let $t(x) = f(x)\cdot g(x)$, and t_n the number of nodes at level n in the generating tree of $\Omega \otimes \Omega'$. The statement clearly holds for $n = 0$, $t_0 = 1 = f_0g_0$, and for $n = 1$, $t_1 = f_1g_0 + f_0g_1 = a + b$. Figure 1 shows the generating tree for the rule $\Omega \otimes \Omega'$; the number of nodes at level n of the tree can be considered as a sum of $n+1$ terms, which are as follows:
- $g_n = f_0g_n$; indeed this is the number of nodes at level n in Ω' generating tree, which is a proper subtree of $\Omega \otimes \Omega'$ having the axiom as its root.
- f_1g_{n-1}; indeed, by construction, the generating tree of $\Omega \otimes \Omega'$ has f_1 subtrees, isomorphic to Ω', the roots of which lie at level 1; each of them has g_{n-1} nodes at level n of $\Omega \otimes \Omega'$ generating tree. Generally, at level n of $\Omega \otimes \Omega'$ generating tree, there are f_i times the nodes at level $n-i$ of Ω' generating tree (that is g_{n-i} nodes), $0 \leq i \leq n$.

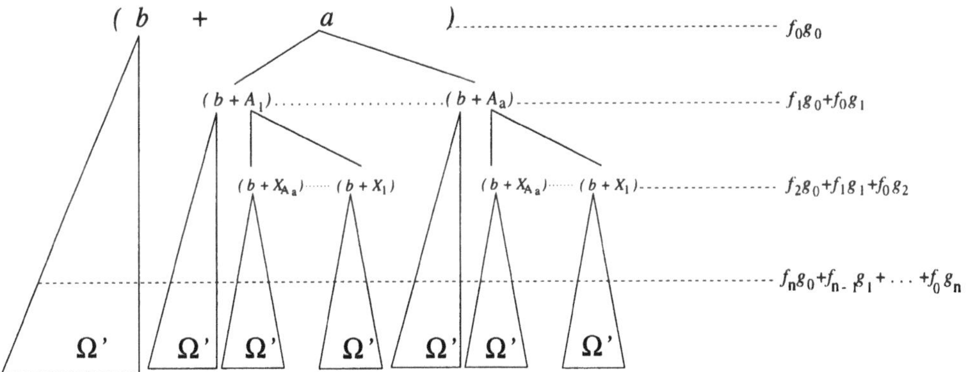

Figure 1: The generating tree for $\Omega \otimes \Omega'$.

Thus the total number of nodes at level n of $\Omega \otimes \Omega'$ generating tree is $t_n = f_0g_n + \ldots + f_ng_0$. As the product is commutative, $\Omega\otimes\Omega'$ and $\Omega'\otimes\Omega$ are equivalent rules but of a different shape.

Example 2.2 i) *Product of Catalan and Fibonacci numbers.* The succession rule obtained by applying the operation $\otimes$ to the rules for Catalan and Fibonacci numbers is as follows:

$$\Omega_C \otimes \Omega_{\mathcal{F}} \qquad \begin{cases} (4) \\ (k+2) \rightsquigarrow (1)(2)(4)(5)...(k)(k+1) \\ (1) \rightsquigarrow (2) \\ (2) \rightsquigarrow (1)(2), \end{cases}$$

and it defines the number sequence $1, 4, 12, 35, 95, ...$

ii) *The rule for the square Catalan numbers.* The rule obtained is:

$$\Omega_C^2 : \qquad \begin{cases} (4) \\ (k+2) \rightsquigarrow (\bar{2})(\bar{3})(4)(5)...(k)(k+1)(k+2)(k+3) \\ (\bar{k}) \rightsquigarrow (\bar{2})(\bar{3})...(\bar{k})(\overline{k+1}). \end{cases}$$

which can be easily simplified as the following nice rule:

$$\Omega_C^2 : \qquad \begin{cases} (4) \\ (k) \rightsquigarrow (2)(3)\dots(k)(k+1). \end{cases}$$

2.3 The Star of a succession rule

The *star* of the succession rule Ω, denoted as Ω^*, is the rule defining the number sequence having $\frac{1}{1-f_0(x)} = 1 + f_0(x) + f_0^2(x) + \dots + f^n(x) + \dots = \sum_{n\geq 0} f_0^n(x)$ as generating function, where $f_0(x) = f(x) - 1$. Let (a) be the axiom and $(a) \rightsquigarrow A_1 \dots A_a$ the production of (a). The rule Ω^* is:

$$\Omega^* : \begin{cases} (a) \\ (a) \rightsquigarrow (A_1 + a)\dots(A_a + a) \\ (h+a) \rightsquigarrow (A_1 + a)\dots(A_a + a)(e_1(h) + a)(e_2(h) + a)\dots(e_h(h) + a). \end{cases}$$

The proof is similar to the one given for the product of two succession rules.

Example 2.3 *The star of Schröder numbers.* We start from the rule Ω_S:

$$\begin{cases} (2) \\ (2k) \rightsquigarrow (2)(4)^2\dots(2k)^2(2k+2), \text{and obtain } \Omega_S^* : \end{cases}$$

$$\begin{cases} (2) \\ (2) \rightsquigarrow (4)(6) \\ (2k+2) \rightsquigarrow (4)^2(6)^3\dots(2k+2)^2(2k+4). \end{cases}$$

2.4 An application of rule operations to enumerative combinatorics

A Grand Dyck path is a a sequence of rise and fall steps ($(1,1)$ and $(1,-1)$ respectively) in the plane $\mathbb{N} \times \mathbb{Z}$, running from $(0,0)$ to $(2n,0)$.
Let us determine a succession rule that enumerates Grand Dyck paths according to their semilenght, by applying some operations to succession rules. Grand Dyck paths are in bijection with Grand Dyck words, which are generated by the following non-ambiguous grammar:

$$\begin{cases} S \to aAbS|bBaS|\epsilon \\ A \to aAbA|\epsilon \\ B \to bBaB|\epsilon, \end{cases}$$

where a encodes a rise step, and b a fall step; its generating function is $f(x) = \frac{1}{\sqrt{1-4x}}$. Let $f_D(x)$ be the generating function for Dyck words, enumerated by Catalan numbers. We can write $f(x)$ as:

$$f(x) = \frac{1}{1 - [(f_D(x) - 1) + (f_D(x) - 1)]}.$$

Thus the rule Ω for Grand Dyck paths can be obtained as $\Omega = (\Omega_C \oplus \Omega_C)^*$, where Ω_C represents the rule for Catalan numbers:

$$\Omega_C : \quad \begin{cases} (1) \\ (h) \rightsquigarrow (2)(3)\dots(h)(h+1). \end{cases} \tag{3}$$

By applying sum and star operations, we obtain the following rule:

$$\Omega : \quad \begin{cases} (2) \\ (h) \rightsquigarrow (3)(3)(4)\dots(h)(h+1). \end{cases} \tag{4}$$

It should be noticed that this is the same rule found in [8] and which recursively constructs the class of Grand Dyck paths according to the ECO method.

2.5 Partial sum of a succession rule

Let Ω be a succession rule, defining the sequence $\{f_n\}_n$ and having $f(x)$ as generating function. The ***partial sum*** $\Sigma\Omega$, is the rule defining the sequence $\{F_n\}_n = \left\{\sum_{j\leq n} f_j\right\}_n$. We can obtain $\Sigma\Omega$ by means of the product operation, since $F(x) = \sum_n F_n x^n = \frac{1}{1-x} \cdot f(x)$. Thus:

$$\Sigma\Omega = \Omega_1 \otimes \Omega,$$

where $\Omega_1 \begin{cases} (1) \\ (1) \rightsquigarrow (1) \end{cases}$ is the natural rule for the sequence $f_n = 1, \forall n$. By applying the product operation we get:

$$\Sigma\Omega : \begin{cases} (a+1) \\ (1) \rightsquigarrow (1) \\ (h+1) \rightsquigarrow (1)(e_1(h)+1)(e_2(h)+1)\dots(e_h(h)+1). \end{cases}$$

For example, the rule Ω_C for Catalan numbers leads to the rule:

$$\Sigma\Omega_C : \begin{cases} (3) \\ (1) \rightsquigarrow (1)(h+1) \rightsquigarrow (1)(3)(4)\dots(h+1)(h+2), \end{cases}$$

giving the sequence $1, 3, 8, 22, 64, \dots$

3 Other operations

Let Ω and Ω' be succession rules, and as usual, $\{f_n\}_n$ and $\{g_n\}_n$ their sequences, with their respective generating functions $f(x)$ and $g(x)$. The *Hadamard product* of Ω and Ω', denoted as $\Omega \odot \Omega'$, is the rule defining the sequence $\{f_n g_n\}_n$. Generally, it is not so simple to determine the generating function $f(x) \odot g(x)$, but the Hadamard product of two $\mathbb{N}$-rational series was proved to be $\mathbb{N}$-rational ([9]).

We start by giving an example of how to construct the rule $\Omega \odot \Omega'$ in the case of finite rules. Let Ω be the rule for the Pell numbers, $\{1, 2, 5, 12, 29, \dots\}$, and Ω' the rule for the Fibonacci numbers having an odd index, $\{1, 2, 5, 13, 34, \dots\}$,

$$\Omega : \quad \begin{cases} (2) \\ (2) \rightsquigarrow (2)(3) \\ (3) \rightsquigarrow (2)(2)(3), \end{cases} \qquad \Omega' : \quad \begin{cases} (\bar{2}) \\ (\bar{2}) \rightsquigarrow (\bar{2})(\bar{3}) \\ (\bar{3}) \rightsquigarrow (\bar{2})(\bar{3})(\bar{3}). \end{cases}$$

For each label (h) of Ω and $(\bar{k})$ of Ω', $(h \cdot k)$ is a label of the rule $\Omega \odot \Omega'$, and it is coloured only if there is already another label having the same value; the axiom is $(a \cdot b)$, where (a) and (b) are the axioms of the rules; if the productions of (h) and $(\bar{k})$ are:

$$(h) \rightsquigarrow (c_1)\dots(c_h)$$

$$(\bar{k}) \rightsquigarrow \overline{(e_1)}\dots\overline{(e_k)},$$

then the production of $(h \cdot k)$ is:

$$(h \cdot k) \rightsquigarrow (c_1 \cdot e_1)\dots(c_1 \cdot e_k)\dots(c_h \cdot e_1)\dots(c_h \cdot e_k).$$

Going back to our example, the labels of $\Omega \odot \Omega'$ are $(2 \cdot \bar{2}) = (4)$, $(2 \cdot \bar{3}) = (6)$, $(3 \cdot \bar{2}) = (\bar{6})$, $(3 \cdot \bar{3}) = (9)$. For instance, the production for the label (4) is:

$$(4) = (2 \cdot \bar{2}) \rightsquigarrow (2 \cdot \bar{2})(2 \cdot \bar{3})(3 \cdot \bar{2})(3 \cdot \bar{3}) = (4)(6)(\bar{6})(9).$$

In the same way we obtain:

$$\Omega \odot \Omega' : \quad \begin{cases} (4) \\ (4) \rightsquigarrow (4)(6)(\bar{6})(9) \\ (6) \rightsquigarrow (4)(6)(6)(\bar{6})(9)(9) \\ (\bar{6}) \rightsquigarrow (4)(4)(6)(6)(\bar{6})(9) \\ (9) \rightsquigarrow (4)(4)(6)(6)(6)(6)(\bar{6})(9)(9). \end{cases}$$

The rule $\Omega \odot \Omega'$ has ij labels, being i and j the numbers of labels of Ω and Ω' respectively. The method we described above enables us to obtain the product $\Omega \odot \Omega'$ for finite rules, and it also proves that *the Hadamard product of two finite rules is a finite rule.* Some problems arise when attempting to find a general formula for the Hadamard product of two rules Ω and Ω' having each an infinite number of labels. Generally speaking, the best we can do is to write a finite rule Ω^k which *approximates* the rule $\Omega \odot \Omega'$ with the required precision, depending on the parameter $k \in \mathbb{N}^+$ (see [8]).

Neverthless there are some cases when the application of the previously defined operation on succession rules becomes particularly easy and helpful.

(1) Let $k \in \mathbb{N}$; the rule Ω_k is a rule for the sequence $\{F_n\}_n = \{k^n f_n\}_n$; since the generating function $F(x) = \sum_n F_n x^n = f(x) \odot \frac{1}{1-kx}$, we have:

$$\Omega_k = \Omega \odot \Omega_1,$$

where Ω_1 is the rule for $\{k^n\}_n$, that is:

$$\begin{cases} (k) \\ (k) \rightsquigarrow (k)^k; \end{cases}$$

by applying the operation $\odot$ we get:

$$\Omega_k : \quad \begin{cases} (ka) \\ (kh) \rightsquigarrow (e_1(h))^k (e_2(h))^k \dots (e_h(h))^k. \end{cases}$$

For example, let us take into consideration the rule Ω for Motzkin numbers $\{1, 1, 2, 4, 9, 21, \dots, M_n, \dots\}$:

$$\Omega : \quad \begin{cases} (1) \\ (h) \rightsquigarrow (1)(2)\dots(h-1)(h+1); \end{cases} \tag{5}$$

for $k = 2$ we get the rule Ω_2 giving the sequence $\{1, 2, 8, 32, 144, \dots, 2^n M_n, \dots\}$:

$$\Omega_2 : \quad \begin{cases} (2) \\ (2h) \rightsquigarrow (2)^2(4)^2 \dots (2h-2)^2(2h+2)^2. \end{cases}$$

(2) The rule $[n+1]\Omega$ defines the sequence $F_n = (n+1)f_n$, $n \in \mathbb{N}$. As the generating function is $F(x) = \sum_n F_n x^n = f(x) \cdot \frac{1}{(1-x)^2}$, we have:

$$[n+1]\Omega = \Omega \odot \Omega',$$

where Ω' is the rule for the sequence $n+1$, that is:

$$\Omega' : \quad \begin{cases} (2) \\ (2) \rightsquigarrow (1)(2) \\ (1) \rightsquigarrow (1), \end{cases}$$

so we obtain:

$$[n+1]\Omega : \quad \begin{cases} (\overline{2a}) \\ (h) \rightsquigarrow (e_1(h))(e_2(h)) \dots (e_h(h)) \\ (\overline{2h}) \rightsquigarrow (e_1(h))(e_2(h)) \dots (e_h(h))(\overline{2e_1(h)}) \dots (\overline{2e_h(h)}). \end{cases}$$

For example, let Ω be the rule for Pell numbers, then:

$$[n+1]\Omega : \quad \begin{cases} (4) \\ (2) \rightsquigarrow (2)(3) \\ (3) \rightsquigarrow (2)(2)(3) \\ (4) \rightsquigarrow (2)(3)(4)(6) \\ (6) \rightsquigarrow (2)(2)(3)(4)(4)(6). \end{cases}$$

Let G_n be the number of Grand Dyck paths having semilength n and C_n the nth Catalan number. As usual, let Ω_C be the rule (3) for Catalan numbers, and Ω_G the rule defining the sequence $\{G_n\}$. From the combinatorial identity $G_n = (n+1)C_n$, we have:

$$\Omega_G = [n+1]\Omega_C,$$

and thus

$$\Omega_G : \quad \begin{cases} (\overline{2}) \\ (h) \rightsquigarrow (2)(3) \dots (h)(h+1) \\ (\overline{2h}) \rightsquigarrow (2)(3) \dots (h)(h+1)(\overline{4})(\overline{6}) \dots (\overline{2h})(\overline{2h+2}), \end{cases}$$

is a rule counting $\{G_n\}$, equivalent to (4).

Moreover it is easy to prove the following property.

Proposition 3.1 *Let Ω be a rule defining the sequence $\{f_n\}_n$. Then a rule Ω' defining a sequence $\{g_n\}_n$, such that $f_n = g_n - rg_{n-1}$, for $n > 1$, exists:*

$$\Omega' : \quad \begin{cases} (a+r) \\ (r) \rightsquigarrow (r)^r \\ (h+r) \rightsquigarrow (r)^r(e_1(h)+r)(e_2(h)+r) \dots (e_h(h)+r). \end{cases}$$

3.1 Open problems

There are several open problems related to the definition of an algebra on succession rules and arising from the operations we have introduced. Below the most interesting problems are mentioned:

- *Equivalence.* Is there a criterion allowing us to establish whether two given succession rules are equivalent simply by working on their labels, that is, with no need to determine the corresponding generating functions?

- *Subtraction.* Given two rules Ω and Ω', defining the sequences $\{f_n\}$ and $\{g_n\}$ respectively, such that $f_n > g_n$ for each $n > 0$, let $\Omega \ominus \Omega'$ be the rule defining the sequence $\{h_n\}_n$ such that $h_n = \begin{cases} 1 & \text{if } n = 0 \\ f_n - g_n & \text{otherwise.} \end{cases}$

  The construction of the rule $\Omega_1 \ominus \Omega_2$ constitutes an open problem.

- *Inversion.* Let $\{f_n\}_n$ be a non decreasing sequence of positive integers. Is there a method allowing us to decide whether a succession rule defining the sequence $\{f_n\}_n$ exists and, in this case, to find it? We remark that this problem can be solved for finite rules.

3.2 A Conjecture

Conjecture: *if a succession rule has a rational generating function, then it is equivalent to a finite succession rule.* It is sufficient to prove that each rational generating function of a succession rule satisfies the properties of the generating functions of finite rules established in Section 1. If the conjecture proves true, rational functions such as (2) cannot be the generating functions of any succession rule. For example, let Ω be the rule, studied in [1], whose set of labels is the whole set of prime numbers:

$$\Omega : \quad \begin{cases} (2) \\ (p_n) \rightsquigarrow (p_{n+1})(q_n)(r_n)(2)^{p_n-3}, \end{cases}$$

where p_n denotes the nth prime number, and q_n and r_n are two primes such that $2p_n - p_{n+1} + 3 = q_n + r_n$; its generating function is rational, $f(x) = \frac{1-2x}{1-4x+3x^2}$, thus, according to our conjecture, a finite succession rule Ω' equivalent to Ω can be found:

$$\Omega' : \quad \begin{cases} (2) \\ (2) \rightsquigarrow (2)(3) \\ (3) \rightsquigarrow (2)(3)(4) \\ (4) \rightsquigarrow (2)(3)(4)(4). \end{cases}$$

It should be noticed that the rule Ω' was further exploited in [8], being the 4-approximating rule for Catalan numbers, and it describes a recursive construction for Dyck paths whose maximal ordinate is 4.

References

[1] C. Banderier, M. Bousquet-Mélou, A. Denise, P. Flajolet, D. Gardy and D. Gouyou-Beauchamps, On generating functions of generating trees, *Proceedings of 11th FPSAC,* (1999) 40-52.

[2] S. Corteel, Problèmes énumératifs issus de l'Informatique, de la Physique et de la Combinatorie, *Phd Thesis, Université de Paris-Sud,* 6082, (2000).

[3] E. Barcucci, A. Del Lungo, A. Frosini, S. Rinaldi, From rational functions to regular languages, *Proceedings of 12th FPSAC, Moscow* (2000).

[4] E. Barcucci, A. Del Lungo, E. Pergola, R. Pinzani, ECO: a methodology for the Enumeration of Combinatorial Objects, *Journal of Difference Equations and Applications,* Vol.5 (1999) 435-490.

[5] E. Barcucci, A. Del Lungo, E. Pergola, R. Pinzani, Some combinatorial interpretations of q-analogs of Schröder numbers, *Annals of Combinatorics,* 3 (1999) 173-192.

[6] F. R. K. Chung, R. L. Graham, V. E. Hoggatt, M. Kleimann, The number of Baxter permutations, *J. Combin. Theory Ser. A,* 24 (1978) 382-394.

[7] D. Merlini, C. Verri, Generating trees and proper Riordan Arrays, *Discrete Mathematics,* 218 (2000) 107-122.

[8] E. Pergola, R. Pinzani, S. Rinaldi, ECO-Approximation of algebraic functions, *Proceedings of 12th FPSAC, Moscow* (2000).

[9] A. Salomaa and M. Soittola, *Automata-theoretic aspects of formal power series*, Springer-Verlag, New York, (1978).

[10] N. J. A. Sloane and S. Plouffe, *The encyclopedia of integer sequences*, Academic press, (1996).

[11] J. West, Generating trees and the Catalan and Schröder numbers, *Discrete Mathematics,* 146 (1995) 247-262.

Part III

Algorithms and Optimization

Trends in Mathematics, © 2000 Birkhäuser Verlag Basel/Switzerland

Convergence of a Genetic Algorithm with finite population

J. BÉRARD AND A. BIENVENÜE *Laboratoire de Probabilités, Combinatoire et Statistique; Université Claude Bernard, Lyon I.*
[berard,bienvenu]@jonas.univ-lyon1.fr

Abstract. *We study the asymptotic behaviour of a mutation-selection genetic algorithm on the integers with finite population, defined by a simple random walk and the fitness function* $f(x) = x$. *We prove the convergence in law of the normalized population and a large deviations principle.*

Key words. *genetic algorithm, convergence in law, large deviations, population dynamics, random walks, interacting particle systems.*

1 Introduction

1.1 Motivation

Genetic algorithms were formally introduced by Holland [6] in 1975 as an optimization method based on a biological analogy with the natural mechanisms of evolution, and they are now a very popular tool for practically solving hard combinatorial optimization problems. Broadly speaking, a genetic algorithm describes a finite population of individuals evolving under the effect of several "genetic" operators:

- selection: the population is randomly resampled from the previous one. Individuals with high fitness are more likely to be selected, whereas individuals with low fitness tend to be eliminated;
- mutation: the individuals are subject to random mutations;
- mating: a new population of "offsprings" is created from pairs of individuals of the previous population.

In the combinatorial optimization setting, individuals are feasible solutions, and the fitness of an individual is measured by the function to be maximized. Thus, the selection operator directs the evolution towards a fitness increase, while mutation and mating preserve the population diversity, and allow the algorithm to visit large parts of the space of solutions. However, despite the success encountered by genetic algorithms in practical applications, and the numerous experimental studies devoted to them, few rigorous results are available about their behaviour.

In [3], R. Cerf obtained asymptotic convergence results for genetic algorithms with rare transitions.

In [8], Y. Rabinovich and A. Wigderson studied the convergence speed of several genetic algorithms defined on binary strings, and the divergence speed of some algorithms defined on the integers. The algorithm we study belongs to the latter category. We are here in a very simplified context, so our results do not apply to complex optimization situations. However, in this setting, we can get a detailed understanding of the effects of selection through a rigorous mathematical treatment.

Moreover, this model has been used in biology to study the evolution of population of viruses (cf. L. Tsimring, H. Levine and D. Kessler [9]). In that context, D. Bonnaz applied theoretical physics methods to get predictions concerning the behaviour of the model (asymptotics of mean and variance of the population), supported by numerical simulations and biological experiments. Our results confirm these predictions mathematically.

1.2 Description of the model

The main object of our study is a mutation-selection genetic algorithm on the integers with finite population, defined by a simple random walk and the fitness function $f(x) = x$.

Let $p \in \mathbb{N}^*$. Let $X_n = (X_n^{(i)})_{1 \leqslant i \leqslant p}$ be the Markov chain with state space $\mathbb{N}^p$ starting from $X_0 = (1, \dots, 1)$ and defined by the following transitions:

1. **selection step:** $X_n \longrightarrow X'_n$

   If $X_n = (0, \dots, 0)$, then $X'_n = (1, \dots, 1)$. Else, the $X_n'^{(i)}$, $1 \leqslant i \leqslant p$, are chosen randomly and independently among the $\left\{X_n^{(i)},\ 1 \leqslant i \leqslant p\right\}$'s according to the probability law

$$\frac{1}{S_n} \sum_{i=1}^{p} X_n^{(i)} \delta_{X_n^{(i)}}, \quad \text{where} \quad S_n = \sum_{i=1}^{p} X_n^{(i)}.$$

2. **mutation step:** $X'_n \longrightarrow X_{n+1}$

   The p particles $X_n'^{(i)}$ evolve independently, each of them performs one step of a simple random walk on $\mathbb{Z}$ (symmetric, to the nearest neighbours) and their new positions are the $X_{n+1}^{(i)}, 1 \leqslant i \leqslant p$.

The population at time n is formed by the p individuals $X_n^{(1)}, \dots, X_n^{(p)}$. Thus, the number of individuals in a generation is kept fixed during the time-evolution of the population. Note that, at a given time, there may be several individuals taking the same value. The choice $X_n = (0, \dots, 0) \longrightarrow X'_n = (1, \dots, 1)$ is arbitrary and the point here is to prevent individuals from taking negative values. We could have suitably modified the mutation operator instead.

We use the following notations:

$$\mathcal{F}_n = \sigma\left(X_0, X'_0, \dots, X_{n-1}, X'_{n-1}, X_n\right), \quad F_n = S_n/p, \quad \text{and} \quad \Delta F_n = F_{n+1} - F_n.$$

We define the diameter of $x \in \mathbb{N}^p$ as $d(x) = \sup_{1 \leqslant i,j \leqslant p} |x_i - x_j|$.

1.3 Statement of the results

In this paper, we focus on the asymptotic behaviour of (X_n) as n goes to infinity, with fixed p.

Theorem *$(X_n^{(1)}, \dots, X_n^{(p)})/\sqrt{n}$ converges in law to $(1, \dots, 1)\gamma_p$, where γ_p stands for a random variable whose law is the same as the law at time $t = 1$ of a $(2p-1)$-dimensional Bessel process starting from zero.*

Theorem *For all $1 \leqslant i \leqslant p$, the sequence $(X_n^{(i)}/n)$ satisfies a large deviations principle on $[0, 1]$, with the same rate function as in the case of a simple reflected random walk:*

$$I(a) = \frac{1}{2}\left[(1-a)\log(1-a) + (1+a)\log(1+a)\right].$$

These results clearly show the effect of selection: when we normalize the population by $\sqrt{n}$, it behaves like a random point in $\mathbb{R}_+$ whose limiting law diverges from the reflected normal law (that we would get without selection) as the population size increases, giving more and more weight to large values. However, when we normalize by n, we get the same large deviations result as in the case of a simple reflected random walk, so the effect of selection cannot be seen on that scale.

We point out that these results differ strikingly from those of C. Mazza and D. Piau [7] about the infinite-population version of the model. Indeed, they found a linear growth rate of average fitness with time, as opposed to our $\sqrt{n}$ scaling factor. A satisfactory explanation of this phenomenon is still lacking.

1.4 Contents

In section 2, we give two preliminary results about the behaviour of the cloud of particles:

- a stochastic minorization of the process by a coupled process that tends to infinity in probability,
- a stochastic majorization of the diameter of the cloud.

In section 3, we give several technical lemmas.

Section 4 is devoted to the detailed exposition of the convergence in law of the differences (Theorem 1) and of the normalized cloud (Theorem 2 and Corollary 7). The large deviations principle (Theorem 3) is presented in section 5. For the sake of brevity, we just sketch the proofs here, and we refer to [1] for detailed arguments.

2 Two preliminary results

The following lemma shows that the cloud of particles X_n is minorized stochastically by a mutation-selection process Y_n with uniform selection. We note that, for fixed n, each $Y_n^{(i)}$ follows the same probability law as the position at time n of a simple reflected random walk. However, for fixed i, the process $(Y_n^{(i)})_n$ is not a simple reflected random walk.

Lemma 1 *There is a process $Y_n = (Y_n^{(i)})_{1 \leqslant i \leqslant p}$ such that:*

- *for all $n \geqslant 1$ and $1 \leqslant i \leqslant p$, we have $Y_n^{(i)} \leqslant X_n^{(i)}$*
- *for all $n \geqslant 1$ and $1 \leqslant i \leqslant p$, $Y_n^{(i)}$ follows the same probability law as the position at time n of a simple symmetric reflected random walk on $\mathbb{N}$ with transition probabilities $p_{0\to 0} = p_{0\to 1} = 1/2$.*

Corollary 2 *For all $1 \leqslant i \leqslant p$, $Y_n^{(i)}$ tends to infinity in probability, that is:*

$$\forall K > 0, \qquad \mathrm{P}\left[Y_n^{(i)} \geqslant K\right] \xrightarrow[n\to+\infty]{} 1.$$

Proof

We construct a coupling between X and Y, where Y is a mutation-selection process with constant fitness function (uniform selection), whose mutation steps are given by a simple symmetric reflected random walk on $\mathbb{N}$ with transition probabilities $p_{0\to 0} = p_{0\to 1} = 1/2$. We check that we can couple pathwise the mutation and selection steps of X and Y so as to constantly preserve the relation $Y_n^{(i)} \leqslant X_n^{(i)}$.

□

The following lemma provides a stochastic majorization of the diameter of the cloud.

Lemma 3 *For all $n \in \mathbb{N}^*$, there is a geometric random variable G_n, with parameter p^{1-p}, such that $d(X_n) \leqslant 2G_n$.*

Proof

We note that the conditional probability q_n knowing $\mathcal{F}_n$ for the selection step between X_n and X'_n to give the same value to all the $X_n'^{(i)}$'s is given by

$$q_n = \begin{cases} S_n^{-p} \sum_{k=1}^{p} \left(X_n^{(k)}\right)^p & \text{if } X_n \neq (0,\dots,0), \\ 1 & \text{if } X_n = (0,\dots,0). \end{cases}$$

Hence we have: $q_n \geqslant p^{1-p}$.

We can then construct a geometric random variable G_n with parameter p^{1-p} that dominates the time interval between n and the last time before n when the selection step gives the same value to all the $X_n'^{(i)}$'s. Noting that a mutation step cannot increase the diameter by more than 2, the diameter $d(X_n)$ cannot exceed $2G_n$.

□

3 Technical lemmas

The following result stems from Corollary 2 and Lemma 3, and will be of constant use in the sequel: the transition probabilities of the selection step can be approximated by those corresponding to a uniform selection.

Lemma 4 *For $1 \leqslant a, b \leqslant p$, we have :*

$$\mathbf{1}_{\{X_n \neq 0\}} \left| \frac{X_n^{(a)} X_n^{(b)}}{S_n^2} - \frac{1}{p^2} \right| \leqslant \mathbf{1}_{\{X_n \neq 0\}} A_n(a,b,p)/F_n,$$

and

$$\mathbf{1}_{\{X_n \neq 0\}} \left| \frac{X_n^{(a)}}{S_n} - \frac{1}{p} \right| \leqslant \mathbf{1}_{\{X_n \neq 0\}} B_n(a,p)/F_n,$$

where the sequences $(A_n(a,b,p))_n$ and $(B_n(a,p))_n$ are bounded in every L^s-space, $1 \leqslant s < \infty$.

We shall use the next two lemmas for deducing the asymptotic behaviour of some moments from approximate recursion relations.

Lemma 5 *If $(u_n)_{n \geqslant 0}$ is a sequence of complex numbers satisfying a recursion relation of the form:*

$$u_{n+1} = \alpha u_n + \beta n^q + o(n^q),$$

where $|\alpha| < 1$, $\beta \in \mathbb{R}$ and $q \in \mathbb{N}$, then

$$u_n = \frac{\beta}{1-\alpha} n^q + o(n^q).$$

Lemma 6 *If $(u_n)_{n \geqslant 0}$ is a sequence of real numbers satisfying a recursion relation of the form:*

$$u_{n+1} = a n^\alpha + u_n + o(n^\alpha),$$

where $a \in \mathbb{R}$ and $\alpha \geqslant 0$, then

$$u_n = \frac{a}{1+\alpha} n^{\alpha+1} + o(n^{\alpha+1}).$$

4 Convergence in law

We start with a result about the behaviour of the unnormalized differences of particles positions:

Theorem 1 *For all $1 \leqslant i \neq j \leqslant p$, the sequence $X_n^{(i)} - X_n^{(j)}$ converges in law to a random variable Δ whose law is given by*

$$(1-\alpha)\delta_0 + \sum_{n \geqslant 1} (1-a)^{n-1} a(\delta_{2n} + \delta_{-2n})\alpha/2,$$

where $\alpha = \sqrt{p}/(\sqrt{p}+1)$ *and* $a = 2/(\sqrt{p}+1)$. *The variable* Δ *may be written as* $\Delta = 2ZG$, *where the law of the random variable* Z *is*

$$(\delta_1 + \delta_{-1})\alpha/2 + (1-\alpha)\delta_0,$$

and G *is a geometric random variable independent from* Z: $\mathrm{P}[G = n] = (1-a)^{n-1}a$ *for* $n \geqslant 1$.

Proof

Let $t \in \mathbb{R}$ and $1 \leqslant i \neq j \leqslant p$.

Let

$$\phi_n(t) = \mathrm{E}\left(e^{it(X_n^{(i)} - X_n^{(j)})}\right).$$

Using Lemma 4, we approximate the transition probabilities of the selection step by uniform ones, and we get a recursion relation of the form:

$$\phi_{n+1}(t) = \cos^2 t\left(\frac{1}{p} + \frac{p-1}{p}\phi_n(t)\right) + l_n(t)$$

where, for fixed t, $l_n(t) \to 0$ as $n \to +\infty$.

This relation, together with Lemma 5 (setting $\alpha = \cos^2 t.(p-1)/p$, $\beta = p^{-1}\cos^2 t$ and $q = 0$), shows that, for fixed t,

$$\phi_n(t) \xrightarrow[n\to+\infty]{} \frac{\cos^2 t}{p - (p-1)\cos^2 t}.$$

According to Lévy's Theorem (cf. [5]), the result follows.

□

Theorem 2 $F_n/\sqrt{n}$ *converges in law to* γ_p, *where* γ_p *stands for a random variable whose law is the same as the law at time* $t = 1$ *of a* $(2p-1)$*-dimensional Bessel process starting from zero.*

The use of characteristic functions does not seem to be appropriate here, so we use the method of moments: we estimate recursively the even-order moments of F_n.

Proof

We shall prove by induction on h that $\mathrm{E}((F_n/\sqrt{n})^{2h})$ tends to $\gamma_{p,h}$ as n goes to infinity, for $h \in \mathbb{N}$. This is obvious for $h = 0$. Let $h \geqslant 1$, and suppose the result established for all $q \leqslant h-1$. We have to show that $\mathrm{E}((F_n/\sqrt{n})^{2h})$ converges. To see this, we shall show that $(\mathrm{E}(F_{n+1}^{2h}) - \mathrm{E}(F_n^{2h}))/n^{h-1}$ converges.

$$\begin{aligned}
\mathrm{E}\left(F_{n+1}^{2h} - F_n^{2h} \mid \mathcal{F}_n\right) &= \mathrm{E}\left((F_n + \Delta F_n)^{2h} - F_n^{2h} \mid \mathcal{F}_n\right) \\
&= \binom{2h}{1} F_n^{2h-1} \mathrm{E}\left(\Delta F_n \mid \mathcal{F}_n\right) \\
&\quad + \binom{2h}{2} F_n^{2h-2} \mathrm{E}\left((\Delta F_n)^2 \mid \mathcal{F}_n\right) \\
&\quad \underbrace{+ \sum_{l=3}^{2h} \binom{2h}{l} F_n^{2h-l} \mathrm{E}\left((\Delta F_n)^l \mid \mathcal{F}_n\right).}_{(\star)}
\end{aligned} \tag{1}$$

The induction hypothesis and the preceding lemmas imply that the expectation of the term $(\star)$ is $o(n^{h-1})$, so we only have to evaluate the first two terms. To achieve this, we write explicitly the conditional expectations $\mathrm{E}\left(\Delta F_n \mid \mathcal{F}_n\right)$ and $\mathrm{E}\left((\Delta F_n)^2 \mid \mathcal{F}_n\right)$. Then, approximating the transition probabilities of two consecutive selection steps thanks to Lemma 4, and using the induction hypothesis, we note that the expectations we have to evaluate satisfy recursion relations similar to those of Lemma 5, so we finally get:

$$\mathrm{E}\left(F_{n+1}^{2h} - F_n^{2h}\right) = \left[2(p-1)h + h(2h-1)\right] \gamma_{p,h-1} n^{h-1} + o(n^{h-1}).$$

From Lemma 6, we then deduce that $\mathrm{E}((F_n/\sqrt{n})^{2h})$ converges to

$$\gamma_{p,h} = \left[2(p-1) + (2h-1)\right] \gamma_{p,h-1},$$

so the proof by induction is complete. Moreover, the sequence $(\gamma_{p,h})_{h \geqslant 0}$ satisfies Carleman's condition (cf. [5]):

$$\sum_{h=1}^{+\infty} \gamma_{p,h}^{-1/2h} = +\infty,$$

hence, the convergence of moments gives us the convergence in law of $F_n/\sqrt{n}$.

We now identify the limiting law γ_p. We have:

$$\gamma_{p,h} = \mathrm{E}\gamma_p^{2h} = \prod_{i=1}^{h} \left(2(p-1) + (2i-1)\right).$$

These are the even-order moments of the symmetric law with density

$$\bar{f}_p(x) = \frac{1}{\sqrt{2\pi}\; 1 \times 3 \times 5 \times \cdots \times (2p-3)} x^{2(p-1)} e^{-x^2/2}.$$

Hence, the symmetrization of the law of γ_p is the law with density $\bar{f}_p$. Noting that the law of γ_p is supported by $\mathbb{R}_+$, it has a density given by

$$f_p(x) = 2\, \mathbf{1}_{\mathbb{R}_+}(x)\, \bar{f}_p(x).$$

□

We check that for $p = 1$, γ_p follows a normal law.

Corollary 7 *$(X_n^{(1)}, \dots, X_n^{(p)})/\sqrt{n}$ converges in law to $(1, \dots, 1)\gamma_p$.*

5 Large Deviations

Theorem 3 *For all $1 \leqslant i \leqslant p$, the sequence $(X_n^{(i)}/n)$ satisfies a large deviations principle on $[0,1]$ with the same rate function as in the case of a simple reflected random walk:*

$$I(a) = \frac{1}{2}\left[(1-a)\log(1-a) + (1+a)\log(1+a)\right].$$

Proof

We shall prove that, for all $t \geqslant 0$, we have:

$$\lim_{n \to +\infty} \frac{1}{n} \log \mathrm{E}(e^{tX_n^{(1)}}) = \log\cosh(t),$$

and the result will be a consequence of the Gärtner-Ellis theorem (cf. [4]).

We note that:

$$\mathrm{E}(e^{t(X_{n+1}^{(1)})}) = \cosh(t)\mathrm{E}(e^{t(X_n^{(1)})})(1 + m_n(t)) \tag{2}$$

where

$$\begin{aligned} m_n(t) &= \frac{1}{\mathrm{E}(e^{t(X_n^{(1)})})}\mathrm{E}\left(\mathbf{1}_{\{X_n \neq 0\}} \sum_{1 \leqslant a \leqslant p} \frac{X_n^{(1)} - X_n^{(a)}}{S_n} e^{t(X_n^{(1)})}\right) \\ &+ \frac{1}{\mathrm{E}(e^{t(X_n^{(1)})})} P(X_n = 0)(e^t - 1). \end{aligned} \tag{3}$$

We first show that $0 \leqslant m_n(t) \leqslant 3$ for $n \geqslant n_0$, so we get the following bounds:

$$(\cosh(t))^n \leqslant \mathrm{E}(e^{t(X_n^{(1)})}) \leqslant (\cosh(t))^{n-n_0} 3^{n-n_0} e^{n_0 t} \qquad \forall n \geqslant n_0.$$

These estimates together with Lemma 4 and the Hölder inequality allow us to prove that $m_n(t) \to 0$ as $n \to +\infty$.

□

6 Conclusion

This paper confirms mathematically the results of [2], and provides much more detailed information. Several important aspects of the asymptotic behaviour of the algorithm studied are now understood (role of selection, of population size). However, the question of non-asymptotic behaviour (when the number of iterations depends in some way of the population size) remains open, and further investigation in this direction would certainly lead us to a better understanding of the difference we observed between the finite and infinite-population models. The law of $(X_n)_{n\geqslant 0}$ as a stochastic process is also an interesting object.

Moreover, one can think of many variants and generalizations of the algorithm we presented (using other fitness functions, other types of mutation, or including mating), and the results we obtained incite us to go on studying simplified models of genetic algorithms in order to improve our understanding of their behaviour, and especially the respective roles of parameters and genetic operators.

Acknowledgements

We gratefully acknowledge C. Mazza and D. Piau for their suggestions and comments.

References

[1] Jean Bérard and Alexis Bienvenüe. Comportement asymptotique d'un algorithme génétique. Preprint, 2000.

[2] Didier Bonnaz. *Modèles stochastiques de dynamique des populations.* Thèse de doctorat, Université de Lausanne, February 1998.

[3] Raphaël Cerf. Asymptotic convergence of genetic algorithms. *Adv. in Appl. Probab.*, 30(2):521–550, 1998.

[4] Amir Dembo and Ofer Zeitouni. *Large deviations techniques and applications.* Springer-Verlag, New-York, second edition, 1998.

[5] Richard Durrett. *Probability: theory and examples.* Duxbury Press, Belmont, CA, second edition, 1996.

[6] John H. Holland. *Adaptation in natural and artificial systems.* University of Michigan Press, Ann Arbor, Mich., 1975. An introductory analysis with applications to biology, control, and artificial intelligence.

[7] Christian Mazza and Didier Piau. On the effect of selection in genetic algorithms. Submitted, 2000.

[8] Yuri Rabinovich and Avi Wigderson. Techniques for bounding the convergence rate of genetic algorithms. *Random Structures Algorithms*, 14(2):111–138, 1999.

[9] Lev S. Tsimring, Herbert Levine and David A. Kessler. RNA virus evolution via a fitness-space model. *Phys. Rev. Lett.*, 76:4440, 1996.

Trends in Mathematics, © 2000 Birkhäuser Verlag Basel/Switzerland

Complexity issues for a redistribution problem

MOSHE DROR *Management Information System Department, College of Business and Public Administration, University of Arizona, Tucson, Arizona, 85721, USA.* mdror@mail.bpa.arizona.edu
DOMINIQUE FORTIN *INRIA, Domaine de Voluceau, Rocquencourt, B.P. 105, 78153 Le Chesnay Cedex, France.* Dominique.Fortin@inria.fr
CATHERINE ROUCAIROL *Laboratoire PRiSM, Université de Versailles-St-Quentin-en-Yvelines, 45, av. des Etats-Unis, 78035 Versailles Cedex, France.* Catherine.Roucairol@prism.uvsq.fr

Abstract. *Given a single discrete and non dedicated commodity, we consider the deterministic problem of transporting such commodities from a set of suppliers to a set of customers with one vehicle of limited capacity Q. We give a VRP-like formulation, dimension of associated polytope as well as complex analysis of practical complexity in terms of number of feasible solutions. From a heuristic viewpoint, a dynamic programming algorithm is given to retrieve in polynomial time, the best solution in an exponential neighborhood.*

Key words. *redistribution, routing problem, dynamic programming*

1 Introduction

Given a single discrete and non dedicated commodity, we consider the deterministic problem of transporting such commodities from a set of suppliers (or surplus nodes) to a set of customers (or shortage nodes) with one vehicle of limited capacity Q. The problem is to find a minimal length feasible tour that pickups commodities at supplier points and delivers them at customer points without violating the capacity of the vehicle. The tour must begin and end at a depot, and could visit each point (supply or demand) more than once except the depot (exactly one). Without loss of generality, we consider only unit surplus/shortage nodes in order to avoid multiple visits of arcs and nodes in underlying graph; upto a pseudo-polynomial clique expansion of demand nodes, the non unit demand problem turns into a unit demand problem. Furthermore, the overall demand is assumed to be balanced to prevent adding dummy nodes.

Let $\mathrm{RED}_Q(\mathcal{S}_{2n}, D)$ redistribution problem under unit demand assumption, capacity Q and distance matrix D; let $\mathcal{V} = N^+ \cup N^- \cup \{0\}$ the set of surplus, shortage and depot nodes, we assume the underlying graph G to be complete on surplus and shortage nodes while the depot is ingoing (resp. outgoing) to every surplus (resp. shortage) node. $G = (\mathcal{V}, \mathcal{K}_{2n} \times 0)$ where $|N^+| = |N^-| = n$. Sometimes we will use $+1$ (resp. -1) instead of surplus (resp. shortage) naming and refer to parity for membership to one or the other sets.

In section 2 we address complexity issues for $\mathrm{RED}_Q(\mathcal{S}_{2n}, D)$ problem; in section 3 we give dimension of $\mathrm{RED}_Q(\mathcal{S}_{2n}, D)$ polytope and finally in section 4, given z, a feasible solution, we retrieve in polynomial time the best solution within an exponential neighborhood associated with z under relaxation of capacity constraint (it is known from meta heuristics that non feasible solutions may help to escape from a local optimum so that this assumption should be considered in a loosely restrictive sense).

2 Complexity issues under unit demand

2.1 Unit Capacity $Q = 1$

Under unit capacity, the vehicle should go from the depot to a surplus node, then to a shortage node etc until completion of tour; removing the depot from both endpoints then, w.l.o.g. we may fix shortage nodes of $\text{RED}_1(\mathcal{S}_{2n}, D)$ in even location so that the following decision problems are NP-hard.

$$\text{RED}_1(\mathcal{S}_{2n}, D) \leq d \quad \text{iff} \quad \text{TSP}(EO_{2n}, D) \leq d$$

where $\text{TSP}(EO, D)$ stands for the traveling salesman problem over even-odd permutations which is NP-hard [DW97] through the following transformation.

$$D' = \left[\begin{array}{ccccc|ccccc}
\infty & 0 & \infty & \infty & \infty & \infty & \infty & \infty & \infty & \infty \\
\infty & \infty & D[1,2] & \infty & D[1,3] & \infty & D[1,4] & \infty & D[1,5] & \\
\infty & \infty & \infty & 0 & \infty & \infty & \infty & \infty & \infty & \infty \\
D[2,1] & \infty & \infty & \infty & D[2,3] & \infty & D[2,4] & \infty & D[2,5] & \\
\infty & \infty & \infty & \infty & \infty & 0 & \infty & \infty & \infty & \infty \\
\hline
D[3,1] & \infty & D[3,2] & \infty & \infty & \infty & D[3,4] & \infty & D[3,5] & \\
\infty & \infty & \infty & \infty & \infty & \infty & \infty & 0 & \infty & \infty \\
D[4,1] & \infty & D[4,2] & \infty & D[4,3] & \infty & \infty & \infty & D[4,5] & \\
\infty & \infty & \infty & \infty & \infty & \infty & \infty & \infty & \infty & 0 \\
D[5,1] & \infty & D[5,2] & \infty & D[5,3] & \infty & D[4,5] & \infty & \infty &
\end{array}\right]$$

Figure 1: $\text{TSP}(EO_{10}, D') \leq d$ iff $\text{TSP}(\mathcal{S}_5, D) \leq d$

2.2 Capacity $Q > 1$

2.2.1 Dyck excursions

Counting number of circuits amounts to count Dyck words of height at most capacity. Let C be a class then the sequence class $\sigma\{C\}$ is defined as the infinite sum

$$\sigma\{C\} = \{\epsilon\} + C + (C \times C) + (C \times C \times C) + \cdots$$

where $\{\epsilon\}$ is a void structure (or a structure of null size). This sequence operation is analogous to Kleene star operation $C^\star$ for monoid. Let $V = \{-1, +1\}$ be our alphabet of unit demands, then the ordinary generating function to count the number of circuit fulfilling the capacity constraint is given by

$$\begin{aligned} C^q(z) &= \sum_n C^q_{2n} z^n \\ C^q(z) &= \sigma\{(+1)\ C^{q-1}(z)\ (-1)\} = \frac{1}{1 - zC^{q-1}(z)} \end{aligned}$$

where evenness allows to assign only one z for both $-1, +1$ demand nodes. We acknowledge Ph. Flajolet [FS93] for pointing out this elegant fashion for counting those circuits; it's a different instance from the genuine counting of average height in trees by de Bruijn et al. [dBKR72]. For small values of q and n, expansion of continued fraction leads to the recurrence:

$$C^q_{2n} = \sum_{i=1}^{n} C^q_{2(n-i)} C^{q-1}_{2(i-1)}$$

We follow hereafter, the complete counting expansion (see [FS96] pages 51—53) to reach a direct expression in q and n.

The continued fraction suggests to write it as a quotient of 2 polynomials, namely $C^q(z) = \frac{F^{q+1}(z)}{F^{q+2}(z)}$ leading to the recurrence and the associated solution :

$$\begin{aligned} F^{q+2}(z) &= F^{q+1}(z) - zF^q(z) \\ C^q(z) &= 2\frac{(1+\sqrt{1-4z})^{q+1} - (1-\sqrt{1-4z})^{q+1}}{(1+\sqrt{1-4z})^{q+2} - (1-\sqrt{1-4z})^{q+2}} \end{aligned}$$

where $F^q(z)$ are known as the Fibonacci polynomials due to their values for $z = -1$.

The actual values are related to the Catalan's numbers through the generating function of trees with n nodes $C(z) = \text{node} \ \times \sigma\{C(z)\}$

$$\begin{aligned} C(z) &= \sum_{n\geq 1} C_{n-1} z^n = \frac{z}{1 - C(z)} = \frac{1 - \sqrt{1-4z}}{2} \\ C_n &= \frac{1}{n+1}\binom{2n}{n} \end{aligned}$$

Then, both generating functions are related by using the quotient $y(z) = \frac{C(z)}{1-C(z)} = \frac{1-\sqrt{1-4z}}{1+\sqrt{1-4z}}$

$$\begin{aligned} C(z) - zC^{q-2}(z) &= \sqrt{1-4z}\frac{y^q(z)}{1-y^q(z)} = \frac{1-y(z)}{1+y(z)}\frac{y^q(z)}{1-y^q(z)} \\ C_n - C^{q-2}_{2n} &= \sum_{j\geq 1} \binom{2n}{n+1-jq} - 2\binom{2n}{n-jq} + \binom{2n}{n-1-jq} \end{aligned}$$

C_{2n}^{q}	n									
Q	1	2	3	4	5	6	7	8	9	10
1	1	1	1	1	1	1	1	1	1	1
2	1	2	4	8	16	32	64	128	256	512
3	1	2	5	13	34	89	233	610	1597	4181
4	1	2	5	14	41	122	365	1094	3281	9842
5	1	2	5	14	42	131	417	1341	4334	14041
6	1	2	5	14	42	132	428	1416	4744	16016
7	1	2	5	14	42	132	429	1429	4846	16645
8	1	2	5	14	42	132	429	1430	4861	16778
9	1	2	5	14	42	132	429	1430	4862	16795
10	1	2	5	14	42	132	429	1430	4862	16796

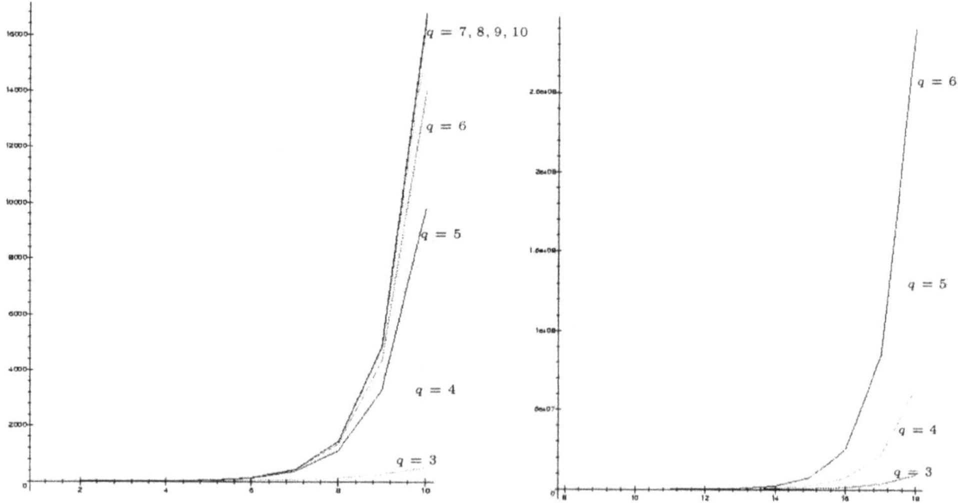

Figure 2: number of tours under small capacity and length

where the second equality is thrown from complex analysis [FS96] and is valid as soon as $n > q - 2$

$$
\begin{aligned}
C_n - C_{2n}^{q-2} &= \frac{1}{2i\pi} \int_{\Gamma} \frac{1 - y(z)}{1 + y(z)} \frac{y^q(z)}{1 - y^q(z)} \frac{dz}{z^{n+2}} = \frac{1}{2i\pi} \int_{\Gamma} \frac{(1-y)^2}{1 - y^q} (1+y)^{2n} \frac{dy}{y^{n+2}} \\
&= [y^{n+1}](1 - 2y + y^2)(1+y)^{2n} \sum_{j \geq 0} y^{jq}
\end{aligned}
$$

Asymptotic behavior comes as $o(\rho^n)$ where ρ is the smallest singularity, here $z = 1/4$, therefore number of circuits tends to $o(4^n)$ as n grows to infinity.

2.2.2 Dyck walks in a strip $[0, Q]$

Let u be a parameter to count the load in vehicle, then the transition from state (z^n, u^q) to $(z^{n+1}, u^{q'})$ comes from either a pickup move (formally u^1) or a delivery move (formally u^{-1}). Assuming c_n counts number of walks from $(0,0)$ to (z^n, u^q) then recurrence is:

$$\begin{aligned} c_{n+1}(u) &= (u+u^{-1})c_n(u) - u^{-1}[u^{-1}](u+u^{-1})c_n(u) - u^{Q+1}[u^{Q+1}](u+u^{-1})c_n(u) \\ &= (u+u^{-1})c_n(u) - u^{-1}[u^0]c_n(u) - u^{Q+1}[u^Q]c_n(u) \end{aligned}$$

where $[u^{-1}]f(u)$ denotes the coefficient of u^{-1} in formal expression $f(u)$ and the negative terms account for walks outside the strip $[0, Q]$. Notice that the strip condition nullifies $[u^{-2}]c_n(u)$ and $[u^{Q+2}]c_n(u)$.

Multiplying by z^{n+1} both sides and summing over every term $z^n c_n(u)$ yields the generating function $C(z,u) = \sum_n z^n c_n(u)$ counting number of walks of a given length to a given load.

$$C(z,u) = c_0(u) + z(u+u^{-1})C(z,u) - zu^{-1}C(z,0) - zu^{Q+1}\frac{1}{Q!}\frac{\partial^Q C}{\partial u^Q}(z,0)$$

$$(u - zu^2 - z)C(z,u) = c_0(u)u - zC(z,0) - zu^{Q+2}\frac{1}{Q!}\frac{\partial^Q C}{\partial u^Q}(z,0)$$

using $[u^k]f(u) = \frac{1}{k!}\frac{\partial^k f}{\partial u^Q}(0)$; any intial load may be considered through $c_0(u)$ while in standard formulation we consider an empty truck $c_0(u) = u^0 = 1$.

Consider $C(z,0)$, $\frac{\partial^Q C}{\partial u^Q}(z,0)$ as unknowns in above equation then solving $u - zu^2 - z = 0$ in u gives 2 solutions

$$\begin{aligned} u_1(z) &= \frac{1-\sqrt{1-4z^2}}{2z} \\ u_2(z) &= \frac{1+\sqrt{1-4z^2}}{2z} \end{aligned}$$

from which we could solve a system of 2 linear equations to retrieve the unknowns in terms of z :

$$\begin{aligned} zC(z,0) &= c_0(u)\frac{u_2^{Q+1}(z) - u_1^{Q+1}(z)}{u_2^{Q+2}(z) - u_1^{Q+2}(z)} \\ z\frac{1}{Q!}\frac{\partial^Q C}{\partial u^Q}(z,0) &= c_0(u)\frac{u_2(z) - u_1(z)}{u_2^{Q+2}(z) - u_1^{Q+2}(z)} \end{aligned}$$

and finally get the formal generating function :

$$C(z,u) = \frac{c_0(u)}{u - zu^2 - z}\left(u - \frac{u_2^{Q+1}(z) - u_1^{Q+1}(z)}{u_2^{Q+2}(z) - u_1^{Q+2}(z)} - u^{Q+2}\frac{u_2(z) - u_1(z)}{u_2^{Q+2}(z) - u_1^{Q+2}(z)}\right)$$

The generating function of walks is given by any u^k , i.e. assigning $u = 1$

$$\begin{aligned} C(z,1) &= \frac{c_0(u)}{1-2z}\left(1 - \frac{u_2^{Q+1}(z) - u_1^{Q+1}(z)}{u_2^{Q+2}(z) - u_1^{Q+2}(z)} - \frac{u_2(z) - u_1(z)}{u_2^{Q+2}(z) - u_1^{Q+2}(z)}\right) \\ &= \frac{c_0(u)}{1-2z}\left(1 - z(1+y(z))\frac{1-y^{Q+1}(z)}{1-y^{Q+2}(z)} - z^{Q+1}(1+y(z))^{Q+1}\frac{1-y(z)}{1-y^{Q+2}(z)}\right) \end{aligned}$$

where $y(z) = \frac{u_1(z)}{u_2(z)} = \frac{1-\sqrt{1-4z^2}}{1+\sqrt{1-4z^2}}$. By grouping terms, it yields

$$\begin{aligned} C(z,1) &= c_0(u)\sum_{i\geq 0} 2^i z^i - c_0(u)\sum_{i\geq 0} 2^i z^{1+i}\frac{(1+y(z))(1-y^{Q+1}(z))}{1-y^{Q+2}(z)} \\ &\quad - c_0(u)\sum_{i\geq 0} 2^i z^{Q+1+i}\frac{(1+y(z))^{Q+1}(1-y(z))}{1-y^{Q+2}(z)} \end{aligned}$$

W_{2n}^q	n									
Q	1	2	3	4	5	6	7	8	9	10
1	1	1	1	1	1	1	1	1	1	1
2	2	4	8	16	32	64	128	256	512	1024
3	2	5	13	34	89	233	610	1597	4181	10946
4	2	6	18	54	162	486	1458	4374	13122	39366
5	2	6	19	61	197	638	2069	6714	21794	70755
6	2	6	20	68	232	792	2704	9232	31520	107616
7	2	6	20	69	241	846	2977	10490	36994	130532
8	2	6	20	70	250	900	3250	11750	42500	153750
9	2	6	20	70	251	911	3327	12190	44744	164407
10	2	6	20	70	252	922	3404	12630	46988	175066

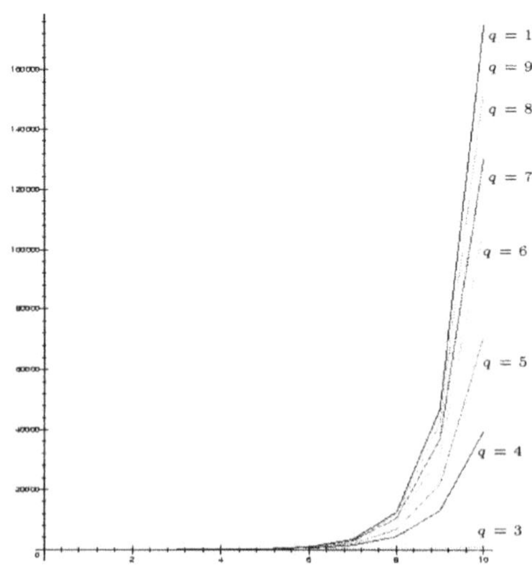

Figure 3: number of walks under small capacity and length

On the other hand, the number of excursions is retrieved through assigning $u = 0$

$$\begin{aligned} C(z,0) &= \frac{c_0(u)}{z}\frac{u_2^{Q+1}(z) - u_1^{Q+1}(z)}{u_2^{Q+2}(z) - u_1^{Q+2}(z)} \\ &= 2c_0(u)\frac{(1+\sqrt{1-4z^2})^{Q+1} - (1-\sqrt{1-4z^2})^{Q+1}}{(1+\sqrt{1-4z^2})^{Q+2} - (1-\sqrt{1-4z^2})^{Q+2}} \end{aligned}$$

Of course, last expression looks like $C^q(z)$ provided we substitute z^2 to z in the latter to relax evenness simplification.

$$C(z,u) = \frac{c_0(u)}{u-u^2z-z}\left(u - 2z\frac{(1+\sqrt{1-4z^2})^{Q+1}-(1-\sqrt{1-4z^2})^{Q+1}}{(1+\sqrt{1-4z^2})^{Q+2}-(1-\sqrt{1-4z^2})^{Q+2}}\right.$$
$$\left. -2^{Q+2}u^{Q+2}z^{Q+1}\frac{\sqrt{1-4z^2}}{(1+\sqrt{1-4z^2})^{Q+2}-(1-\sqrt{1-4z^2})^{Q+2}}\right)$$

3 Dimension of redistribution polytope

For any $V \subseteq \mathcal{V}$ of $\mathrm{RED}_Q(\mathcal{S}_{2n}, D)$, $\overline{V} = \mathcal{V} \setminus V$ will denote the complementary of V in $\mathcal{V}$; then redistribution is formulated as:

$$\min \sum_{i\in\mathcal{V}}\sum_{j\in\mathcal{V}} d_{ij}x_{ij}$$

subject to

$$\sum_{j\in\mathcal{V}} x_{ij} + \sum_{j\in\mathcal{V}} x_{ji} = 2, \quad i \in \mathcal{V} \tag{1}$$
$$\sum_{j\in\mathcal{V}} x_{ij} - \sum_{j\in\mathcal{V}} x_{ji} = 0, \quad i \in \mathcal{V} \tag{2}$$
$$\sum_{\substack{i \in V^+ \subseteq N^+ \\ j \in V^- \subseteq N^-}} x_{ij} \leq 2\min(|V^+|, |V^-|) \tag{3}$$
$$x_{ij} \in \{0,1\}$$

where d_{ij} is the distance between nodes i and j as given by D. For sake of conciseness we denote RED_Q^n the redistribution polytope.

3.1 Dimension of linear relaxation

Let $\mathrm{RED}_{Q/\mathbb{R}}^n$ be the linear relaxation of binary constraints $x_{ij} \in \{0,1\}$. Summing over all $i \in \mathcal{V}$ degree (1) and flow (2) equalities shows there are related by the 2 equalities

$$\sum_{i\in\mathcal{V}}\sum_{j\in\mathcal{V}} x_{ij} + \sum_{i\in\mathcal{V}}\sum_{j\in\mathcal{V}} x_{ji} = 2(2n+1)$$
$$\sum_{i\in\mathcal{V}}\sum_{j\in\mathcal{V}} x_{ij} - \sum_{i\in\mathcal{V}}\sum_{j\in\mathcal{V}} x_{ji} = 0$$

therefore $\dim(\mathrm{RED}_{Q/\mathbb{R}}^n) = |\text{variables}| - 2(2n+1) + 2$ so that $\dim(\mathrm{RED}_Q^n) \leq |\text{variables}| - 4n$.

- unit capacity $Q = 1$; number of variables reduces to $2n(n+1)$ since a surplus (resp. shortage) node is connected to a shortage (resp. surplus) node or the depot; therefore $\dim(\text{RED}_Q^n) \leq 2n(n-1)$.
- capacity $Q > 1$; from $G = (\mathcal{V}, \mathcal{K}_{2n} \times 0)$ we get $|\text{variables}| = 2n(2n-1)+2n = 4n^2$ yielding $\dim(\text{RED}_Q^n) \leq 4n(n-1)$.

3.2 The result

Proposition 3.1 *Redistribution polytope has dimension either* $\dim(RED_1^n) = 2n(n-1)$ *for unit capacity or* $\dim(RED_Q^n) = 4n(n-1)$.

Proof. let $\sum_i \sum_j \alpha_{ij} x_{ij} = \alpha_0$ an equation satisfied by any solution to redistribution problem; we will show that it is a linear combination of degree and flow equalities in order to prove the result.

First, given a sequence of nodes i, j, k, l, m in a solution of redistribution problem, we claim that there exist

1. a solution that contains i, l, k, j, m under mild assumptions (for instance, j, l both belong to N^+ or N^-; then after simplifying remaining terms, we get

$$\alpha_{ij} + \alpha_{jk} + \alpha_{kl} + \alpha_{lm} = \alpha_{il} + \alpha_{lk} + \alpha_{kj} + \alpha_{jm}$$

2. 2 solutions that contain i, l, m, j, k and i, j, m, l, k where memberships afford the pairs $(j, k), (l, m)$ to be swapped, leading to

$$\alpha_{il} + \alpha_{lm} + \alpha_{mj} + \alpha_{jk} = \alpha_{ij} + \alpha_{jm} + \alpha_{ml} + \alpha_{lk}$$

It is easy to check that those 4 solutions exist even for the simplest case $n = 2, Q = 1$.From previous 2 equalities, we derive:

$$2(\alpha_{ij} - \alpha_{il} = (\alpha_{kj} - \alpha_{kl}) + (\alpha_{mj} - \alpha_{ml})$$

implying that α is linear in i and j; therefore exist some β_i, β_j such that

$$\alpha_{ij} = \beta_i + \beta_j$$

Second, using $\sum_j x_{ij} = 1$ from degree and flow constraints fulfilled in any point i, we could rewrite our equality as

$$\sum_i \sum_j \beta_j x_{ij} = \alpha_0 - \sum_i \beta_i$$

Just take the alternating sequence of surplus and shortage nodes as a particular solution and apply it to above equality; then, we obtain

$$\alpha_0 = 2 \sum_i \beta_i$$

Altogether, equality rewrites as $\sum_i \sum_j (\beta_i + \beta_j) x_{ij} = 2 \sum_i \beta_i$, hence the required linear combination in degree and flow equalities. □

4 Searching an exponential neighborhood

4.1 Notations

A feasible solution of $\mathrm{RED}_Q(\mathcal{S}_{2n}, D)$ is denoted $\rho \in \mathcal{S}_{2n}$, its cost $D(\rho)$ and is composed of *excursions* $(\rho_1, \rho_2, \ldots, \rho_c)$; given a point $k \in [1, 2n]$, its image under permutation ρ is k^ρ. A transposition (k_i, k_j) has standard meaning of permutation group theory, while a left circular shift (composition of adjacent transpositions) $(k_i, k_j) \circ (k_j, k_l) \circ (k_l, k_m) \circ (k_m, k_n)$ or in disjoint cycle notation $(k_m, k_n)(k_l, k_m)(k_j, k_l)(k_i, k_j)$ is abbreviated by its cycle representation $(k_i \ldots k_n)$ and correspondingly for a right circular shift $(k_n, k_m) \circ (k_m, k_l) \circ (k_l, k_j) \circ (k_j, k_i)$, in disjoint cycle notation $(k_j, k_i)(k_l, k_j)(k_m, k_l)(k_n, k_m)$ its shortcut $(k_n \ldots k_i)$ consistent with composition of transpositions. Notice, the additive representation of taking the image under a given permutation ρ, $(k_i^\rho)^\rho = k_i^{2\rho}$ so that the reverse permutation is denoted $-\rho$ and identity 0; it will make further formulae more readable than multiplicative representation, in particular we will take benefit from $k_i = k_i^{0\rho}$ to shrink notation.

Provided all points involved in an *independent* (ordering meaningless) composition of consecutive transposition do not overlap, the notation $(k_i, k_j)^l = (k_i, k_j) \circ (k_i^\rho, k_j^\rho) \ldots \circ (k_i^{l\rho}, k_j^{l\rho})$ extends transposition to *chain* of points under ρ. The same extension applies to circular shift in the same fashion e.g. either $(k_n \ldots k_i)^l$ or $(k_n \ldots k_i)^l$ under the non overlapping chain condition.

In the sequel, we consider *dense* sets of consecutive transposition, or otherwise stated, composition of a given permutation ρ through sequence of disjoint circular shifts $\rho \circ (k_1 \ldots k_i) \circ (k_j \ldots k_i^\prime) \circ \ldots$ under the restriction that all excursions occur in the operation; a excursion occurs either once when its selected point appears within one shift only, or twice if two distinct points occur at any endpoint of 2 consecutive shifts like in above example (2 consecutive shifts may have same or reverse direction). Burkard and Deineko [BD94] name this set *dense* since it amounts to extend composition of ρ by either left side or right side while consecutively choosing points in increasing order of excursions $\rho_1, \rho_2 \ldots$ This neighborhood comes from the nice properties

$$
\begin{aligned}
(k_i \ldots k_{j-1}) \circ (k_{j-1}, k_j) &= (k_i \ldots k_j) \\
(k_{j-1}, k_j) \circ (k_{j-1} \ldots k_i) &= (k_j \ldots k_i)
\end{aligned}
$$

together with their chain counterpart (under non overlapping condition).

$$
\begin{aligned}
(k_i \ldots k_{j-1})^l \circ (k_{j-1}, k_j)^l &= (k_i \ldots k_j)^l \\
(k_{j-1}, k_j)^l \circ (k_{j-1} \ldots k_i)^l &= (k_j \ldots k_i)^l
\end{aligned}
$$

4.2 Patching on same parity points

We assume that only points with same parity (member of either N^+ or N^-) are ripple shifted to discard feasibility checking on the corresponding transformation.

Let us consider the sequence of differential cost associated with a left shift

$$\begin{aligned}
D(\rho\circ(k_i\dots k_{j-1})\circ(k_{j-1},k_j)) &- D(\rho\circ(k_i\dots k_{j-1}))\\
&= \mathbf{d_{k_{j-1}k_j^\rho}} - d_{k_{j-1}k_i^\rho} + \overline{d_{k_jk_i^\rho} - d_{k_jk_j^\rho}}\\
D(\rho\circ(k_i\dots k_{j-2})\circ(k_{j-2},k_{j-1})) &- D(\rho\circ(k_i\dots k_{j-2}))\\
&= d_{k_{j-2}k_{j-1}^\rho} - d_{k_{j-2}k_i^\rho} + d_{k_{j-1}k_i^\rho} - \mathbf{d_{k_{j-1}k_{j-1}^\rho}}\\
\dots &= \dots\\
D(\ rho\ \ circ(k_i,k_{i+1})) - D(\rho)\\
&= \underline{d_{k_ik_{i+1}^\rho} - d_{k_ik_{i+1}^\rho}} + d_{k_{i+1}k_i^\rho} - d_{k_{i+1}k_{i+1}^\rho}
\end{aligned}$$

where opening, closing and incremental terms are emphasized (underlined, overlined and in bold face respectively). Hereafter, we will keep emphasized notation for ease of reading but no meaning is attached to it.

Introducing $\delta_{k_jk_i} = d_{k_jk_i^\rho} - d_{k_jk_j^\rho}$ then a separation property appears while summing up consecutive lines above, after cancellation of terms with opposite sign and using $\Delta_{(k_i\dots k_{j-1})}$ for partial sums

$$\begin{aligned}
\Delta_{(k_i\dots k_j)} &= \Delta_{(k_i\dots k_{j-1})} + \delta_{\mathbf{k_{j-1}k_j}} &(4)\\
D(\rho\circ(k_i\dots k_j)) - D(\rho) &= \Delta_{(k_i\dots k_j)} + \overline{\delta_{k_jk_i}} &(5)\\
D(\rho\circ(k_{j-1},k_j)) - D(\rho) &= \underline{\delta_{k_{j-1}k_j}} + \delta_{k_jk_{j-1}} &(6)
\end{aligned}$$

Correspondingly, right shift yields

$$\begin{aligned}
\Delta_{(k_j\dots k_i)} &= \Delta_{(k_{j-1}\dots k_i)} + \delta_{\mathbf{k_jk_{j-1}}} &(7)\\
D(\rho\circ(k_j\dots k_i)) - D(\rho) &= \Delta_{(k_j\dots k_i)} + \overline{\delta_{k_ik_j}} &(8)\\
D(\rho\circ(k_j,k_{j-1})) - D(\rho) &= \underline{\delta_{k_jk_{j-1}}} + \delta_{k_{j-1}k_j} &(9)
\end{aligned}$$

Then the dynamic programming applies in a straigthforward fashion by introducing minimization in the course of summation. First, let $\Delta_{k_jk_j}$ define the best differential cost obtained by patching excursion ρ_1 upto excursion ρ_j; second, let $\Delta_{k_{j-1}k_j}$ be the best differential cost obtained by using transposition (k_{j-1},k_j) between excursions ρ_{j-1} and ρ_j, it comes either by ending a shift or breaking the dense set and restarting a new shift.

$$\Delta_{(k_i\dots k_j)} = \min_{k_{j-1}\in\rho_{j-1}} \Delta_{(k_i\dots k_{j-1})} + \delta_{\mathbf{k_{j-1}k_j}} \tag{10}$$

$$\Delta_{(k_j\dots k_i)} = \min_{k_{j-1}\in\rho_{j-1}} \Delta_{(k_{j-1}\dots k_i)} + \delta_{\mathbf{k_jk_{j-1}}} \tag{11}$$

$$\Delta_{k_{j-1}k_j} = \begin{cases} \underline{\delta_{k_{j-1}k_j}} + \min\limits_{k'_{j-1}\neq k_{j-1}\in\rho_{j-1}} \Delta_{k'_{j-1}k'_{j-1}} \\ \\ \min\limits_{k_i\in\rho_i, i=1\dots j-2} \Delta_{(k_{j-1}\dots k_i)} + \overline{\delta_{k_ik_j}} \end{cases} \tag{12}$$

$$\Delta_{k_j k_{j-1}} = \begin{cases} \underline{\delta_{k_j k_{j-1}}} + \min_{k'_{j-1} \neq k_{j-1} \in \rho_{j-1}} \Delta_{k'_{j-1} k'_{j-1}} \\ \\ \min_{k_i \in \rho_i, i=1\ldots j-2} \Delta_{(k_i \ldots k_{j-1})} + \overline{\delta_{k_j k_i}} \end{cases} \quad (13)$$

$$\Delta_{k_j k_j} = \min_{k_{j-1} \in \rho_{j-1}} \begin{cases} \Delta_{k_{j-1} k_j} + \delta_{\mathbf{k_j k_{j-1}}} \\ \Delta_{k_j k_{j-1}} + \delta_{\mathbf{k_{j-1} k_j}} \end{cases} \quad (14)$$

where the only subtlety relies on the exchange of the last increment and opening/closing terms between $\Delta_{k_j k_j}$ and $\Delta_{k_{j-1} k_j} (\Delta_{k_j k_{j-1}}$ resp.) compared to original recurrence : in the latter, we dealt with excursions upto $k-1$ so that the last increment (closing term resp.) should be introduced in the former for a closing (single transposition (k_{j-1}, k_j) resp.) in order to fulfill recurrence/separation property .

At end, the best differential cost from a given permutation ρ is retrieved from $\min_{k_c \in \rho_c} \Delta_{k_c k_c}$ and the actual shifted points along excursions ρ_1 to ρ_c through tracing back the best differential cost.

Proposition 4.1 *Best differential cost for point patching along excursion ρ_1 to excursion ρ_c can be solved in $O(n^2 |\rho_{\max}|)$ time and $\Omega(4n^2)$ space.*

Proof. There are $4n^2$ elements $\Delta_{k_i k_j}$ where the actual shift is retrieved from the relative position of k_i and k_j; from which $O(n^2)$ require at most $|\rho_{\max}| = \max_1^c |\rho_j|$. Only closing recurrences $\Delta_{k_{j-1} k_j}$ pay more and at most $\sum_1^{c-1} |\rho_i| \leq 2n$, but since we consider only same parity exchange, they are no more than $2(\sum_2^c |\rho_{i-1}|/2 |\rho_i|/2) \leq |\rho_{\max}|/2 \sum_1^c |\rho_i| \leq n |\rho_{\max}|$. □

Proposition 4.2 *Best differential cost for point patching along excursion ρ_1 to excursion ρ_c is retrieved within an exponential neighborhood of size $\frac{1}{4} |\rho_1| |\rho_2| \prod_{k=3}^{c} (|\rho_{k-1}| + 1) |\rho_k|$*

Proof. Let s_k be the number of shifts examined when point patching along excursion ρ_1 to excursion ρ_k; clearly $s_2 = |\rho_1||\rho_2|/4$. Let us suppose s_{k-1} to be known, there are $(|\rho_{k-1}| - 1)|\rho_k|$ different opening ways to compose excursion ρ_{k-1} with ρ_k and $2|\rho_k|$ different ways to increment or to close current shift. Altogether, it yields $s_k = (|\rho_{k-1}| + 1)|\rho_k| s_{k-1}$ □

It is worth noting that Burkard and Deineko requirement of patching all excursions in order, comes from the TSP they are dealing with; but for excursion patching, it is compulsory, so that the dynamic programming recurrence becomes

$$\Delta_{(k_i \ldots k_j)} = \min_{k_{j-1} \in \rho_{j-1}} \Delta_{(k_i \ldots k_{j-1})} + \delta_{\mathbf{k_{j-1} k_j}} \quad (15)$$

$$\Delta_{(k_j \ldots k_i)} = \min_{k_{j-1} \in \rho_{j-1}} \Delta_{(k_{j-1} \ldots k_i)} + \delta_{\mathbf{k_j k_{j-1}}} \quad (16)$$

$$\Delta_{k_{j-1} k_j} = \begin{cases} \underline{\delta_{k_{j-1} k_j}} + \min_{k'_i \in \rho_i, i=1\ldots j-1} \Delta_{k'_i k'_i} \\ \\ \min_{k_i \in \rho_i, i=1\ldots j-2} \Delta_{(k_{j-1} \ldots k_i)} + \overline{\delta_{k_i k_j}} \end{cases} \quad (17)$$

$$\Delta_{k_j k_{j-1}} = \begin{cases} \underline{\delta_{k_j k_{j-1}}} + \min\limits_{k'_i \in \rho_i, i=1\ldots j-1} \Delta_{k'_i k'_i} \\ \\ \min\limits_{k_i \in \rho_i, i=1\ldots j-2} \Delta_{(k_i \ldots k_{j-1})} + \overline{\delta_{k_j k_i}} \end{cases} \tag{18}$$

$$\Delta_{k_j k_j} = \min_{k_{j-1} \in \rho_{j-1}} \begin{cases} \Delta_{k_{j-1} k_{j-1}} \\ \Delta_{k_{j-1} k_j} + \delta_{\mathbf{k_j k_{j-1}}} \\ \Delta_{k_j k_{j-1}} + \delta_{\mathbf{k_{j-1} k_j}} \end{cases} \tag{19}$$

where skipping implicitly underlies the disjoint condition $k'_{j-1} \neq k_{j-1}$ whenever best patching happens to be up to ρ_{j-1} and where possibly no transposition occurs between ρ_{j-1} and ρ_j. Clearly, time complexity remains in the same bound.

Notice that from feasibility assumption, endpoints of any excursion may be shifted since they all have +1 parity. However, cumulative load along patching may be addressed in the same way to keep track of feasibility interchange...

5 Concluding remarks

We addressed a generic redistribution problem from a complexity point of view; both polyhedral dimension and dynamic programming approach were set in a usual framework (VRP or TSP-like) and are appealing to study further facet defining equalities. In the past, we already experiment a combined transportation and TSP formulation whose lagrangian relaxation leads to difficult step length adjustment (resulting in a low speed of convergence even with small sized instances). With this formulation, a Branch-and-Cut strategy seems more promising on the one hand; an efficient dynamic programming upper bound was given on the other hand, to fill the gap.

References

[BD94] R. E. Burkard and V. Deĭneko. Polynomially solvable cases of the traveling salesman problem and a new exponential neighborhood. Technical Report 1, Technische Universitt Graz, Austria, July 1994.

[dBKR72] N. G. de Bruijn, D. E. Knuth, and S. O. Rice. *The average height of planted trees*, pages 15–22. Academic press, 1972.

[DFNR99] M. Dror, D. Fortin, E. Naudin, and C. Roucairol. Heuristics for the redistribution of a single commodity. In *MIC'99*, July 1999. Angras de Reis, Brazil.

[DW97] V. Deĭneko and G. J. Woeginger. A study od exponential neighborhoods for the travelling salesman problem and for the quadratic assignment problem. Technical Report 05, Technische Universitt Graz, Austria, July 1997.

[FS93] Ph. Flajolet and R. Sedgewick. The average case analysis of algorithms: Counting and generating functions. Technical Report RR1888, INRIA Rocquencourt, France, April 1993.

[FS96] Ph. Flajolet and R. Sedgewick. The average case analysis of algorithms: Mellin transfoem asymptotics. Technical Report RR2956, INRIA Rocquencourt, France, August 1996.

Trends in Mathematics, © 2000 Birkhäuser Verlag Basel/Switzerland

On the rate of escape of a mutation-selection algorithm

CHRISTIAN MAZZA AND DIDIER PIAU *LaPCS, UFR de Mathématiques, Université Claude Bernard Lyon-I, 43 Bd du 11-Novembre-1918, 69622 Villeurbanne Cedex, France.* [mazza,piau]@jonas.univ-lyon1.fr

Abstract. *We consider a genetic algorithm associated with mutation and selection, modeled as a measure valued dynamical system on the integers. A simple symmetric random walk induces the mutation and the fitness is linear. We prove that the rate of escape of the fitness is of order the number of iteration steps, in a strong sense, since a.s. convergence and L^p convergence hold. Furthermore, the normalized algorithm converges in law, and a large deviations principle hold.*

Key words. *Genetic algorithms, dynamical systems.*

1 Introduction

Finite population genetic algorithms, introduced in Holland [8], are widely used in applications. They are relevant in many areas, for instance biology, computer science, optimisation and signal processing. Mathematically speaking, genetic algorithms are Markov chains on product spaces E^p, where E is the state space and $p \geq 1$ is the size of the running population. Generic genetic algorithms are the various combination of three basic operators: mutation, mating and selection.

Despite the successes of these algorithms in the applications and numerous experimental studies, few rigorous results on their behaviour are available.

Del Moral and Guionnet [3, 4, 5] study the infinite population limit of mutation-selection algorithms, in connection with nonlinear filtering. In particular, these authors show that the empirical law on E of the population $x(n)$ at a discrete time $n \geq 0$ converges, when p goes to infinity, toward the solution of a dynamical system with values on the space of probability measures on E. This dynamical system is the composition of a selection operator W and a mutation operator M. For a given fitness $f \geq 0$ and a law μ such that $\mu(f) \in (0, +\infty)$, the selection replaces μ by the law $W(\mu)$ such that, for any bounded g,

$$W(\mu)(g) := \mu(f\, g)/\mu(f). \tag{1}$$

Hence, $W(\mu)$ favors, more than μ, the states of high fitness. An equivalent description of W is that, if X is a random variable of law μ, then $W(\mu)$ is the law of a random variable $\widehat{X}$ such that

$$E(g(\widehat{X})) = E(f(X)\, g(X))/E(f(X)),$$

for any bounded function g. The mutation replaces μ by $M(\mu) := \mu\, Q$, where Q is the mutation kernel. The dynamical system $(\mu_n)_{n\geq 0}$ takes then the form $\mu_{n+1} := W(M(\mu_n))$.

Similar dynamical systems are considered in computer science [12], and in biology [7]. In the mating step, one can replace the linear kernel K by a quadratic operator, but the dynamical system is then much harder to analyze from the mathematical point of view. Rabinovich and Wigderson [11] get significant results about several examples of this combination of selection and mating. We refer the reader to Mazza and Piau [10], where similar results are proved for a new class of dynamical systems, generated by Mandelbrot martingales.

In this paper, we consider a mutation-selection operator on the integer line, with fitness $f(x) = |x|$. This model is the infinite population version of models used in biology to describe the evolution of populations of RNA viruses, see [9] and the references therein. Eq. (1) defines the selection W, and a random walk induces the mutation M by the convolution of a given probability measure ν. That is, one sets $M(\mu) = \mu * \nu$. Obviously, this is a much simplified setting, and our aim is to understand the role of selection in the convergence to equilibrium, i.e., in this case, to infinity.

Assuming, for further simplicity, that $\nu = \frac{1}{2}(\delta_{-1} + \delta_{+1})$, and that the dynamics starts from δ_1, we prove that the mean fitness at epoch n behaves like $2n/\pi$. The convergence holds a.s. and in every L^p space, $p \geq 1$. In addition, we prove that a large deviations principle and a central limit theorem hold, see Theorem 1.

Notice, about the linear growth these results demonstrate, that the mean fitness of the random walk, without selection, is of order $\sqrt{n}$. The best possible fitness that the random walk can reach after n steps being $n + 1$, our mutation-selection algorithm has the best possible order. Furthermore, we mention that Bérard and Bienvenüe [2] study the finite population version of this algorithm. They prove that, in this case, the growth is of order $\sqrt{n}$ (in a precise sense, see [2]). The transition finite population/infinite population is not fully understood yet.

We state our results in Section 2 as Theorem 1. Section 3 proves some technical lemmas, which we use in Section 4 for the proof of Theorem 1.

2 Results

We study the measure valued dynamical system

$$p_{n+1} = W(p_n * \nu), \quad \text{with} \quad f(x) = |x|,$$

where $*$ is the convolution of probability measures. We focus on the case of the simple symmetric random walk on the integers, where $\nu = \frac{1}{2}(\delta_{-1} + \delta_{+1})$. As stated in the introduction, this is the infinite population version of a simple model for the evolution of populations of RNA viruses considered in [9], where mutants move in fitness space according to a reflected random walk on $\{0, 1, 2, \ldots\}$. Our study sheds some light on the heuristic and experimental results presented in [9]. We are interested in the mean fitness at epoch n, i.e.

$$E(f(X_n)) = p_n(f).$$

Recall that $(X_n)_n$ is any sequence of random variables, such that p_n is the law of X_n. We prove the following results.

Theorem 1 *Consider the mutation-selection dynamics, where the mutation is based on the simple symmetric random walk starting at* $+1$*, and the selection on the fitness function* $f(x) = |x|$.

(1) The mean fitness at epoch n *increases linearly in the following sense:* $E(f(X_n))/n$ *converges to* $2/\pi$.

(2) The sequence $(X_n/n)_n$ *satisfies a large deviations principle of good rate function*

$$I(a) := \sup_{u \in (0,1)} [\log(2u) - a \log \tan(\pi u/2)].$$

(3) Since $I(a) > 0$ *for every* $a \neq 2/\pi$*, a consequence is that* X_n/n *converges to* $2/\pi$ *a.s. Furthermore,* X_n/n *converges to* $2/\pi$ *in every space* L^p, $p \geq 1$*, and the reduced random variable* $(X_n - (2/\pi)\, n)/\sqrt{n}$ *converges in law to a centered Gaussian law of variance* $4/\pi^2$.

Thus, selection has a rather strong effect, since, by the central limit theorem, the mean fitness of the classical (reflected) random walk, which corresponds to mutations without selection, is of order $\sqrt{n}$.

3 Lemmata

3.1 Multiplicative functionals of a Markov chain

Recall that, for a given fitness f, W is defined by (1). Let Q be an irreducible stochastic matrix. For a given measure ν, we consider the dynamical system defined by $p_0 := \nu$ and $p_{n+1} := W(p_n Q)$. The following classical lemma relates the value p_n of the dynamical system to the law of the Markov chain of transition matrix Q.

Lemma 1 *Let* $(Y_n)_{n \geq 0}$ *be a Markov chain of transition matrix* Q *and initial law* ν*, and* $\mathbf{1}(A)$ *be the indicator function of the set* A*. Then,*

$$p_n(x) = \frac{E_\nu(\mathbf{1}(Y_n = x)\, Z_n)}{E_\nu(Z_n)}, \quad \textit{where} \quad Z_n := \prod_{k=1}^{n} f(Y_k).$$

We introduce i.i.d. random variables $(\varepsilon_k)_k$ of law ν. We assume that the random walk starts from $S_0 = +1$, and that the initial law p_0 is δ_1. We set

$$S_n := 1 + \sum_{k=1}^{n} \varepsilon_k,$$

and we let X_n be any random variable of law p_n. Hence, $X_0 = 1$ a.s., and Lemma 1 yields

$$p_n(x) = \frac{E(\mathbf{1}(S_n = x)\, Z_n)}{E(Z_n)} \quad \text{with} \quad Z_n := \prod_{k=1}^{n} |S_k|. \tag{2}$$

3.2 A decomposition

Since the random walk starts from $S_0 = +1$, one can drop the absolute value signs in (2). Since ε_{n+1} is centered and independent of everything else in the expectations of (2), one can add it to S_n in the numerator of (2). This yields

$$p_n(f) = v_{n+1}/v_n, \qquad v_n := E(V_n), \qquad V_n := \prod_{k=1}^{n} S_k.$$

The following decomposition of V_n is our starting point.

Lemma 2 *(1) For $n \geq 0$, call Φ_n the set of functions*

$$\varphi : \{0, 1, \ldots, n\} \to \{0, 1, \ldots, n\},$$

such that $\varphi(k) \leq k$ for every k. For $\varphi \in \Phi_n$, let φ_i be the cardinal of $\varphi^{-1}(i)$. Then,

$$V_n = \sum_{\varphi \in \Phi_n} \varepsilon(\varphi), \quad \text{where} \quad \varepsilon(\varphi) := \varepsilon_1^{\varphi_1} \cdots \varepsilon_n^{\varphi_n}.$$

(2) Moreover, v_n is the cardinal of Ψ_n, where $\Psi_n \subset \Phi_n$ is the set of functions φ such that φ_i is even for every $i \geq 1$.
(3) The sequence $(v_n)_n$ is uniquely defined by

$$1/(1 - \sin s) = \sum_{n \geq 0} v_n \, s^n/n!, \quad |s| < \pi/2.$$

We state separately a consequence of this lemma. Recall that Bernoulli numbers $(B_n)_{n\geq 0}$ and Euler numbers $(E_n)_{n\geq 0}$ can be defined as the coefficients of the expansions, see p. 804 of [1]:

$$s/(\exp(s) - 1) =: \sum_{n \geq 0} B_n \, s^n/n! \quad \text{and} \quad 1/\cosh s =: \sum_{n \geq 0} E_n \, s^n/n!.$$

One can write v_{2n} as a multiple of B_{2n+2} and v_{2n+1} as a multiple of E_{2n+2}. Since precise estimates of Bernoulli and Euler numbers are known, one gets:

Lemma 3 *Let $v_n^0 := 2\,(2/\pi)^{n+2}\,(n+1)!$. Then, $v_n = v_n^0\,(1 + w_n)$, where $w_n \sim (-1)^n/3^{n+2}$.*

4 Proof of Theorem 1

Lemma 3 yields an equivalent of the mean fitness $p_n(f)$, i.e. Part (1) of Theorem 1, since

$$p_n(f) = v_{n+1}/v_n \sim v_{n+1}^0/v_n^0 \sim (2/\pi)\,n.$$

We now compute the Laplace transform of p_n, defined, for any t, by

$$\xi_n(t) := E(\exp(t\,X_n)) = E(V_n \, \exp(t\,S_n))/v_n.$$

Using the expansion of V_n in Lemma 2, one has to evaluate, for $\varphi \in \Phi_n$,

$$E(\varepsilon(\varphi)\exp(t\,S_n)) = \mathrm{e}^t\,(\sinh t)^{d(\varphi)}\,(\cosh t)^{n-d(\varphi)},$$

where $d(\varphi)$ is the number of indices $i \in \{1, 2, \dots, n\}$ such that φ_i is odd. Hence, $\xi_n(t)$ is a simple function of $g_n(\tanh t)$, where g_n is the generating function of $d(\cdot)$. It turns out that g_n may be computed with the help of the function ℓ introduced in Lemma 2. One gets

$$\begin{aligned} \xi_n(t) &= \mathrm{e}^t\,(\cosh t)^n\,g_n(\tanh t)/v_n \\ &= (\cosh t)^{-1}\,\ell^{(n)}(s)/v_n = (\cos s)\cdot \ell^{(n)}(s)/\ell^{(n)}(0), \end{aligned}$$

where $\sin s = \tanh t$. Here, s is a function of t, whose sign is the sign of t, and which may be defined by $|s| < \pi/2$ and by any of the following three relations:

$$\tanh t = \sin s, \quad \cosh t = 1/\cos s, \quad \sinh t = \tan s. \tag{3}$$

Lemma 3 then yields an equivalent of $\xi_n(t)$ (we skip the details and refer the reader to [10]):

$$\xi_n(t) \sim (\cos s)\cdot(1 - 2s/\pi)^{-(n+2)},$$

for any t, when n goes to infinity. Setting

$$\lambda(t) := \lim_{n\to+\infty} n^{-1}\,\log \xi_n(t) = -\log(1 - 2s/\pi),$$

one sees that $\lambda(t)$ is always finite. Secondly, since the support of X_n is a subset of $[1, n+1]$, the sequence of the laws of X_n/n is exponentially tight. Finally, since $t \mapsto s$ is smooth, the function λ itself is smooth. Hence, a Gärtner–Ellis type theorem, see Corollary 4.6.14 of [6], shows that X_n/n satisfies the large deviations principle

$$\begin{aligned} \limsup n^{-1}\,\log P(X_n \in n\,F) &\leq -\inf\{I(a)\,;\, a \in F\}, \\ \liminf\, n^{-1}\,\log P(X_n \in n\,U) &\geq -\inf\{I(a)\,;\, a \in U\}, \end{aligned}$$

for any closed set F and any open set U of the real line. The action I is a good rate function and I is the Fenchel–Legendre transform of λ, i.e.

$$I(a) := \sup\,(at - \lambda(t)) = \sup\,(at + \log(1 - 2s/\pi))\,,$$

where the supremum is over all the real numbers t, and where s is still related to t through (3).

One gets the convergence in law of $X'_n := n^{-1/2}\,((\pi/2)\,X_n - n)$ with the same methods, through the estimation of

$$\eta_n(t) := E(\exp(t\,X'_n)) = \mathrm{e}^{-t\sqrt{n}}\,\xi_n(\tau/\sqrt{n}),$$

with $\tau := \pi\,t/2$. Then, $\eta_n(t)$ converges to $\exp(t^2/2)$, a fact which ends the proof of Theorem 1.

References

[1] M. Abramowitz and I. A. Stegun, editors (1992). *Handbook of mathematical functions with formulas, graphs, and mathematical tables.* Dover, New York. Reprint of the 1972 edition.

[2] J. Bérard and A. Bienvenüe (2000). Convergence d'un algorithme génétique en population finie. *Preprint.*

[3] P. Del Moral and A. Guionnet (1999a). Central limit theorem for nonlinear filtering and interacting particle systems. *Ann. Appl. Probab.*, 9(2): 275–297.

[4] P. Del Moral and A. Guionnet (1999b). On the stability of measure valued processes with applications to filtering. *C. R. Acad. Sci. Paris Sér. I Math.*, 329(5): 429–434.

[5] P. Del Moral and A. Guionnet (1999c). On the stability of measure valued processes. Applications to non linear filtering and interacting particle systems. *Preprint.*

[6] A. Dembo and O. Zeitouni (1998). *Large deviations techniques and applications.* Springer-Verlag, New York, second edition.

[7] J. Hoffbauer and K. Sigmund (1988). *The theory of evolution and dynamical systems.* London Math. Society, Students Texts 7. Cambridge Univ. Press.

[8] J. Holland (1975). *Adaptation in natural and artificial systems.* Univ. of Michigan Press.

[9] D. Kessler, H. Levine, D. Ridgway and L. Tsimring (1997). Evolution on a smooth landscape. *J. of Statist. Physics*, 87(3/4): 519-544.

[10] C. Mazza et D. Piau (2000). On the effect of selection in genetic algorithms. *Preprint.*

[11] Y. Rabinovich and A. Wigderson (1999). Techniques for bounding the convergence rate of genetic algorithms. *Random Structures Algorithms*, 14(2): 111–138.

[12] M. Vose and A. Wright (1995). Stability of vertex fixed points and applications. In *Foundations of genetic algorithms* 3. L. Whitley and M. Vose Eds. Morgan Kaufmann.

Trends in Mathematics, © 2000 Birkhäuser Verlag Basel/Switzerland

Randomized Rendezvous

YVES MÉTIVIER, NASSER SAHEB AND AKKA ZEMMARI *LaBRI, Université Bordeaux I - ENSERB, 351 cours de la Libération 33405 Talence, France.*

Abstract. *In this paper we propose and analyze a randomized algorithm to get rendezvous in an anonymous graph. We examine in particular the probability to obtain at least one rendezvous and the expected number of rendezvous. We study the rendezvous number distribution in the cases of chain graphs, rings and complete graphs.*

Key words. *randomized algorithm, analysis of algorithm, graph, rendezvous.*

1 Introduction

We consider an asynchronous distributed network of anonymous processors with an arbitrary topology; processors communicate by exchanging messages. It is represented as a connected graph where vertices represent processors, and two vertices are connected by an edge if the corresponding processors have a direct communication link. We consider systems with asynchronous message passing: a process sends a message to another processor by depositing the message in the corresponding channel, and there is no fixed upper bound on how long it takes for the message to be delivered.

In synchronous message passing, the sender and the receiver must both be ready to communicate. A communication takes place only if the participant processors are waiting for the communication: this is termed a rendezvous.

Angluin [1] proved that there is no deterministic algorithm to implement synchronous message passing in an anonymous network that passes messages asynchronously (see [16]).

In this paper, we consider the following distributed randomized procedure. Every message will be a single bit:

Each vertex v repeats forever the following actions:
the vertex v selects one of its neighbours $c(v)$ chosen at random;
the vertex v sends 1 *to* $c(v)$;
the vertex v sends 0 *to its neighbours different from* $c(v)$;
the vertex v receives messages from all its neighbours.
(There is a rendezvous between v and $c(v)$ if v receives* 1 *from* $c(v)$ **)*

Our analysis is based on the consideration of *rounds*: in order to measure the performance of the algorithm in terms of the number of rendezvous taking place, we assume that at some instant each node sends and receives messages. Thus this parameter of interest, which is the (random) number of rendezvous, is the maximal number (i.e. under the assumption that all nodes are active) authorized by the algorithm. We believe that in many real cases similar approaches and computations may be applied to measure the complexity and performances of randomized distributed algorithms.

The first investigations, related to the number of rendezvous, are carried out on the properties of the expected number of rendezvous. We get the asymptotic lower bound $1-e^{-1/2}$ for the probability of a success in a round. It follows that the algorithm is a Las Vegas one. Sharper lower bounds are obtained for the classes of graphs with bounded degrees. As a direct consequence of the definitions, we compute easily the probability of a rendezvous for vertices, from which we derive the expected waiting time for a vertex to get a rendezvous. Elementary computations provides the expected waiting time between two rendezvous for edges. We also study the rather surprising effect of adding a new edge on the number of rendezvous. It is shown that the impact is not monotone. In some cases it is positive in others negative, both on the expected number of rendezvous and on the probability of a success. The asymptotic distribution of the rendezvous number is fully characterized for the class of complete graphs and that of ring graphs. Further investigations, being the subject of forthcoming studies, show that the introduced randomized algorithm has a good performance whenever it is applied in lowly linked networks. In the case of highly linked ones, it can be improved if one introduces messages using more than one bit.

Many problems have no solution in distributed computing [11]. The introduction of randomization makes possible tasks that admit no deterministic solutions; for instance, the election problem in an anonymous network. The impossibility result on the election problem comes from the fact that the symmetry between the processors cannot be broken in an anonymous network that passes messages asynchronously.

Many papers and results are based on the same model. During a basic computation step, two adjacent vertices exchange their labels and then compute new ones: for example, in [1] an election algorithm is given for complete graphs or in [5, 12] election algorithms are given for prime rings (rings having a prime size). In these cases, our randomized algorithm may be considered as a basic step for the implementation of these algorithms in an anonymous asynchronous system where processors communicate with asynchronous message passing.

General considerations about randomized distributed algorithms may be found in [16] and some techniques used in the design for the analysis of randomized algorithms are presented in [14, 8, 9].

Our paper is organized as follows. Section 2 contains basic notions. Section 3 gives general results. Section 4 studies the probability to get at least one rendezvous in particular cases. Section 5 provides a uniform lower bound for the success probability. Section 6 is devoted to the rendezvous number distribution on special classes of graphs. We only sketch the proofs and we refer the reader to [13] for more details.

2 Basic Notions and Notation on Graphs

We use standard terminology of graph theory [3]. A *simple graph* $G = (V, E)$ is defined as a finite set V of *vertices* together with a set E of *edges* which is a set of pairs of different vertices, $E \subseteq \{\{v, v'\} \mid v, v' \in V,\ v \neq v'\}$. The cardinality of V is called the *size* of the graph. A *tree* is a connected graph containing no cycles. A *forest* is a graph whose connected components are trees. A *quasi-tree* τ is a connected graph containing precisely one cycle. A *quasi-forest* is a graph whose connected components are quasi-trees. In this paper we consider only simple (i.e. without loop) and connected graphs.

3 First Results

Let $G = (V, E)$ be a connected simple graph of size $n > 1$. The purpose of this section is to provide a formal basis for the randomized procedure described in the introduction and to give simple general results on its analysis.

3.1 Definition and Characterization of a Call

Definition 1. *Let $G = (V, E)$ be a graph. A call over $G = (V, E)$ is a function c from V into itself which maps each $v \in V$ to one of its neighbours.*

Let c be a call, according to the definition, there is a rendezvous if and only if there exist two vertices v and w such that $c(v) = w$ and $c(w) = v$. A call c over $G = (V, E)$ will be a *success*, if there is at least one rendezvous. Otherwise it will be a *failure*. It is convenient to represent a call c over G by a directed graph $G_c = (V, A)$, where A contains an arc from v to w if and only if $w = c(v)$. Clearly G_c is a simple graph whose vertices are all of outer degree 1. It has, therefore, $n = |V|$ arcs. Moreover it is easy to see that:

Lemma 1. *Let c be a call over the graph G. Then c is a failure if and only if G_c has no cycle of length 2.*

Lemma 2. *If $G = (V, E)$ is a tree then any call over G is a success.*

If we consider the case of K_n the complete graph of size n, a call corresponds to the combinatorial notion of endofunctions [4].

Obviously there is a one-to-one correspondence between the set of calls over K_n and the set of endofunctions on $\{1, ..., n\}$ without fix-point. A call is a success (resp. a failure) if and only if it corresponds to an endofunction without a fix-point containing at least one cycle of length 2 (resp. without cycles of length 2).

It is also easy to see that a failure call corresponds to a quasi-forests with cycles of length greater than 2, see [13].

3.2 Probability of Success on a Graph

We assume that all the adjacent vertices to v have the same chance equal to $\frac{1}{d(v)}$ to be chosen, where $d(v)$ is the degree of the vertex v. Thus any edge $e = \{v, w\} \in E$ has a probability $\frac{1}{d(v)}$ to be the bearer of the unique message of v to w. The adjacent vertices v and w are said to *meet* each other, if v and w contact one other: there is a rendezvous. Throughout this study it is assumed that each vertex behaves independently in a memoryless manner.

Each vertex v has $d(v)$ possible choices, consider now the probability measure, which assigns to each call over G the probability $\alpha(G)$ equal to:

$$\alpha(G) = \prod_{v \in V} \frac{1}{d(v)}.$$

Let $s(G)$ be the probability of a success and $f(G)$ that of a failure. From Lemma 1, we deduce:

Lemma 3. *We have $f(G) = \alpha(G)N(G)$ and $s(G) = 1 - \alpha(G)N(G)$, where $N(G)$ is the number of calls c over G for which G_c has no cycle of length 2.*

The probability $f(G)$ may be obtained using quasi-forests. Let $F(G)$ be the set of spanning quasi-forests of G, if ϕ is a spanning quasi-forest of G then $|\phi|$ denotes the number of quasi-trees of ϕ. With these notations and using the characterization of failures by means of quasi-forests we obtain

$$f(G) = \prod_{v \in V} \frac{1}{d(v)} \Big(\sum_{\phi \in F(G)} 2^{|\phi|} \Big).$$

In order to get an exact expression for the probability distribution of the number of rendezvous for a random call, we consider matchings. A matching over $G = (V, E)$ is a subset M of E such that for any pair e and e' in M, $e \cap e' = \emptyset$. We associate to a matching M the rendezvous corresponding to meetings between end-points of edges of the matching, this set of rendezvous is by definition the rendezvous over the matching M. Let $e = \{v, w\}$ be an edge, $e^{(1)}$ denotes the event of a rendezvous over e and $e^{(0)}$ the complementary. The probability of a rendezvous over e is $Pr(e^{(1)}) = Pr(\{v, w\}^{(1)}) = \frac{1}{d(v)d(w)}$. Let $M = \{e_1, ..., e_k\}$ be a matching, in the same manner the probability $Pr(M)$ of the rendezvous over M is

$$Pr(M) = Pr(e_1^{(1)} \wedge e_2^{(1)} \wedge ... \wedge e_k^{(1)}) = \prod_{\{v,w\} \in M} \frac{1}{d(v)d(w)} = \prod_{e \in M} Pr(e^{(1)}).$$

For the integer k, a *k-matching* over G is a matching of cardinality k. Let $\mathcal{M}_k$ denote the set of all k-matchings. Let $q_k = \sum_{M \in \mathcal{M}_k} Pr(M),\ k = 0, 1, ..., \lfloor n/2 \rfloor$. According to this definition, it should be noted that $q_0 = 1$. By a straightforward application of the principle of inclusion-exclusion, we have:

Proposition 1. *Let the sequence $q_k, k = 0, 1, ..., \lfloor n/2 \rfloor$ be defined as above for the connected graph G of size n. Then, for the integer l in the above stated range, the probability of having* at least *l rendezvous over G is $P_l = \sum_{0 \le i \le \lfloor n/2 \rfloor - l} (-1)^i q_{l+i}$. In particular the probability of a success is $s(G) = P_1 = \sum_{0 \le i \le \lfloor n/2 \rfloor - 1} (-1)^i q_{i+1}$.*

It is possible also to derive rather simple expressions for the probability $s(G)$ of success and subsequently that of the expected number of necessary calls in the case of special classes of graphs. For instance:

Example 1. Let G be a ring graph (cycle) of size $n \geq 2$. The number $N(G)$, used in Lemma 3, is equal to 2. Hence $f(G) = \frac{1}{2^{n-1}}$ and $s(G) = 1 - \frac{1}{2^{n-1}}$. The expected number of necessary calls to get a success is thus $\frac{2^{n-1}}{2^{n-1}-1}$.

The impact of the addition of an edge on the probability of success is not monotone.

- If we add an edge to a tree the probability of the success decreases.
- The graph G of Figure 1, due to H. Austinat and V. Diekert [2], shows that the addition of an edge may increase this probability. In fact, for this graph, we have $s(G) = 1156/1600 = 0.7225$. Let G' denote the graph obtained from G by adding the edge $\{1, 2\}$. Then we have $s(G') = 4742/6400 = 0.7409....$

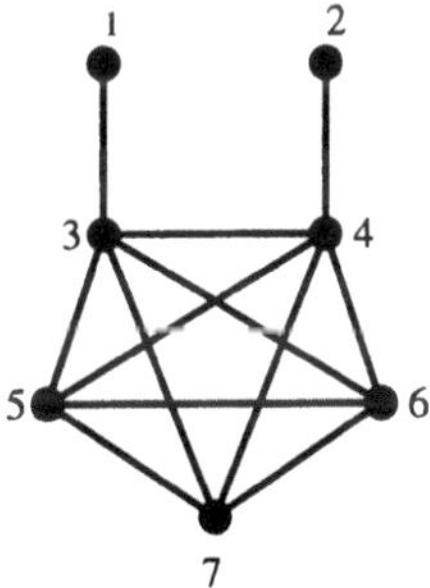

Figure 1

3.3 Expected Time Between Successive Rendezvous

For a vertex v the probability $p(v)$ of a rendezvous involving this vertex can be computed easily thanks to the independence of the choice for vertices and thanks to the fact that events associated to rendezvous over incident edges are disjoint:

$$p(v) = \sum_{e \text{ incident with } v} Pr(e^{(1)}) = \frac{1}{d(v)} \sum_{w \text{ adjacent to } v} \frac{1}{d(w)}.$$

¿From this formula it is clear that $p(v) = 1$ if and only if all w adjacent to v are leaves. We get the expected number of calls to get a rendezvous for v:

$$\frac{d(v)}{\displaystyle\sum_{w \text{ adjacent to } v} \frac{1}{d(w)}}.$$

Thus, for a graph of degree at most d, the expected time between two successive rendezvous for a vertex is bounded by d.

The expected time between two successive rendezvous on an edge $e = \{v, w\}$ is $d(v)d(w)$. Once more, if we consider a graph of bounded degree this value is bounded by d^2.

3.4 Expected Number of Rendezvous

Let X be the number of rendezvous of a call over G, the expected number of rendezvous over G, denoted $\overline{M}(G)$, is $E(X)$, the expected value of X. This parameter may be considered as a measure of the degree of parallelism of the rendezvous algorithm. We have:

Proposition 2. *The expected number of rendezvous over the graph G is:*

$$\overline{M}(G) = \sum_{\{v,w\}\in E} \frac{1}{d(v)d(w)}.$$

Consider the following particular cases.

Example 2. If G is a complete graph of size $n \geq 2$, we have $\overline{M}(G) = \binom{n}{2}\frac{1}{(n-1)^2} = \frac{n}{2(n-1)}$.

Example 3. If G is a cycle of size $n \geq 2$, we have $\overline{M}(G) = \frac{n}{4}$.

Example 4. If $G = (V, E)$ has a degree bounded by d then $\overline{M}(G) \geq \frac{|E|}{d^2}$.

If we consider the case of a tree T of size n with degrees bounded by d, we get $\overline{M}(T) \geq \frac{n-1}{d^2}$. In the case of regular graphs of size n and of degree d, we have $\overline{M}(T) = \frac{n}{2d}$.

We are interested by the impact of the addition of an edge on $\overline{M}(G)$. The above examples illustrate the fact that the number of edges does not necessarily favour the events of rendezvous. Nevertheless, Figure 2 and Figure 3 show that the expected number of rendezvous is not monotone with respect to the additions of new edges.

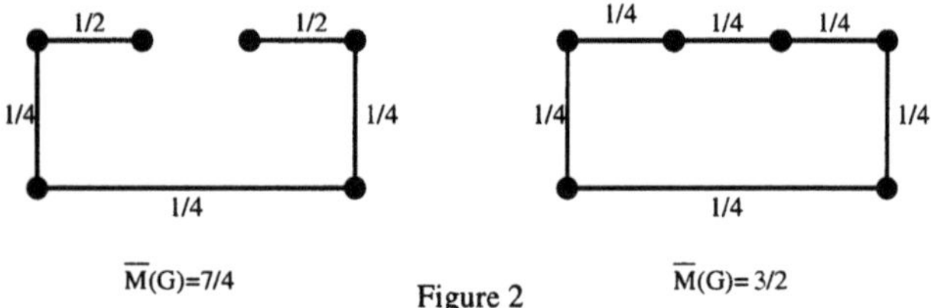

Figure 2

Proposition 3 gives a lower bound for the number of rendezvous expectation.

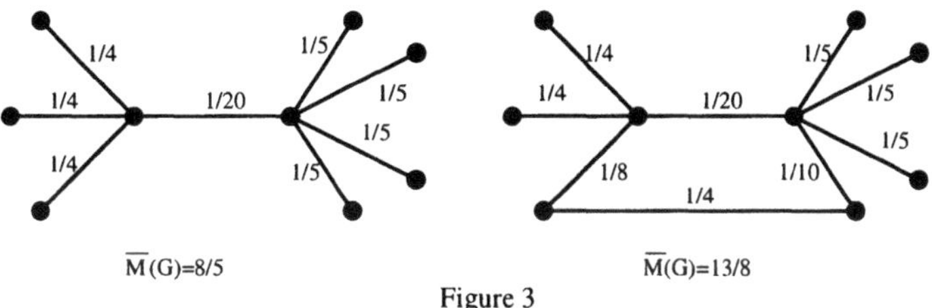

Figure 3

Proposition 3. *For a given fixed positive integer n, the complete graph K_n minimizes the expected number of rendezvous over graphs of size n. The minimal expected value realized by K_n is $\frac{n}{2(n-1)}$.*

Proof. Given a graph $G = (V, E)$, if we denote by $\overline{M}(G)$ the expected number of rendezvous in G, we have $\overline{M}(G) = \frac{1}{2}\sum_{v\in V} p(v)$, where $p(v)$ is as in Section 3.3. Since $p(v) = \frac{1}{d(v)} \sum\limits_{\{w,v\}\in E} \dfrac{1}{d(w)}$, and $d(w) \leq n-1$, we have $p(v) \geq \frac{1}{n-1}$. Summing on all vertices, we get $\overline{M}(G) \geq \frac{n}{2(n-1)}$. By Example 2, if G is the complete graph of size n, $\overline{M}(G) = \frac{n}{2(n-1)}$. □

4 Probability of Success in Particular Cases

In this section we study the probability of getting at least one rendezvous over the graph. The considered graphs are special classes of graphs. Let e be an edge, we recall that $e^{(1)}$ denotes the event of a rendezvous over e and $e^{(0)}$ the complementary event.

4.1 Graphs with Bounded Degrees

We start by the case where $G = (V, E)$ is a graph of degree at most d. The following proposition gives a lower bound for the failure probability.

Proposition 4. *Let $G = (V, E)$ be a d-bounded degree graph, and $s(G)$ denote its success probability. Then we have*

$$s(G) \geq 1 - (1 - \frac{1}{d^2})^{|E|}.$$

Proof. A straightforward computation based on conditional probabilities, (see [13]).

□

The above bound becomes very interesting if the ratio $|E|$ over d is important. Indeed, the above formula shows that $f(G) \leq e^{\frac{-|E|}{d^2}}$.

In particular in the case of $d-$regular graphs G, we have $|E| = \frac{nd}{2}$ and therefore:

Corollary 1. *Let G be a $d-$regular graph, the failure probability $f(G)$ satisfies $f(G) \leq e^{\frac{-n}{2d}}$.*

4.2 Complete Graphs

For this class of graphs we have:

Proposition 5. *Let K_n be the complete graph of size n, then:*

- $s(K_n) = \sum_{k\geq 1}(-1)^{k+1}\frac{n!}{k!2^k(n-2k)!}\frac{1}{(n-1)^{2k}}$,
- *$s(K_n)$ is asymptotically $1 - e^{-1/2}$,*
- *and the expected number of necessary calls to get a success is asymptotically $\sqrt{e}/(\sqrt{e}-1)$.*

Proof. For a fixed k, the k-matchings of K_n are all of the same probability $q_k = \frac{1}{(n-1)^{2k}}$. On the other hand, an easy computation yields their number which is $\frac{n!}{k!2^k(n-2k)!}$. We have thus:

$$s(K_n) = \sum_{k\geq 1}(-1)^{k+1}\frac{n!}{k!2^k(n-2k)!}\frac{1}{(n-1)^{2k}}.$$

The expected number of necessary calls to get a success is $\frac{1}{s(K_n)}$. The above expression is difficult to compute. Nevertheless, if we use a combinatorial reasoning, we can estimate it asymptotically.

As we saw in Section 3.1, a call c over G is a failure, if G_c is without cycle of length 1 or 2. A translation of specifications of types shows that the exponential generating function $F(z)$ of the number of such graphs is $F(z) = c(t(z))$, where $t(z) = ze^{t(z)}$ is the EGF(exponential generating function) of number of labeled trees and $c(z) = \frac{1}{1-z}e^{-z-z^2/2}$ that of the number of cycles of length at least 3, see [7]. The unique singularity of $F(z) = c(t(z))$ is $z_0 = 1/e$, since $t(z_0) = 1$ iff $z_0 = 1/e$. In [6, 7], the authors show that $t(z) \sim 1 - 2^{1/2}\sqrt{1-ez}$. Hence $F(z) \sim \frac{1}{\sqrt{2e^3}}(1-ez)^{-1/2}$. From which, we get easily $[z^n]F(z) \sim \frac{1}{\sqrt{2e^3}}e^n\frac{1}{\sqrt{\pi n}}$. This yields the number of failure calls $N(K_n)$ on the complete graph K_n. In order to get the failure probability over K_n, we have to divide $N(K_n)$ by $(n-1)^n$ which is the total number of calls over K_n. Using the Stirling formula, we derive $f(K_n) \sim e^{-1/2}$, or $s(K_n) \sim 1 - e^{-1/2}$. And the expected number of necessary calls to get a success is asymptotically $\frac{\sqrt{e}}{\sqrt{e}-1}$. □

5 Lower Bound for the Probability of Success

Proposition 4 gives a lower bound for the success probability if the graph is of maximum degree d. Corollary 1 shows how this bound is important if d is small enough in comparison with n. But this bound becomes uninteresting if d is too large and $|E|$ is not large enough. It is therefore interesting to find a uniform bound which does not depend on d or on $|E|$. The goal of this section is to give such a bound. Indeed, we have the following theorem:

Theorem 1. *The probability $s(G)$ of a success in a call over any graph $G = (V, E)$ is bounded from below by $1 - e^{-\overline{M}(G)}$, where $\overline{M}(G)$ denotes the expected number of rendezvous in G.*

Proof. It is easy to see that $f(G) \leq \prod_{i=1}^{m}(1 - Pr(e_i^{(1)}))$. By virtue of Proposition 2, we have $\sum_{i=1}^{m} Pr(e_i^{(1)}) = \overline{M}(G)$.
Thus, the bound on $f(G)$ is maximal when $Pr(e_i^{(1)}) = \frac{\overline{M}(G)}{m}$, $\forall i = 1, \ldots, m$.
Hence,

$$f(G) \leq \prod_{i=1}^{m}(1 - \frac{\overline{M}(G)}{m}) = (1 - \frac{\overline{M}(G)}{m})^m \leq e^{-\overline{M}(G)}.$$

□

Corollary 2.(Robson[1]) *The probability $s(G)$ of a success call over any graph $G = (V, E)$ is lower bounded by $1 - e^{-1/2}$.*

Remark 1. In spite of this corollary and Proposition 5, which asserts that $f(K_n) \leq e^{-1/2}$, it is not known whether or not the complete graph K_n minimizes the probability of success on graphs of size n.

Remark 2. The above theorem shows that the algorithm is a Las Vegas one.

6 Rendezvous Number Distribution

The previous study can be refined by determining the distribution of the rendezvous number in a given graph. It seems also interesting to evaluate the asymptotic behaviour of this random variable for graphs of large size. Let us consider for instance star, complete and chain graphs. For the first class of graphs the number of rendezvous is always 1. For the second class it takes value in the integer interval $[0, \frac{n}{2}]$ and its mathematical expectation is $\frac{n}{2(n-1)}$. For the third class it takes value in $[1, \frac{n}{2}]$ with the mathematical expectation $\frac{n+1}{4}$. Although the computation of the distribution is feasible in principle, no simple method is available and a standard technique based on a direct numbering is quite complicated.

As a first attempt in this direction we calculate the asymptotic distribution of the rendezvous number for the two extreme cases of complete graphs and chain graphs (which is the same as for ring graphs). We show in the case of complete graphs, as n grows to infinity, this random number remains an integer finite valued random variable with a distribution which will be determined in the sequel; the same study for chain graphs reveals a quite different behaviour: the expected

[1]A first direct proof of the corollary was provided by J. M. Robson [15]

rendezvous number grows and the normalized random variable tends in law to a normal distribution.

6.1 Rendezvous Number in Ring and Chain Graphs

We start with the study of chain graphs for which the rendezvous number distribution can easily be computed. We then prove that this number for the ring graphs is asymptotically the same as in the case of chain graphs. Let $G = (V, E)$ be a chain of size n and let X_n denote the random variable which counts the number of rendezvous in G. In this section we are interested in the asymptotic distribution of this r.v. It is supposed that all vertices are active. We first prove the following lemma which provides an exact expression for the probability of exactly k rendezvous on the graph.

Lemma 5. *For any integer k, the probability of having exactly k rendezvous is*

$$Pr(X_n = k) = \frac{1}{2^{n-2}}\binom{n-1}{2k-1}.$$

Proof. Let $\phi_n(x)$ be the ordinary generating function for the r.v. X_n, i.e. $\phi_n(x) = \sum_{k=0}^{\infty} Pr(X_n = k)x^k$. It is technically convenient to consider also generating function $\psi_{n-1}(x)$ for another r.v. which counts the rendezvous number on the chain graph whenever one of the endpoints is passive. A combinatorial reasoning shows that we have the following recurrences

$$\begin{cases} \phi_n(x) &= \frac{1}{2}\phi_{n-1}(x) + \frac{1}{2}x\psi_{n-1}(x), \quad \forall n \geq 2 \\ \psi_n(x) &= \frac{1}{2}\psi_{n-1}(x) + \frac{1}{2}\phi_{n-1}(x), \quad \forall n \geq 2 \end{cases}$$

with $\phi_1(x) = \psi_1(x) = 1$. A straightforward technique provides the solution

$$\phi_n(x) = \frac{2\sqrt{x}}{1+\sqrt{x}}\left(\frac{1+\sqrt{x}}{2}\right)^n - \frac{2\sqrt{x}}{1-\sqrt{x}}\left(\frac{1-\sqrt{x}}{2}\right)^n,$$

from which we get easily $[x^k]\phi_n(x) = \frac{1}{2^{n-2}}\binom{n-1}{2k-1}$. The lemma follows. □

The above generating function can be used to compute interesting parameters of X_n. We have in particular:

Corollary 3. *The variance of the random variable X_n defined above is equal to $\frac{n-1}{16}$.*

More interesting, the generating function $\phi_n(x)$ can be used to show that asymptotic behaviour of this r.v. is normal. Indeed, we have the following theorem.

Theorem 2. *The normalized variable defined by $Y_n = (4X_n - n)/\sqrt{n}$, has a distribution which tends to the normal distribution $\mathcal{N}(0, 1)$, i.e. for any real interval*

$[a, b]$

$$Pr(a < Y_n \leq b) \to \frac{1}{\sqrt{2\pi}} \int_a^b e^{-x^2/2} dx, \quad as\ n \to \infty.$$

Proof. For any integer k let $j = 2k - 1$, and $l = n - 2k + 1$. Lemma 5 shows that

$$Pr(X_n = k) = \frac{1}{2} \frac{n!}{j!l!} \left(\frac{1}{2}\right)^j \left(\frac{1}{2}\right)^l.$$

By the Stirling formula we have $Pr(X_n = k) \sim \frac{1}{\sqrt{2\pi n}} \left(\frac{n}{2j}\right)^j \left(\frac{n}{2l}\right)^l$ and $\ln \left(\frac{n}{2j}\right)^j \left(\frac{n}{2l}\right)^l$ is equivalent to $-\frac{x^2}{2}$, where $x = \frac{4k-n}{\sqrt{n}}$. The theorem is now proved by the same reasoning as [10], p. 22. □

The case of ring graphs has been the subject of numerous studies in distrbutive algorithms. We prove the following theorem on the asymptotic behaviour of the rendezvous number over a ring graph, by using the fact that this number is asymptotically the same as the number of rendezvous over a chain graph, see [13]. Let the r.v. Z_n denote the rendezvous number in a ring graph of size n.

Theorem 3. *Define the normalized r.v. by* $V_n = \frac{4Z_n - n}{\sqrt{n}}$. *Then, as* $n \to \infty$*:*

$$Pr(a < V_n \leq b) \to \frac{1}{\sqrt{2\pi}} \int_a^b e^{-x^2/2} dx, \quad \text{for any real interval } [a, b].$$

6.2 Rendezvous Number in Complete Graphs

In this section, we are interested in the asymptotic behaviour of the rendezvous number in complete graphs. Here $G = (V, E)$ is a complete graph, X_n the random variable which counts the rendezvous number on G. We first prove:

Lemma 6. *Let* m *be any nonnegative integer. As* $n \to \infty$*, the probability of having at least* m *rendezvous over* G*,* $Pr(X_n \geq m)$ *tends to* $\sum_{k \geq m} (-1)^{k+m} \frac{1}{k!2^k}$.

Proof. From Proposition 1, and using the same argument as in the proof of the first point of Proposition 5, the probability of having at least m rendezvous over G is

$$Pr(X_n \geq m) = \sum_{k \geq m} (-1)^{k+m} \frac{n!}{k!2^k (n-2k)!} \frac{1}{(n-1)^{2k}}.$$

Using the Stirling formula, we get easily the assertion. □

It is now easy to derive a simple characterization of the asymptotic distribution:

Theorem 4. *For every positive integer m, the probability for X_n to be equal to m tends to* $2\frac{(-1)^m}{\sqrt{e}} - \frac{1}{m!2^m} - 2(-1)^m \sum_{k<m}(-1)^k \frac{1}{k!2^k}$ *and* $Pr(X_n = 0)$ *to* $1/\sqrt{e}$, *as* $n \to \infty$.

Proof. Let X denote $\lim_{n\to\infty} X_n$. Then $Pr(X = m) = Pr(X \geq m) - Pr(X \geq m+1)$. Using Lemma 6, the theorem follows. □

References

[1] D. Angluin, *Local and Global Properties in Networks of Processors,* In Proceedings of the 12*th* Symposium on Theory of Computing (1980), 82-93.

[2] H. Austinat and V. Diekert, Private communication.

[3] C. Berge, *Graphes et Hypergraphes,*

[4] F. Bergeron, G. Labelle, and P. Leroux, *Combinatorial Species and Tree-like Structures,* Cambridge University Press (1998).

[5] J. E. Burns and J. Pachl, *Uniform self-stabilizing rings,* ACM Trans. Program. Lang. Syst. 11, (1989), 330-344.

[6] P. Flajolet, and A. Odlyzko, *Random Mapping Statistics,* in Advances in Cryptology, Lecture Notes in Computer Science vol 434, (1990), 329-354.

[7] R. Sedgewick and P. Flajolet, *An Introduction to the Analysis of Algorithms,* Addison Wesley (1996).

[8] *Probabilistic Methods for Algorithmic Discrete Mathematics,* Editors: M. Habib, C. McDiarmid, J. Ramirez-Alfonsin and B. Reed, Springer (1998).

[9] R. Gupta, S. A. Smolka, and S. Bhaskar, *On randomization in sequential and distributed algorithms,* ACM Comput. surv. 26, 1 (Mar. 1994) 7–86.

[10] M. Loève, *Probability Theory,* (1955).

[11] N. Lynch, *A hundred impossibility proofs for distributed computing,* proceedings of the eighth annual ACM symposium on Principles of Distributed Computing, (1989), 1–28.

[12] A. Mazurkiewicz, *Solvability of the asynchronous ranking problem,* Inf. Processing Letters 28, (1988), 221–224.

[13] Y. Métivier, N. Saheb, A. Zemmari, *Randomized Rendezvous,* Research Report 1228-00, January 2000, LaBRI, Université Bordeaux 1, 351, Cours de la Libération, 33405, Talence Cedex.

[14] R. Motwani and P. Raghavan, *Randomized algorithms,* Cambridge University Press (1995).

[15] J. M. Robson, Private communication

[16] G. Tel, *Introduction to distributed algorithms,* Cambridge University Press (1994).

Part IV

Performance Evaluation

Trends in Mathematics, © 2000 Birkhäuser Verlag Basel/Switzerland

Computing Closed-Form Stochastic Bounds on the Stationary Distribution of Markov Chains

MOUAD BEN MAMOUN *PRiSM, Université Versailles St Quentin, 45, Av. des États-Unis,78035 Versailles France.* mobe@prism.uvsq.fr
NIHAL PEKERGIN *PRiSM, Université Versailles St Quentin, 45, Av. des États-Unis,78035 Versailles France, and CERMSEM, Université Paris 1, 106-112, Bld de l'Hopital, 75013 Paris France.* nih@prism.uvsq.fr

Abstract. *In this paper, we define a particular class of transition probability matrices for discrete time Markov chains, and demonstrate that there exists a closed form solution to compute the steady state distribution for this class of Markov chains. We give an algorithm to compute a monotone and bounding matrices in the sense of the $\leq_{st}$ ordering that belongs to this particular class, for a given stochastic matrix P. Therefore, it is possible to compute the bounds in the sense of the $\leq_{st}$ ordering on the stationary distribution of P through the closed-form solution applied to the bounding matrix.*

Key words. *Markov chains, Stochastic Comparison Method, $\leq_{st}$ ordering, stationary distribution.*

1 Introduction

In performance evaluation studies, it is usual to have Markovian models in order to be able to compute numerically transient and steady-state distributions. Therefore the considered performance measures defined as functionals of these distributions can be quantitatively evaluated. However the state space size of Markovian models grows exponentially with the size of model parameters. Because of this state space explosion problem, the numerical solution of the Markovian models are generally limited to the small values of model parameters. The stochastic comparison approach is proposed in order to overcome this problem. Intuitively, this approach consists in analyzing "simpler" models than the underlying one in order to compute bounds on the performance measures. In fact "simpler" models in the context of the Markov chains may be the models on a reduced state space size, or models having special numerical solutions. In [8], [1] an algorithm to construct a monotone bounding Markov chain on a reduced size state space is given. In [6], the state space reduction by taking into account the dynamic of the underlying model to compute stochastic bounds is explained. The numerical solutions of the Nearly Completely Decomposable (NCD) Markov chains by the stochastic comparison approach is considered in [9],[2]. In [4], [5] some of the proposed Markovian bounding models have special structures letting to have matrix-geometric solutions. In this paper we define a class of Markov chains having a closed form solution to compute steady-state distribution. The application of the stochastic comparison approach is to construct a bounding Markov chain having this special form for a given matrix. Therefore the steady-state of the underlying Markov chain is not computed by applying classical numerical methods, but its bounding distribution is computed through the closed-form solution of the bounding chain.

The stochastic comparison method has been largely applied in different areas of the applied probability: economics, reliability, queueing networks. There exist several stochastic ordering, and in this paper we apply the *strong* ($\leq_{st}$) ordering which is also known as the first order stochastic dominance, sample-path stochastic ordering ($\leq_d$) [7].

The paper is organized as follows: in section 2, we give some preliminaries on the stochastic comparison method. In section 3, we define the class of Markov chains having a closed form solution and give the algorithm to compute bounding Markov chains having this special structure. Finally, we conclude by giving some perspectives to extend this work.

2 Preliminaries

A time-homogeneous discrete time Markov chains ($\{X_n, n \geq 0\}$) may be defined by its one-step probability transition matrix, P and its initial state (X_0). The stationary probability distribution (probability vector), π is computed by resolving the following linear equation system:

$$\pi P = \pi, \quad \text{where} \quad ||\pi||_1 = 1 \tag{1}$$

The complexity order to compute the stationary vector is then $\theta(N^3)$ where N is the state space size of the Markov chain.

Here we state the basic definitions and theorems for further informations we refer to [7].

Definition 1 *Let X and Y be two random variables taking values on a totally ordered space E.*

$$X \leq_{st} Y \iff Ef(X) \leq Ef(Y)$$

$\forall$ nondecreasing function $f : E \to R$ whenever the expectations exist.

In fact, in the case of the finite state space this is equivalent to the following definition.

Definition 2 *Let X and Y be two random variables taking values on the finite state space $E = \{1, 2, \cdots, n\}$, and $\mathbf{p} = [p_1 \cdots p_i \cdots p_n]$, $\mathbf{q} = [q_1 \cdots q_i \cdots q_n]$ be probability vectors such that*

$$p_i = Prob(X = i) \quad and \quad q_i = Prob(Y = i) \quad for \quad 1 \leq i \leq n.$$

$X \leq_{st} Y$ ($\mathbf{p} \leq_{st} \mathbf{q}$) if and only if

$$\sum_{i=j}^{n} p_i \leq \sum_{i=j}^{n} q_i \quad for \quad j = n, n-1, \cdots, 1.$$

We use the following definition to compare Markov chains (definition 4.1.2 of [[7], p.59]).

Definition 3 *Let $\{X_n, n \geq 0\}$ and $\{Y_n, n \geq 0\}$ be two discrete time Markov chains. Then $\{X_n, n \geq 0\}$ is smaller than $\{Y_n, n \geq 0\}$ in the sense of the strong stochastic order, symbolically $\{X_n\} \preceq_{st} \{Y_n\}$ if*

$$X_n \preceq_{st} Y_n, \qquad \forall\, n \geq 0$$

It is shown in Theorem 4.2.5 of [[7], p.65]) that monotonicity and comparability of the probability transition matrices of time-homogeneous Markov chains yield sufficient conditions to compare them stochastically.

Theorem 1 *Let P and Q be stochastic matrices which are respectively the probability transition matrices of time-homogeneous Markov chains $\{X_n, n \geq 0\}$ and $\{Y_n, n \geq 0\}$. Then $\{X_n\} \preceq_{st} \{Y_n\}$ if*

- $X_0 \leq_{st} Y_0$
- *monotonicity of at least one of the probability transition matrices holds, that is,*

$$either P[i,*] \leq_{st} P[j,*] \;\; or \;\; Q[i,*] \leq_{st} Q[j,*] \quad \forall i,j \;\; such \;\; that \;\; i \leq j,$$

- *comparability of the transition matrices holds ($P \leq_{st} Q$), that is,*

$$P[i,*] \leq_{st} Q[i,*] \quad \forall i.$$

The following lemma let to compare the stationary distributions of Markov chains which can be considered as the limiting case ($\Pi_P = \lim_{n\to\infty} X(n)$).

Lemma 1 *Let Q be a $\leq_{st}$-monotone, upper bounding matrix for P,*

$$\pi_P \leq_{st} \pi_Q$$

if the steady-states (π_P, π_Q) exist.

3 Class of Markov chains having a closed-form solution to compute the stationary distribution

Let us first define a class of time-homogeneous Markov chains (probability transition matrices) that will be called class C.

Definition 4 *A stochastic matrix $P = (p_{i,j})_{1\leq i,j\leq n}$ belongs to class C , if for each column j there exists a real constant c_j satisfying the following conditions:*

$$p_{i+1,j} = p_{i,j} + c_j, \qquad 1 \leq i \leq n-1,$$

which is equivalent to

$$p_{i,j} = p_{1,j} + (i-1)\, c_j, \qquad 1 \leq i,j \leq n. \tag{2}$$

In fact, the stochastic matrices of class C are defined by their first row and a set of real constants c_j, $1 \leq j \leq n$. Since P is a stochastic matrix, the sum of elements in each row must be equal to 1, thus $\sum_{j=1}^{n} c_j = 0$. The regular form of class C matrices lets to have a closed-form solution to compute the stationary distribution. In the following theorem we give this closed-form solution for class C matrices.

Theorem 2 *Let $P = (p_{i,j})_{1 \leq i,j \leq n}$ be a stochastic matrix in class C with $p_{1,1} \neq 1$. If the stationary vector $\pi = (\pi_1, \pi_2, \ldots, \pi_n)$ exists, then for each j, $1 \leq j \leq n$*

$$\pi_j = p_{1,j} + c_j \frac{\sum_{j=1}^{n} j\ p_{1,j} - 1}{1 - \sum_{j=1}^{n} j\ c_j} \tag{3}$$

Proof. Since the stationary vector exists, and P belongs to class C

$$\pi_j = \sum_{i=1}^{n} \pi_i\ p_{i,j} = p_{1,j} \sum_{i=1}^{n} \pi_i + c_j \sum_{i=1}^{n} (i-1)\ \pi_i = p_{1,j} + c_j \sum_{i=1}^{n} (i-1)\ \pi_i, \quad 1 \leq j \leq n$$

Let $E[\pi]$ denote the expectation of the stationary distribution, thus

$$\pi_j = p_{1,j} + c_j\ (E[\pi] - 1), \qquad 1 \leq j \leq n$$

and

$$E[\pi] = \sum_{j=1}^{n} j\ \pi_j = \sum_{j=1}^{n} j\ p_{1,j} + (E[\pi] - 1) \sum_{j=1}^{n} j\ c_j \tag{4}$$

Let us now show by contradiction that the denominator of the closed-form solution equation 3 is not equal to zero. If $\sum_{j=1}^{n} j c_j = 1$, it follows from equation 4 that $\sum_{j=1}^{n} j p_{1,j} = 1$. We rewrite this equation as $\sum_{j=2}^{n} (j-1) p_{1,j} + \sum_{j=1}^{n} p_{1,j} = 1$. Since the row sum is equal to 1, $\sum_{j=2}^{n} (j-1) p_{1,j} = 0$. Thus for each j, $2 \leq j \leq n$ $(j-1)\ p_{1,j} = 0$, which implies that $p_{1,j} = 0$, $2 \leq j \leq n$ and $p_{1,1} = 1$. That contradicts with the hypothesis that $p_{1,1} \neq 1$.

Therefore it follows from equation 4 that

$$E[\pi] = \frac{\sum_{j=1}^{n} j\ (p_{1,j} - c_j)}{1 - \sum_{j=1}^{n} j\ c_j} \quad \text{and} \quad \pi_j = p_{1,j} + c_j \frac{\sum_{k=1}^{n} k\ p_{1,k} - 1}{1 - \sum_{k=1}^{n} k\ c_k}, \quad 1 \leq j \leq n$$

Let us remark that theorem 2 can be applied when matrix P is irreducible, or there is only one closed irreducible class. Moreover the case $p_{1,1} \neq 1$ is not restrictive, since this case can be avoided in the construction of the upper bound (see section 4).

We now give the following proposition to define the conditions in terms of c_j for a class C matrix to be monotone in the sense of the $\leq_{st}$ ordering.

Proposition 1 *Let P be a stochastic matrix belonging to class C . P is $\leq_{st}$-monotone if and only if* $\sum_{k=j}^{n} c_k \geq 0, \quad \forall j \in \{1, \ldots, n\}$.

Proof. The $\leq_{st}$-monotonicity of matrix P is equivalent to $P[i,*] \preceq_{st} P[i+1,*], \ \forall\, i \in \{1,\ldots,n-1\}$. Since $P \in \mathcal{C}$ it follows from equation 2

$$\begin{aligned} P \leq_{st} \textit{monotone} \quad & \iff \quad \forall\, i, \ \sum_{k=j}^{n} p_{i,k} \leq \sum_{k=j}^{n} p_{i+1,k}, \quad 1 \leq j \leq n \\ & \iff \quad \forall\, i, \ \sum_{k=j}^{n} p_{i,k} \leq \sum_{k=j}^{n} (p_{i,k} + c_k), \quad 1 \leq j \leq n \\ & \iff \quad \sum_{k=j}^{n} c_k \geq 0, \quad 1 \leq j \leq n. \end{aligned}$$

4 Algorithm to compute a monotone upper bounding class $\mathcal{C}$ matrix

In fact a large number of the probability transition matrices that arising from real applications do not belong to class $\mathcal{C}$. However the closed-form solution for the stationary distribution of this class makes it interesting for the resolution of large size Markov chains. Therefore, we give the following algorithm to construct a $\leq_{st}$-monotone, upper bounding matrix Q which belongs to class $\mathcal{C}$ for an irreducible matrix P. The stationary distribution of Q, Π_Q can be computed through the closed-form solution (eq. 2) and provides $\leq_{st}$ upper bounds on the stationary distribution of P ($\Pi_P \leq_{st} \Pi_Q$).

Notice that since the upper bounding matrix Q belongs to class $\mathcal{C}$, we must determine its first row $q_{1,j}, \quad 1 \leq j \leq n$, and the coefficients for the columns $c_j, \ 1 \leq j \leq n$. In the following algorithm, the construction is done from the right to the left to satisfy the $\preceq_{st}$ comparability conditions ($P \leq_{st} Q$), and from up to down to satisfy $\leq_{st}$- monotonicity conditions on Q.

Algorithm

column n: determine $q_{1,n}$, and c_n:

$$\begin{aligned} q_{1,n} &= \max_{1 \leq i \leq n-1} \left[\frac{1}{n-i} \left((n-1)p_{i,n} - (i-1) \right) \right] \\ c_n &= \left[\max_{2 \leq i \leq n} \left(\frac{p_{i,n} - q_{1,n}}{i-1} \right) \right]^+ \end{aligned}$$

column j: For $j := n-1$ *downto* 2 Determine $q_{1,j}$, and c_j :

$$\begin{aligned} q_{1,j} &= \left[\max_{1 \leq i \leq n-1} g(i) \right]^+ \\ c_j &= \max\left(\frac{-q_{1,j}}{n-1}, \ \alpha_j^+ - \sum_{k=j+1}^{n} c_k\right) \end{aligned}$$

where,

$$g(i) = \frac{n-1}{n-i}\left[\sum_{k=j}^{n} p_{i,k} - \sum_{k=j+1}^{n} q_{i,k}\right] + \frac{i-1}{n-i}\left[\sum_{k=j+1}^{n} q_{n,k} - 1\right]$$

$$\alpha_j = \max_{2\leq i\leq n}\left[\frac{\sum_{k=j}^{n} p_{i,k} - \sum_{k=j}^{n} q_{1,k}}{i-1}\right]$$

column 1: Normalize the elements of the first column to have sum rows=1:

$$q_{i,1} = 1 - \sum_{j=2}^{n} q_{i,j}; \quad 1 \leq i \leq n$$

Let us give the following example to illustrate the computing of the upper bound. Let P be a matrix which does not belong to class $\mathcal{C}$, and Q its upper bounding matrix computed through the proposed algorithm.

$$P = \begin{pmatrix} 0.5 & 0.1 & 0.4 \\ 0.7 & 0.1 & 0.2 \\ 0.3 & 0.2 & 0.5 \end{pmatrix} \qquad Q = \begin{pmatrix} 0.5 & 0.1 & 0.4 \\ 0.4 & 0.15 & 0.45 \\ 0.3 & 0.2 & 0.5 \end{pmatrix}$$

Since $c_3 = 0.05,\ c_2 = 0.05\ c_1 = -0.1$, Q belongs to class $\mathcal{C}$. The corresponding stationary distributions are :

$$\pi_P = (0.4456; 0.1413; 0.4130) \qquad \pi_Q = (0.3941; 0.1529; 0.4529)$$

$$\pi_P \leq_{st} \pi_Q$$

We first give the following lemma to state the properties of the entries of matrix Q and c_j, whose proof is given in appendix.

Lemma 2 *Let $q_{1,j},\ c_j \quad 1 \leq j \leq n$ be computed from the previous algorithm for an irreducible stochastic matrix P. Then the following conditions are satisfied:*

1. $\sum_{k=j}^{n} c_k \geq 0, \quad 1 \leq j \leq n,$
2. *for each* $i, \quad \sum_{k=j}^{n} p_{i,k} \leq \sum_{k=j}^{n} q_{i,k}, \quad 1 \leq j \leq n,$
3. *for each* $i, \quad q_{i,j} \geq 0, \quad 1 \leq j \leq n,$
4. *for each* $i, \quad \sum_{k=j}^{n} q_{i,k} \leq 1, \quad 2 \leq j \leq n.$

Let us give the main theorem on the properties of the bounding matrix Q computed from the given algorithm.

Theorem 3 *Let $P = (p_{i,j})_{1\leq i,j\leq n}$ be an irreducible stochastic matrix and $Q = (q_{i,j})_{1\leq i,j\leq n}$ be the bounding matrix of P computed through the previous algorithm. Then the following assertions hold:*

1. *Q is a stochastic matrix,*
2. *Q is a class C matrix,*
3. *$P \preceq_{st} Q$,*
4. *Q is $\leq_{st}$ monotone.*

Proof.

1. Q is a stochastic matrix $\iff \begin{cases} 1 \geq q_{i,j} \geq 0, & 1 \leq i,j \leq n \\ \sum_{j=1}^{n} q_{i,j} = 1, & 1 \leq i \leq n \end{cases}$

   The proof follows from the third and the fourth condition of lemma 2, and the construction of the first column of the algorithm to normalize the sum row of the bounding matrix.

2. It must be proven that equations 2 are satisfied to have a class C matrix. These are satisfied by construction of columns j $(2 \leq j \leq n)$. By construction of the first column, for each $i \in \{1, \cdots, n\}$:

$$q_{i,1} = 1 - \sum_{j=2}^{n} q_{i,j} = 1 - \sum_{j=2}^{n} (q_{1,j} + (i-1)\, c_j) = q_{1,1} - (i-1) \sum_{j=2}^{n} c_j$$

   thus $c_1 = -\sum_{j=2}^{n} c_j$.

3. $P \preceq_{st} Q \iff P[i,*] \leq_{st} Q[i,*] \quad \forall i$. This follows from the second condition of lemma 2.

4. The $\leq_{st}$ *–monotonicity* of class C matrix which is given in proposition 1 follows from the first condition of lemma 2.

Let us remark that a $\leq_{st}$-monotone lower bounding matrix can be constructed in a similar manner. Moreover the given upper bounding algorithm can be also used to compute lower bound by permuting state i to state $n - i + 1$.

4.1 Complexity of the algorithm

The stationary distribution of a class C matrix is computed with a complexity $O(n)$, where n is the state space size. However the construction of the $\leq_{st}$-monotone upper bounding matrix is $O(n^2)$ in the worst case without taking into account any optimization in the case of sparse matrices. Therefore, computing stochastic bounds through class C monotone, upper bounding matrix time complexity has a complexity $O(n^2)$. Thus the complexity is considerably reduced comparing to the classical complexity $O(n^3)$ (eq. 1).

On the other hand, by construction of monotone upper bounding matrices, the computed bounding matrix may not be irreducible even the original matrix is irreducible [1],[2] . However the bounding matrices have one essential class of states, so the computed bounding stationary distribution have a sense. We omit here the detailed proof, but emphasize that by construction of the bounding matrix, it can be shown easily that for each $i < n$, there exists a transition to greater states, i.e. $\sum_{j=i+1}^{n} q_{ij} > 0$. Therefore there is only one essential class for the bounding matrix Q.

5 Conclusion

In this paper, we define a class of Markov chains having a closed-form solution to compute the stationary distribution. The transition probability matrices of this class can described through their first row and a constant c_j for each column j, and the stationary distribution is computed through these parameters with a complexity $O(n)$, where n is the state space size. In order to apply the nice property of this class, we give an algorithm to construct a $\leq_{st}$-monotone, upper bounding class C matrix. Therefore, the stochastic bounds on the stationary distribution can be computed with a complexity of $O(n^2)$. The extension of this work is two-fold: first we consider different stochastic ordering relations, second we are looking for other class of Markov chains having "simpler" solution on the stationary distribution.

References

[1] O. Abu-amsha, J. M. Vincent, An algorithm to bound functionals of Markov chains with large state space, *Technical report IMAG-Grenoble, France, RR Mai-25*, (1996).

[2] T. Dayar and N. Pekergin, Stochastic comparison, reorderings, and nearly completely decomposable Markov chains, in the *Proceedings of the International Conference on the Numerical Solution of Markov Chains (NSMC'99)*, pp. 228-246, (Ed. B. Plateau, W. Stewart), Prensas universitarias de Zaragoza, (1999).

[3] J. Keilson, A. Kester, Monotone Matrices and Monotone Markov Processes, *Stochastic Processes and their Applications*, vol.5, pp. 231-241, (1977).

[4] J. Lui, R. Muntz, D. Towsley, Bounding the mean response time of the minimum expected delay routing policy: an algorithmic approach, *IEEE Trans. on Compt.*, vol. 44, no. 12, (Dec. 1995).

[5] J. Lui, R. Muntz, D. Towsley, Computing performance bounds of Fork-Join parallel programs under a multiprocessing environment, *IEEE Trans. Paral. and Dist. Comp.* , vol. 9, no.3, (march 1998). vol. 44, no. 12, (Dec. 1995).

[6] N. Pekergin, Stochastic performance bounds by state space reduction, *Performance Evaluation*, 36-37, pp.1-17, (August 1999).

[7] D. Stoyan, *Comparison Methods for Queues and Other Stochastic Models*, J. Wiley and Sons, 1976.

[8] M. Tremolieres, J.M. Vincent and B. Plateau, Determination of the optimal upper bound of a Markovian Generator, *Technical Report 106, LGI-IMAG, Grenoble, France*, (1992).

[9] L.Truffet, Near complete decomposability: bounding the error by stochastic comparison method, *Adv. Appl. Prob.* , no.29, pp. 830-835, (1997).

Appendix

Proof of lemma 2: The proof is done in two steps. First, we show by induction on $j \in \{2,\ldots,n\}$ that the conditions of the lemma hold, then include the case $j = 1$.

basic step: $k = n$ The first condition ($c_n \geq 0$) follows from the construction of column n. Let us now show that the second condition is satisfied ($q_{i,n} \geq p_{i,n}, \quad 1 \leq i \leq n$)

$$c_n = \left[\max_{2\leq i\leq n}\left(\frac{p_{i,n} - q_{1,n}}{i-1}\right)\right]^+ \geq \frac{p_{i,n} - q_{1,n}}{i-1}, \qquad 2 \leq i \leq n$$

Thus, $q_{i,n} = q_{1,n} + (i-1)\, c_n \geq p_{i,n}, \quad 2 \leq i \leq n$.

Now let us consider the first row, $f(i) = \frac{1}{n-i}\,((n-1)p_{i,n} - (i-1))$, so

$$q_{1,n} = \max_{1\leq i\leq n-1} f(i) \;\geq f(1) = p_{1,n} \geq 0 \tag{5}$$

which completes the proof for the second condition. The third condition ($q_{i,n} \geq 0, \quad 1 \leq i \leq n$) holds since $q_{1,n} \geq 0$ and $c_n \geq 0$. Since $p_{i,n} \leq 1, \quad 1 \leq i \leq n$ then $f(i) \leq 1, \quad 1 \leq i \leq n$. It follows from equation 5 that $q_{1,n} \leq 1$. On the other hand, we have

$$\begin{array}{lll}
 & q_{1,n} \geq \frac{1}{n-i}\,((n-1)p_{i,n} - (i-1)), & 1 \leq i \leq n-1 \\
\Longleftrightarrow & \frac{n-i}{n-1}\, q_{1,n} \geq p_{i,n} - \frac{i-1}{n-1}, & 1 \leq i \leq n-1 \\
\Longleftrightarrow & (1 - \frac{i-1}{n-1})\, q_{1,n} \geq p_{i,n} - \frac{i-1}{n-1}, & 1 \leq i \leq n-1 \\
\Longleftrightarrow & \frac{i-1}{n-1}(1 - q_{1,n}) \geq p_{i,n} - q_{1,n}, & 1 \leq i \leq n-1 \\
\Longleftrightarrow & \frac{1-q_{1,n}}{n-1} \geq \frac{p_{i,n}-q_{1,n}}{i-1}, & 2 \leq i \leq n-1
\end{array}$$

Moreover,

$$\frac{p_{n,n} - q_{1,n}}{n-1} \leq \frac{1 - q_{1,n}}{n-1}$$

thus

$$\max_{2\leq i\leq n}\left(\frac{p_{i,n} - q_{1,n}}{i-1}\right) \leq \frac{1 - q_{1,n}}{n-1} \tag{6}$$

Since $q_{1,n} \leq 1$ we have

$$0 \leq \frac{1 - q_{1,n}}{n-1} \tag{7}$$

¿From equations 6 and 7

$$c_n = \left[\max_{2\leq i\leq n}\left(\frac{p_{i,n} - q_{1,n}}{i-1}\right)\right]^+ \leq \frac{1 - q_{1,n}}{n-1} \leq \frac{1 - q_{1,n}}{i-1}, \qquad 2 \leq i \leq n$$

which implies that $q_{i,n} \leq 1, \quad 2 \leq i \leq n$ and completes the proof for the fourth condition.

induction step : Suppose that the conditions hold for $k, \quad n \geq k \geq j+1$ and let us show that they hold also for $k = j$.

1. $c_j \geq \alpha_j^+ - \sum_{k=j+1}^n c_k \Longrightarrow \sum_{k=j}^n c_k \geq \alpha_j^+ \geq 0$
2.
$$\begin{array}{lcl} c_j & \geq & \alpha_j^+ - \sum_{k=j+1}^n c_k \geq \alpha_j - \sum_{k=j+1}^n c_k \\ \Longrightarrow \sum_{k=j}^n c_k & \geq & \frac{\sum_{k=j}^n p_{i,k} - \sum_{k=j}^n q_{1,k}}{i-1}, \qquad 2 \leq i \leq n \\ \Longrightarrow \sum_{k=j}^n q_{1,k} + (i-1)\sum_{k=j}^n c_k & \geq & \sum_{k=j}^n p_{i,k}, \qquad 2 \leq i \leq n \\ \Longrightarrow \sum_{k=j}^n q_{i,k} & \geq & \sum_{k=j}^n p_{i,k}, \qquad 2 \leq i \leq n \end{array}$$

   On the other hand,

$$\begin{array}{lcl} q_{1,j} & \geq & \max_{1 \leq i \leq n-1} g(i) \\ & \geq & g(1) = \left[\sum_{k=j}^n p_{1,k} - \sum_{k=j+1}^n q_{1,k}\right] \\ \Longrightarrow \sum_{k=j}^n q_{1,k} & \geq & \sum_{k=j}^n p_{1,k} \end{array}$$

3.
$$\begin{array}{ll} & c_j = \max(-q_{1,j}, \ \alpha_j^+ - \sum_{k=j+1}^n c_k) \geq -q_{1,j} \\ \Longrightarrow & c_j \geq \frac{-q_{1,j}}{i-1}, \qquad 2 \leq i \leq n \\ \Longrightarrow & q_{i,j} \geq 0, \qquad 2 \leq i \leq n \end{array}$$

   Moreover, by construction $q_{1,j} \geq 0$.

4. It suffices to show that the following inequality holds

$$c_j \leq \frac{1 - q_{1,j} - \sum_{k=j+1}^n q_{n,k}}{n-1} \tag{8}$$

   Indeed, this inequality implies $\sum_{k=j}^n q_{n,k} \leq 1$. On the other hand we have shown that $\sum_{k=j}^n c_k \geq 0$, so for all $i \in \{1, \ldots, n-1\}$.

$$\sum_{k=j}^n q_{i,k} = \sum_{k=j}^n q_{n,k} - (n-i)\sum_{k=j}^n c_k \leq \sum_{k=j}^n q_{n,k} \leq 1$$

   Let now show inequality 8. For all $i \in \{2, \ldots, n-1\}$, we have
   $q_{1,j} \geq \frac{n-1}{n-i}\left(\sum_{k=j}^n p_{i,k} - \sum_{k=j+1}^n q_{i,k}\right) + \frac{i-1}{n-i}\left(\sum_{k=j+1}^n q_{n,k} - 1\right)$

   $\Longrightarrow \frac{n-i}{n-1} q_{1,j} - \frac{i-1}{n-1} \sum_{k=j+1}^n q_{n,k} \geq p_{i,j} + \sum_{k=j+1}^n p_{i,k} - \sum_{k=j+1}^n q_{i,k} - \frac{i-1}{n-1}$

   $\Longrightarrow \left(1 - \frac{i-1}{n-1}\right) q_{1,j} - \frac{i-1}{n-1} \sum_{k=j+1}^n q_{n,k}$

$$\geq p_{i,j} + \textstyle\sum_{k=j+1}^{n}(p_{i,k} - q_{1,k}) - (i-1)\sum_{k=j+1}^{n} c_k - \frac{i-1}{n-1}$$

$$\Longrightarrow q_{1,j} + \tfrac{i-1}{n-1}(1 - q_{1,j} - \textstyle\sum_{k=j+1}^{n} q_{n,k})$$

$$\geq p_{i,j} + \textstyle\sum_{k=j+1}^{n}(p_{i,k} - q_{1,k}) - (i-1)\sum_{k=j+1}^{n} c_k - \frac{i-1}{n-1}$$

$$\Longrightarrow \tfrac{(1-q_{1,j}-\sum_{k=j+1}^{n} q_{n,k})}{n-1} \geq \tfrac{\sum_{k=j}^{n}(p_{i,k}-q_{1,k})}{i-1} - \textstyle\sum_{k=j+1}^{n} c_k$$

So,

$$\max_{2\leq i\leq n-1}\left(\frac{\sum_{k=j}^{n}(p_{i,k} - q_{1,k})}{i-1}\right) - \sum_{k=j+1}^{n} c_k \leq \frac{1 - q_{1,j} - \sum_{k=j+1}^{n} q_{n,k}}{n-1}$$

On the other hand we can easily prove that

$$\frac{\sum_{k=j}^{n}(p_{n,k} - q_{1,k})}{n-1} - \sum_{k=j+1}^{n} c_k \leq \frac{1 - q_{1,j} - \sum_{k=j+1}^{n} q_{n,k}}{n-1}$$

Hence,

$$\alpha_j - \sum_{k=j+1}^{n} c_k \leq \frac{1 - q_{1,j} - \sum_{k=j+1}^{n} q_{n,k}}{n-1} \tag{9}$$

We also show that

$$-\sum_{k=j+1}^{n} c_k \leq \frac{1 - q_{1,j} - \sum_{k=j+1}^{n} q_{n,k}}{n-1} \tag{10}$$

to conclude that

$$\alpha_j{}^{+} - \sum_{k=j+1}^{n} c_k \leq \frac{1 - q_{1,j} - \sum_{k=j+1}^{n} q_{n,k}}{n-1} \tag{11}$$

We can easily verified that equation 10 is equivalent to

$$q_{1,j} \leq 1 - \sum_{k=j+1}^{n} q_{1,k} \tag{12}$$

Let show this inequality by contradiction. Suppose that $\exists\, i \in \{1, \ldots, n-1\}$ such that $g(i) > 1 - \sum_{k=j+1}^{n} q_{1,k}$. Then $g(i) > 1 - \sum_{k=j+1}^{n} q_{1,k}$

$$\Longleftrightarrow \tfrac{n-1}{n-i}\left[\textstyle\sum_{k=j}^{n} p_{i,k} - \sum_{k=j+1}^{n} q_{i,k}\right] + \tfrac{i-1}{n-i}\textstyle\sum_{k=j+1}^{n} q_{n,k}$$

$$> \tfrac{n-1}{n-i} - \textstyle\sum_{k=j+1}^{n} q_{1,k}$$

$$\Longleftrightarrow \sum_{k=j}^{n} p_{i,k} - \sum_{k=j+1}^{n} q_{i,k} + \tfrac{i-1}{n-1} \sum_{k=j+1}^{n} q_{n,k} > 1 - \tfrac{n-i}{n-1} \sum_{k=j+1}^{n} q_{1,k}$$

$$\Longleftrightarrow \sum_{k=j}^{n} p_{i,k} - \sum_{k=j+1}^{n} (q_{1,k} + (i-1)\ c_k) + \tfrac{i-1}{n-1} \sum_{k=j+1}^{n} (q_{1,k} + (n-1)\ c_k)$$

$$> 1 - \tfrac{n-i}{n-1} \sum_{k=j+1}^{n} q_{1,k}$$

$$\Longleftrightarrow \sum_{k=j}^{n} p_{i,k} + \tfrac{i-n}{n-1} \sum_{k=j+1}^{n} q_{1,k} > 1 - \tfrac{n-i}{n-1} \sum_{k=j+1}^{n} q_{1,k}$$

$$\Longleftrightarrow \sum_{k=j}^{n} p_{i,k} > 1$$

which contradicts with the fact that P is stochastic.

It follows from definition of c_j and inequality 11 that to complete the proof we must show the following

$$-q_{1,j} \leq \frac{1 - q_{1,j} - \sum_{k=j+1}^{n} q_{n,k}}{n-1} \tag{13}$$

In fact, this inequality is equivalent to

$$(2-n)\ q_{1,j} \leq 1 - \sum_{k=j+1}^{n} q_{n,k} \tag{14}$$

By induction hypothesis $0 \leq 1 - \sum_{k=j+1}^{n} q_{n,k}$ and we have already shown that $q_{1,j} \geq 0$ so $(2-n)\ q_{1,j} \leq 0$.

Now consider the case $j = 1$. The first and second conditions follows from $\sum_{k=1}^{n} c_k = 0$. For the last two conditions, since $0 \leq \sum_{k=2}^{n} q_{i,k} \leq 1, \quad 1 \leq i \leq n$ and $q_{i,1} = 1 - \sum_{k=2}^{n} q_{i,k}$, we have $0 \leq q_{i,1} \leq 1 \ \ \forall\, 1 \leq i \leq n$.

Effects of Reordering and Lumping in the Analysis of Discrete-Time SANs

TUĞRUL DAYAR[1] *Department of Computer Engineering, Bilkent University, 06533 Bilkent, Ankara, Turkey.* tugrul@cs.bilkent.edu.tr

Abstract. *In a recent paper [13], it is shown that discrete-time stochastic automata networks (SANs) are lumpable under rather general conditions. Therein, the authors present an efficient iterative aggregation-disaggregation (IAD) algorithm geared towards computing the stationary vector of discrete-time SANs that satisfy the conditions of lumpability. The performance of the proposed IAD solver essentially depends on two parameters. The first is the order in which the automata are lined up, and the second is the size of the lumped matrix. Based on the characteristics of the SAN model at hand, the user may have some flexibility in the choice of these two parameters. In this paper, we give rules of thumb regarding the choice of these parameters on a model from mobile communications.*

Key words. *Stochastic automata networks, discrete-time Markov chains, lumpability, iterative aggregation-disaggregation*

1 Introduction

Stochastic Automata Networks (SANs) [19, 20, 22, 21, 23, 11, 25, 12, 26, 6, 2, 4, 9, 27, 28, 3, 10, 13] comprise a methodology for modeling Markovian systems that have interacting components. The methodology is based on decomposing the system to be investigated into its components and modeling each component independently. Afterwards, interactions and dependencies among components are introduced and the model finalized. The two advantages of SANs that result from this divide-and-conquer approach are the following. Each component can be modeled much easier compared to the global system due to state space reduction. Storage space allocated for components is minimal compared to the case in which transitions from each global state are stored explicitly. However, all this comes at the expense of longer analysis time [12, 26, 2, 6, 4, 9, 27, 3, 13].

A discrete-time system of N components can be modeled by a single stochastic automaton for each component. With this decompositional approach, the global system ends up having as many states as the product of the number of states of the individual components. See [25, Ch. 9] for detailed information regarding SANs. When there are E synchronizing events in the system, automaton k denoted by $\mathcal{A}^{(k)}$ has the corresponding transition probability matrix $P_e^{(k)}$ that represents the contribution of $\mathcal{A}^{(k)}$ to synchronization $e \in \{0, 1, \ldots, E-1\}$ (see [10, p. 333]). For convenience, we number the automata and synchronizing events starting from 0. The underlying discrete-time MC (DTMC) corresponding to the global system can be obtained from

$$P = \sum_{e=0}^{E-1} \bigotimes_{k=0}^{N-1} P_e^{(k)}. \tag{1}$$

[1]This work is supported by TÜBİTAK-CNRS grant.

We refer to the tensor representation in equation (1) associated with the DTMC as the descriptor of the SAN. Assuming that $\mathcal{A}^{(k)}$ has n_k states, the global system has $n = \prod_{k=0}^{N-1} n_k$ states. When there are transition probabilities in $P_e^{(k)}$ that are functions of the global state of the system rather than only $\mathcal{A}^{(k)}$, tensor products become generalized tensor products [23]. We consider the form of the descriptor in equation (1) rather than the one in [19] since it is compact and easier to work with.

The difficulty associated with discrete-time SANs is that the matrices $P_e^{(k)}$ are relatively dense compared to their continuous-time counterparts implying a larger number of floating-point arithmetic operations in the generalized descriptor-vector multiply algorithm (see [9, p. 404]) used in iterative solvers. Generally, the underlying DTMC of a discrete-time SAN is a dense matrix as opposed to the generator corresponding to a continuous-time SAN. Therefore, discrete-time SANs can be used to tackle even small systems composed of interacting components but that have dense DTMCs since the underlying MC is not generated and stored during SAN analysis.

In the next section, we provide information about the wireless ATM model that is investigated. In the third section, we discuss the results in [13] related to ordering and lumping of automata. In the fourth section, we provide results of numerical experiments, and in the fifth section, we conclude.

2 The model

The application considered in [13] is a multiservices resource allocation policy (MRAP) that integrates three types of service over a time division multiple access (TDMA) system in a mobile communication environment. We have the constant bit rate (CBR) service for two types of voice calls (i.e., handover calls from neighboring cells and new calls), the variable bit rate (VBR) service for two types of calls as in CBR service, and the available bit rate (ABR) service for data transfer. A single cell and a single carrier frequency is considered, and the system is modeled as a discrete-time SAN in which state changes occur at TDMA frame boundaries. Regarding the events that take place in the system, the following assumptions are made. Data packet and call arrivals to the system happen at the beginning of a frame, and data packet transmissions finish and calls terminate at the end of the frame. Since each data packet is small enough to be transmitted in a single slot of a TDMA frame, in a particular state of the system it is not possible to see slots occupied by data packets.

Now, let us move to the parameters of the model. Data is queued in a FIFO buffer of size B and has the least priority. The arrival of data packets is modeled as an on-off process. The process moves from the on state to the off state with probability α and vice versa with probability β. The load offered to the system is defined as $\lambda = \beta/(\alpha + \beta)$. Assuming that the time interval between two consecutive on periods is t, the burstiness of such an on-off process is described by the square coefficient of variation, $S_C = Var(t)/[E(t)]^2$. In terms of λ and S_C, $\beta = 2\lambda(1-\lambda)/(S_C + 1 - \lambda)$ and $\alpha = \beta(1-\lambda)/\lambda$. When the on-off process is in the on state, we assume that $i \in \{0, 1, 2, 3\}$ data packets may arrive with probability p_{di}. The mean arrival rate of data packets is defined as $\rho = \sum_{i=1}^{3} i \times p_{di}$. Hence, the global mean arrival rate of data packets is given by $\Gamma = \lambda\rho$. If the number of arriving data packets exceeds the free space in the buffer plus the number of free slots in the current TDMA frame, the excess packets are dropped.

Handover CBR requests have priority over new CBR calls and they respectively arrive with probabilities p_h and p_n. Similarly, the probabilities of VBR new call and handover arrivals during a TDMA frame have geometrical distributions with parameters p_{vn} and p_{vh}, respectively. We do not consider multiple handover or new CBR/VBR call arrivals during a TDMA frame since the associated probabilities with these events are small. Each CBR/VBR call takes up a single slot of a TDMA frame but may span multiple TDMA frames. When all the slots are full, incoming CBR/VBR calls are rejected. The number of CBR calls that may terminate in a TDMA frame depends on the number of active CBR calls, but can be at most M, and hence is modeled as a truncated binomial process with parameter p_s.

On the other hand, a VBR connection is characterized by a state of high intensity and a state of low intensity. In the former, the VBR source transmits data with its peak rate, whereas in the latter, its transmission rate is lower. The reduced transmission rate in the low intensity state in fact means that at some instances of time the slot in a TDMA frame allocated to the VBR connection is not used. A VBR connection moves from the high intensity state to the low intensity state with probability α_v and vice versa with probability β_v. In terms of $\lambda_v = \beta_v/(\alpha_v + \beta_v)$ and its square coefficient of variation S_{C_v}, we have $\beta_v = 2\lambda_v(1-\lambda_v)/(S_{C_v}+1-\lambda_v)$ and $\alpha_v = \beta_v(1-\lambda_v)/\lambda_v$. When in the low intensity state, the VBR connection moves from the case of a busy slot to the case of an empty slot with probability p_{empty} and vice versa with probability p_{busy}. State changes of a VBR connection after it is set up are assumed to take place at the end of a TDMA frame. We further assume that when a VBR connection is set up as either a new call or a handover, it is in the high intensity state. However, the connection can terminate in any state of the VBR source. We also assume that when a VBR connection changes its state from high intensity to low intensity, it enters the state with a busy slot. The number of VBR calls that may terminate in a TDMA frame depends on the number of active VBR calls and the duration of each VBR call is assumed to be a geometric process with parameter p_{vs}.

Each TDMA frame that gives CBR, VBR, and ABR service consists of C slots reserved for CBR traffic and V slots reserved for VBR traffic. ABR traffic can be pushed into any reserved, but unused slots. Hence, data packets can be transmitted in the idle slots among the C reserved for CBR traffic, in the idle slots among the V reserved for VBR traffic, and in those slots among the V that are in the low intensity state but are empty. The SAN model consists of $(3+V)$ automata and 9 synchronizing events. States of all automata are numbered starting from 0. We denote the state index of automaton k by $s\mathcal{A}^{(k)}$. Automaton $\mathcal{A}^{(0)}$ represents the data arrival process and has two states that correspond to the on and off states of the data source. Transitions in this automaton happen independently of other automata. Automaton $\mathcal{A}^{(1)}$ represents the portion of the TDMA frame reserved for CBR calls and has $(C+1)$ states. Automaton $\mathcal{A}^{(2)}$ represents the data buffer and has $(B+1)$ states, where B is the buffer size. Transitions of this automata depend on $\mathcal{A}^{(0)}$ and $\mathcal{A}^{(1)}$. Automata $\mathcal{A}^{(3)}$ through $\mathcal{A}^{(2+V)}$ represent the V slots reserved for VBR traffic. Each automaton corresponding to VBR traffic has four states. State 0 of the automaton corresponds to the case of an idle slot, i.e., the VBR connection is not active. State 1 corresponds to the state of high intensity state, states 2 and 3 correspond to the state of low intensity. Particularly, state 2 indicates that the slot is busy and state 3 indicates that it is empty. Each automaton $\mathcal{A}^{(k)}$, $k \in \{4, 5, \ldots, V+2\}$, that models VBR traffic depends on the automata $\mathcal{A}^{(3)}$,$\mathcal{A}^{(4)}, \ldots, \mathcal{A}^{(k-1)}$ since we assume that arriving VBR calls are dispatched to slots starting from the smallest indexed VBR automaton. We remark that the automata of the SAN that handle CBR and VBR arrivals are mutually independent. Hence, the set of synchronizing events are given by Cartesian product, and e_{ab} denotes the synchronizing event that is triggered by a CBR arrivals and b VBR arrivals for $a, b \in \{0, 1, 2\}$.

The probability matrices associated with the automata are all relatively dense except the ones that correspond to the data buffer when $s\mathcal{A}^{(0)} = 0$. See [13] for a detailed description of the system. In passing, we remark that the discrete-time SAN model of the system has a global state space size of $n = 2(C+1)(B+1)4^V$. For the problem with $(C, V, B) = (8, 2, 15)$, $\lambda = 0.1$, $S_C = 1$, $(p_{d0}, p_{d1}, p_{d2}, p_{d3}) = (0.05, 0.1, 0.25, 0.6)$ (amounting to an average of $\rho = 2.5$ packet arrivals during a TDMA frame), $(p_n, p_h, p_s) = C(5 \times 10^{-6}, 10^{-5}, 5 \times 10^{-6})$, $\lambda_v = 0.5$, $S_{C_v} = 10$, $(p_{empty}, p_{busy}) = (0.9, 0.1)$, $(p_{vn}, p_{vh}, p_{vs}) = V(5 \times 10^{-6}, 10^{-5}, 5 \times 10^{-6})$, and at most $2 (= M)$ CBR departures during a TDMA frame, we have $n = 4,608$ and $nz = 1,618,620$ (number of nonzeros larger than 10^{-16} is 1,174,657). Here nz denotes the number of nonzeros in the underlying DTMC.

We refer to the rejection of an existing call as dropping and to the rejection of a new call or packet as blocking. The performance measures of interest are the dropping probabilities of handover CBR and handover VBR calls, the blocking probabilities of new CBR calls and new VBR calls, and the blocking probability of data packets. Once the steady state vector of the descriptor is computed, each of the performance measures may be determined [13]. In this work, the aim is not to present values of performance measures for a set of parameters, but rather is to discuss the implications on the IAD solver of reordering the automata and choosing the size of the lumped matrix. Therefore, we constrain ourselves to reporting solution times and iteration counts in section 4. Now, let us summarize the theoretical results in [13].

3 Reordering and Lumping

The automata of some discrete-time SANs can be reordered and then renumbered so that transitions of $\mathcal{A}^{(k)}$ for $k \in \{1, 2, \ldots, N-1\}$ depend (if at all) on the states of the lower indexed automata $\mathcal{A}^{(0)}, \mathcal{A}^{(1)}, \ldots, \mathcal{A}^{(k-1)}$ (the functional dependency being represented by $\mathcal{A}^{(k)}[\mathcal{A}^{(0)}, \mathcal{A}^{(1)}, \ldots, \mathcal{A}^{(k-1)}]$) (see [9] for details). Note that there must exist at least one automaton (in our case, it is $\mathcal{A}^{(0)}$) that is independent of all the other automata for such an ordering to be possible. In our model, such an ordering of automata is possible, and it implies that the automaton $\mathcal{A}^{(2)}$ be placed in the last position and the automata $\mathcal{A}^{(k)}$, $k \in \{3, 4, \ldots, V+2\}$, be placed in any position other than the last as long as they are ordered according to increasing index among themselves. Hence, one possibility is $\mathcal{A}^{(0)}, \mathcal{A}^{(1)}, \mathcal{A}^{(3)}, \mathcal{A}^{(4)}, \ldots, \mathcal{A}^{(V+2)}, \mathcal{A}^{(2)}$.

Now, let us assume that the automata are reordered as described and renumbered from 0 to $(N-1)$. To such an ordering of automata correspond block partitionings of the form

$$P_{n \times n} = \begin{pmatrix} P_{11} & P_{12} & \cdots & P_{1K} \\ P_{21} & P_{22} & \cdots & P_{2K} \\ \vdots & \vdots & \ddots & \vdots \\ P_{K1} & P_{K2} & \cdots & P_{KK} \end{pmatrix} \tag{2}$$

in which all the blocks P_{ij} are square, of order $\prod_{k=m}^{N-1} n_k$, and $K = \prod_{k=0}^{m-1} n_k$ for any $m \in \{1, 2, \ldots, N-1\}$. Hence, for the given ordering of automata, there are $(N-1)$ different such partitionings which vary between one that has n_0 diagonal blocks of order $\prod_{k=1}^{N-1} n_k$ and one that has $\prod_{k=0}^{N-2} n_k$ diagonal blocks of order n_{N-1}. Each of the $(N-1)$ block partitionings is lumpable as stated in the next theorem of [13].

Theorem 1 *A discrete-time SAN of N automata and E synchronizing events whose automata are reordered and renumbered so that $\mathcal{A}^{(k)}[\mathcal{A}^{(0)}, \mathcal{A}^{(1)}, \ldots, \mathcal{A}^{(k-1)}]$, $k \in \{1, \ldots, N-1\}$, and that has the descriptor in equation (1) with equal row sums in each $P_e^{(k)}$ for $k = 0, 1, \ldots, N-1$ and $e = 0, 1, \ldots, E-1$ is lumpable with respect to the partitioning in equation (2) for any $m \in \{1, 2, \ldots, N-1\}$.*

Now, we state the more relaxed version of Theorem 1 in [13] for the case of cyclic functional dependencies.

Definition 1 *Let $G(\mathcal{V}, \mathcal{E})$ be the directed graph (digraph) corresponding to a discrete-time SAN in which the vertex $v_k \in \mathcal{V}$ represents $\mathcal{A}^{(k)}$ and the edge $(v_k, v_l) \in E$ if transitions in $\mathcal{A}^{(k)}$ depend on the state of $\mathcal{A}^{(l)}$ (i.e., $\mathcal{A}^{(k)}[\mathcal{A}^{(l)}]$). Then the SAN is said to contain cyclic functional dependencies if and only if the digraph has at least one strongly connected component (SCC) composed of multiple automata.*

Detailed description of the SCC algorithm for digraphs can be found in [1, pp. 191–197].

Theorem 2 *A discrete-time SAN of N automata $\mathcal{A}^{(k)}$, $k = 0, 1, \ldots, N-1$, and E synchronizing events that contains cyclic functional dependencies among its automata is lumpable if the digraph corresponding to the SAN has more than one SCC and each $P_e^{(k)}$ has equal row sums for $k = 0, 1, \ldots, N-1$ and $e = 0, 1, \ldots, E-1$.*

Assuming that P is lumpable with respect to the partition in (2) and is irreducible, in [13] the following modified form of Koury-McAllister-Stewart's IAD algorithm [15] is proposed for computing the stationary probability vector π (i.e., $\pi P = \pi$, $||\pi|| = 1$). The convergence analysis of the KMS algorithm is based on the concept of near complete decomposability (NCD) [18]. In [16] and [24], it is shown that fast convergence is achieved if the degree of coupling, $||F||_\infty$, is small compared to 1, where $P = F + \text{diag}(P_{11}, P_{22}, \dots, P_{KK})$.

Algorithm 1. *IAD algorithm for discrete-time SANs*

1. Let $\pi^{(0)} = (\pi_1^{(0)}, \pi_2^{(0)}, \dots, \pi_K^{(0)})$ be a given initial approximation of π. Set $it = 1$.

2. Aggregation:

   (a) Compute the lumped matrix L of order K with ijth element $l_{ij} = \max(P_{ij}u)$.

   (b) Solve the singular system $\tau(I - L) = 0$ subject to $||\tau||_1 = 1$ for $\tau = (\tau_1, \tau_2, \dots, \tau_K)$.

3. Disaggregation:

   (a) Compute the row vector

$$z^{(it)} = (\tau_1 \frac{\pi_1^{(it-1)}}{||\pi_1^{(it-1)}||_1}, \tau_2 \frac{\pi_2^{(it-1)}}{||\pi_2^{(it-1)}||_1}, \dots, \tau_K \frac{\pi_K^{(it-1)}}{||\pi_K^{(it-1)}||_1}).$$

   (b) Solve the K nonsingular systems of which the ith is given by

$$\pi_i^{(it)}(I - P_{ii}) = b_i^{(it)}$$

   for $\pi_i^{(it)}$, $i = 1, 2, \dots, K$, where

$$b_i^{(it)} = \sum_{j>i} z_j^{(it)} P_{ji} + \sum_{j<i} \pi_j^{(it)} P_{ji}.$$

4. Test $\pi^{(it)}$ for convergence. If the desired accuracy is attained, then stop and take $\pi^{(it)}$ as the stationary probability vector of P. Else set $it = it + 1$ and go to step 3.

In [17], the convergence of a framework of IAD methods is studied. The IAD algorithm considered is different in that there is no requirement of NCDness on the partitioning. Furthermore, a number of relaxations (i.e., smoothings) of the power method kind is performed at the fine level. The authors prove that the errors at the fine and coarse levels are intimately related, and for a strictly positive initial approximation, the IAD approximation converges rapidly to the stationary vector as long as one is very precise in computing at the coarse level and a sufficiently high number of smoothings is performed at the fine level. Numerical results on randomly generated stochastic matrices with varying degrees of coupling and having equal orders of blocks which are all tridiagonal show that convergence is practically independent of the degree of coupling.

The motivation behind proposing Algorithm 1 rather than block Gauss-Seidel (BGS) for discrete-time SANs is that the partitioning in (2) is a balanced one with equal orders of blocks and the aggregate matrix needs to be formed only once due to lumpability. See [8] for recent results on the computation of the stationary vector of Markov chains.

In Algorithm 1, the lumped matrix L of order $K = \prod_{k=0}^{m-1} n_k$ is computed at the outset and solved once for its stationary vector τ. Note that the lumped matrix L is also lumpable if $m > 1$. These hint at the solver to be chosen at step 2 to compute τ. Assuming that L is dense, one may opt for a direct solver such as Gaussian elimination (GE) (or the method of Grassmann-Taksar-Heyman, GTH, if L is relatively ill-conditioned) when K is on the order of hundreds. Else one may use IAD with a lumped matrix of order $\prod_{k=0}^{m'-1} n_k$, where $1 < m' < m$ (or IAD with an NCD partitioning if L is relatively ill-conditioned). In any case, sufficient space must be allocated to store L. As for the disaggregation phase (i.e., a BGS iteration), the right-hand sides $b_i^{(it)}$ at iteration it may be computed efficiently as shown in [13] at the cost of $E(K-2)$ vectors of length $\prod_{k=m}^{N-1} n_k$, that is roughly E vectors of length n. In summary, the proposed solver is limited by $\max(K^2, (E+2)n)$ amount of double precision storage assuming that the lumped matrix is stored in two dimensions. The 2 vectors of length n are used to store the previous and current approximations of the solution.

So far, we have assumed that the underlying DTMC of the given discrete-time SAN is irreducible. Since this may not be the case, a state classification (SC) algorithm that partitions the global state space of a SAN into essential and transient subsets is implemented [13]. The inhibition of transient states is important in removing redundant computation from the IAD solver. However, the case of more than one partition of essential states is not considered since it hints at a modeling problem.

Now, let us return to the model in section 2 and consider lumpable orderings of the automata. Since $\mathcal{A}^{(2)}$ must be the last automaton in a lumpable ordering, there are $(V+1)!$ ways in which $\mathcal{A}^{(0)}$ may precede $\mathcal{A}^{(1)}$ among the first $(V+2)$ automata. The same argument is true of $\mathcal{A}^{(0)}$ succeeding $\mathcal{A}^{(1)}$ among the first $(V+2)$ automata. Therefore, the requirement in Theorem 1 regarding functional dependencies implies that there are $2(V+1)!$ orderings that may be used. Hence, we have 12 lumpable orderings to choose from when $V = 2$ and 48 lumpable orderings to choose from when $V = 3$. Since these are large numbers of orderings to investigate, we treat the VBR automata as a single entity and concentrate on only the 3! orderings of automata given by $[(3,4,\ldots,V+2),1,0,2]$, $[1,(3,4,\ldots,V+2),0,2]$, $[(3,4,\ldots,V+2),0,1,2]$, $[0,(3,4,\ldots,V+2),1,2]$, $[1,0,(3,4,\ldots,V+2),2]$, and $[0,1,(3,4,\ldots,V+2),2]$. We name these orderings respectively $o_1, o_2, \ldots, o_6$. Regarding the lumping parameter, m, we choose values from the set $\{V-1, V, V+1\}$ which implies that the largest lumped matrix formed in step 2 is of order $2(C+1)4^V$ and the smallest system to solve in step 3 is of order $(B+1)$.

We remark that a reordering of discrete-time automata corresponds to a symmetric permutation of the underlying DTMC. In other words, reordering of automata is equivalent to a renumbering of the global states in a SAN model. Since the objective of this study is to investigate the effects of reordering and the lumping parameter, m, on the convergence of Algorithm 1, we choose the integer parameters (C, V, B) so that the lumped matrix formed in step 2 can be solved accurately and rapidly using the GTH method as discussed in [7] and there is sufficient space to factorize in sparse format the K diagonal blocks in step 3(b) at the outset. Hence, we use sparse forward and back substitutions to solve the K nonsingular systems at each iteration of Algorithm 1 and do not need to employ any smoothings.

4 Numerical results

Algorithm 1 is implemented in C++ as part of the software package PEPS [22]. We time the solver on a Pentium III with 64 MBytes of RAM under Linux. In each experiment, we use a tolerance of 10^{-8} on the approximate error $\|\pi^{(it)} - \pi^{(it-1)}\|_2$ in step 4 of Algorithm 1. The approximate residual $\|\pi^{(it)} - \pi^{(it)}P\|_2$ turns out to be less than the approximate error upon termination in all our experiments.

In the first set of experiments, we consider three variants of the problem $(C, V, B) = (8, 2, 15)$, which has a state space size of $n = 4,608$ (see section 2). We set $(p_{d0}, p_{d1}, p_{d2}, p_{d3}) = (0.05, 0.1, 0.25, 0.6)$, $\lambda_v = 0.5$, $S_{C_v} = 10$, $(p_{empty}, p_{busy}) = (0.9, 0.1)$ in the first variant and $(p_{d0}, p_{d1}, p_{d2}, p_{d3}) = (0.4, 0.3, 0.2, 0.1)$, $\lambda_v = 0.5$, $S_{C_v} = 1$, $(p_{empty}, p_{busy}) \in \{(0.9, 0.1), (0.5, 0.5)\}$ in the last two variants. The other parameters are chosen as $M = V$, $S_C = 1$, $\lambda \in \{0.1, 0.3, 0.5, 0.7, 0.9\}$, $(p_n, p_h, p_s) = C(5 \times 10^{-6}, 10^{-5}, 5 \times 10^{-6})$, and $(p_{vn}, p_{vh}, p_{vs}) = V(5 \times 10^{-6}, 10^{-5}, 5 \times 10^{-6})$. The lumpability parameter assumes the values in $\{2, 3, 4\}$. The degree of coupling, $\|F\|_\infty$, associated with the partitioning in equation (2) for each DTMC is on the order of 10^{-1} and mostly close to 1.0. Hence, the lumpable partitionings we consider in Algorithm 1 for the first set of experiments are not NCD. However, the smallest degree of coupling we find for each DTMC using the algorithm in [5] is on the order of 10^{-5}. All this means that even though the lumpable partitionings we consider are not NCD partitionings, there exist highly NCD partitionings for each DTMC, and therefore they are all very ill-conditioned.

The underlying DTMCs of the SAN models in the first set of experiments are reducible with a single subset of essential states and 224 transient states. It is argued in [13] that when a reducible discrete-time SAN has a single subset of essential states and each subset of the partition in equation (2) includes at least one essential state (which is the case in our experiments), the lumped matrix computed in step 2 of Algorithm 1 is irreducible. With such a partitioning, if one starts in step 1 with an initial approximation having zero elements corresponding to transient states, successive computed approximations will have zero elements corresponding to transient states as well. In our experiments, we start with a positive initial approximation and observe that all elements that correspond to transient states become zero at the second iteration. Once the elements that correspond to transient states in an approximate solution become zero, they remain zero. Hence, there is no need to run the time consuming SC algorithm for each of the experiments, since the matrices in each of the three variants have the same nonzero structure.

In the first variant, Algorithm 1 converges for the orderings o_1 and o_2 within 34 iterations when $m \in \{3, 4\}$, whereas for the other four orderings it converges within 34 iterations only when $m = 4$. In all other cases, Algorithm 1 does not converge within 250 iterations to the prespecified tolerance. A similar observation follows for the second and third variants if we replace the 34 respectively with 65 and 120 (achieved in o_4-o_6). Hence, the three variants of this problem seem to be of increasing difficulty as we move from the first to the third, although each one has highly NCD partitionings with degree of coupling on the same order of 10^{-5}. In Table 1, we present the results of numerical experiments with Algorithm 1 for the three variants using o_1. Note that a smaller number of iterations may not imply a smaller solution time as in the first variant for $\lambda = 0.9$. The best solution times are obtained with o_1 (and o_2) when $m = 3$ is used. For each of the six orderings, $m = 4$ gives a lumped matrix of order $K = 288$. For o_1 and o_2, $m = 3$ gives a lumped matrix of order $K = 144$. For the other four orderings, $m = 3$ gives lumped matrices of orders 32 and 72, which are both smaller than 144. For the six orderings, $m = 2$ gives lumped matrices of orders varying between 8 and 36. It takes nearly 0 seconds to solve the lumped matrix using GTH in all cases. Hence, the timing results in Table 1 are for the iterative part of Algorithm 1.

TABLE 1. Solution times in seconds and # of iterations with o_1 for 1st set of experiments.

Variant	m	$\lambda = 0.1$		$\lambda = 0.3$		$\lambda = 0.5$		$\lambda = 0.7$		$\lambda = 0.9$	
		Time	$\#it$	Time	$\#it$	Time	$\#it$	Time	$\#it$	Time	$\#it$
1	4	25	28	30	34	27	30	23	26	22	24
	3	8	18	9	21	8	17	9	20	12	27
	2	84	250^+	84	250^+	84	250^+	84	250^+	84	250^+
2	4	48	51	52	56	41	44	35	38	61	65
	3	22	48	22	47	17	38	17	37	27	61
	2	84	250^+	84	250^+	84	250^+	84	250^+	84	250^+
3	4	111	118	53	57	51	55	44	47	84	90
	3	51	114	23	52	22	49	20	45	40	89
	2	84	250^+	84	250^+	84	250^+	84	250^+	84	250^+

In the second set of experiments, we consider the problem $(C, V, B) = (3, 3, 15)$, which has a state space size of $n = 8,192$. We set $(p_{d0}, p_{d1}, p_{d2}, p_{d3}) = (0.05, 0.1, 0.25, 0.6)$, $\lambda_v = 0.5$, $S_{C_v} = 10$, $(p_{empty}, p_{busy}) = (0.9, 0.1)$, the other parameters being chosen as in the first set of experiments. The lumpability parameter assumes the values in $\{3, 4, 5\}$. Again, even though the lumpable partitionings we consider are not NCD partitionings, there exist highly NCD partitionings for each DTMC, and therefore they are all very ill-conditioned. The underlying DTMCs of the SAN models in the second set of experiments are irreducible.

Algorithm 1 converges for the orderings o_1 and o_2 within 42 iterations when $m \in \{4, 5\}$, whereas for the other four orderings it converges within 43 iterations (achieved in o_4-o_6) only when $m = 5$. In all other cases, Algorithm 1 does not converge within 250 iterations to the prespecified tolerance. The best solution times are obtained with o_1 and o_2 when $m = 4$ is used.

Regarding the ordering of automata, o_1 and o_2 are more advantageous than the other four orderings since they converge for more values of m. The orderings o_1 and o_2 have $\mathcal{A}^{(0)}$ and $\mathcal{A}^{(2)}$ as the last two automata. These automata have transition probabilities of the same order. Furthermore, when $m = N - 2$ is used with o_1 and o_2, the lumped matrix is larger than one would have with the other four orderings. We believe these two factors influence the behavior of Algorithm 1 for o_1 and o_2 when $m = N-2$. Further experiments must be conducted to improve our confidence in this conjecture.

5 Conclusion

Experiments on a problem from mobile communications indicate that the performance of Algorithm 1 for discrete-time SANs is sensitive to the ordering of automata and the choice of the lumpability parameter. Even though the coarse and fine level solutions are computed with high accuracy in our experiments, there are values of the lumpability parameter for which fast convergence of Algorithm 1 is not witnessed. On the other hand, some partitionings converge in a smaller number of iterations than others. The value of the lumpability parameter determines the order of the coupling matrix. Numerical results imply that convergence may not be observed in a reasonable number of iterations when the order of the coupling matrix is smaller than the squareroot of the state space size. Hence, it is recommended for one to consider the largest permissible value of the lumpability parameter when using Algorithm 1.

Acknowledgment. The author thanks Oleg Gusak for setting up the experimental framework.

References

[1] S. Baase, Computer Algorithms: Introduction to Design and Analysis, Addison-Wesley, Reading, Massachusetts, 1988.

[2] P. Buchholz, An aggregation\disaggregation algorithm for stochastic automata networks, Probability in the Engineering and Informational Sciences 11 (1997) 229–253.

[3] P. Buchholz, Projection methods for the analysis of stochastic automata networks, in: Proceedings of the 3rd International Workshop on the Numerical Solution of Markov Chains, B. Plateau, W. J. Stewart, M. Silva, (Eds.), Prensas Universitarias de Zaragoza, Spain, 1999, pp. 149–168.

[4] R. Chan, W. Ching, Circulant preconditioners for stochastic automata networks, Technical Report 98-05 (143), Department of Mathematics, The Chinese University of Hong Kong, April 1998.

[5] T. Dayar, Permuting Markov chains to nearly completely decomposable form, Technical Report BU-CEIS-9808, Department of Computer Engineering and Information Science, Bilkent University, Ankara, Turkey, August 1998, available via ftp from ftp://ftp.cs.bilkent.edu.tr/pub/tech-reports/1998/BU-CEIS-9808.ps.z.

[6] T. Dayar, O.I. Pentakalos, A.B. Stephens, Analytical modeling of robotic tape libraries using stochastic automata, Technical Report TR-97-189, CESDIS, NASA/GSFC, Greenbelt, Maryland, January 1997.

[7] T. Dayar, W.J. Stewart, On the effects of using the Grassmann-Taksar-Heyman method in iterative aggregation-disaggregation, SIAM Journal on Scientific Computing 17 (1996) 287–303.

[8] T. Dayar, W. J. Stewart, Comparison of partitioning techniques for two-level iterative solvers on large, sparse Markov chains, SIAM Journal on Scientific Computing 21 (2000) 1691–1705.

[9] P. Fernandes, B. Plateau, W.J. Stewart, Efficient descriptor-vector multiplications in stochastic automata networks, Journal of the ACM 45 (1998) 381–414.

[10] J.-M. Fourneau, Stochastic automata networks: using structural properties to reduce the state space, in: Proceedings of the 3rd International Workshop on the Numerical Solution of Markov Chains, B. Plateau, W. J. Stewart, M. Silva, (Eds.), Prensas Universitarias de Zaragoza, Spain, 1999, pp. 332–334.

[11] J.-M. Fourneau, H. Maisonniaux, N. Pekergin, V. Véque, Performance evaluation of a buffer policy with stochastic automata networks, in: IFIP Workshop on Modelling and Performance Evaluation of ATM Technology, vol. C-15, La Martinique, IFIP Transactions North-Holland, Amsterdam, 1993, pp. 433–451.

[12] J.-M. Fourneau, F. Quessette, Graphs and stochastic automata networks, in: Computations with Markov Chains: Proceedings of the 2nd International Workshop on the Numerical Solution of Markov Chains, W. J. Stewart, (Ed.), Kluwer, Boston, Massachusetts, 1995, pp. 217–235.

[13] O. Gusak, T. Dayar, J.-M. Fourneau, Discrete-time stochastic automata networks and their efficient analysis, submitted to Performance Evaluation, June 2000.

[14] J.R. Kemeny, J.L. Snell, Finite Markov Chains, Van Nostrand, New York, 1960.

[15] J.R. Koury, D.F. McAllister, W. J. Stewart, Iterative methods for computing stationary distributions of nearly completely decomposable Markov chains, SIAM Journal on Algebraic and Discrete Methods 5 (1984), 164–186.

[16] D.F. McAllister, G.W. Stewart, W.J. Stewart, On a Rayleigh-Ritz refinement technique for nearly uncoupled stochastic matrices, Linear Algebra and Its Applications 60 (1984), 1–25.

[17] I. Marek, P. Mayer, Convergence analysis of an iterative aggregation-disaggregation method for computing stationary probability vectors of stochastic matrices, Numerical Linear Algebra with Applications 5 (1998), 253–274.

[18] C.D. Meyer, Stochastic complementation, uncoupling Markov chains, and the theory of nearly reducible systems, SIAM Review 31 (1989) 240–272.

[19] B. Plateau, De l'évaluation du parallélisme et de la synchronisation, Thèse d'état, Université Paris Sud Orsay, 1984.

[20] B. Plateau, On the stochastic structure of parallelism and synchronization models for distributed algorithms, in: Proceedings of the SIGMETRICS Conference on Measurement and Modelling of Computer Systems, Texas, 1985, pp. 147–154.

[21] B. Plateau, K. Atif, Stochastic automata network for modeling parallel systems, IEEE Transactions on Software Engineering 17 (1991) 1093–1108.

[22] B. Plateau, J.-M. Fourneau, K.-H. Lee, PEPS: A package for solving complex Markov models of parallel systems, in: Modeling Techniques and Tools for Computer Performance Evaluation, R. Puigjaner, D. Ptier, (Eds.), Spain, 1988, pp. 291–305.

[23] B. Plateau, J.-M. Fourneau, A methodology for solving Markov models of parallel systems, Journal of Parallel and Distributed Computing 12 (1991) 370–387.

[24] G.W. Stewart, W.J. Stewart, D.F. McAllister, A two-stage iteration for solving nearly completely decomposable Markov chains, in: Recent Advances in Iterative Methods, IMA Vol. Math. Appl. 60, G. H. Golub, A. Greenbaum, M. Luskin, (Eds.), Springer-Verlag, New York, 1994, pp. 201–216.

[25] W.J. Stewart, Introduction to the Numerical Solution of Markov Chains, Princeton University Press, Princeton, New Jersey, 1994.

[26] W.J. Stewart, K. Atif, B. Plateau, The numerical solution of stochastic automata networks, European Journal of Operational Research 86 (1995) 503–525.

[27] E. Uysal, T. Dayar, Iterative methods based on splittings for stochastic automata networks, European Journal of Operational Research 110 (1998) 166–186.

[28] V. Vèque, J. Ben-Othman, MRAP: A multiservices resource allocation policy for wireless ATM network, Computer Networks and ISDN Systems 29 (1998) 2187–2200.

Trends in Mathematics, © 2000 Birkhäuser Verlag Basel/Switzerland

Large deviations for polling systems

FRANCK DELCOIGNE *Université Paris 10, UFR SEGMI, 200 av. de la République, 92000 Nanterre, FRANCE.*
ARNAUD DE LA FORTELLE *INRIA — Domaine de Voluceau, Rocquencourt, BP 105, 78153 Le Chesnay Cedex, FRANCE.*

Abstract. *We aim at presenting in short the technical report [5], which states a sample path large deviation principle for a rescaled process $n^{-1}Q_{nt}$, where Q_t represents the joint number of clients at time t in a single server 1-limited polling system with Markovian routing. The main goal is to identify the rate function. A so-called empirical generator is introduced, which consists of Q_t and of two empirical measures associated with S_t, the position of the server at time t. The analysis relies on a suitable change of measure and on a representation of fluid limits for polling systems. Finally, the rate function is solution of a meaningful convex program.*

Key words. *Large deviations, polling system, fluid limits, empirical generator, change of measure, contraction principle, entropy, convex program.*

1 Presentation of the model

Consider a polling system consisting of N nodes attended by a single server and denote by $\mathcal{S} \stackrel{\text{def}}{=} \{1, \dots, N\}$ the set of nodes. At node i, arrivals of clients form a Poisson process with rate λ_i. Each customer at node i requires service, whose duration is exponentially distributed with parameter μ_i. Let $\rho_i \stackrel{\text{def}}{=} \lambda_i/\mu_i$, the intensity factor at node i. When the server arrives at a busy node, say i, it serves one customer and then moves to some node, chosen via some ergodic routing matrix $\mathbf{P} = (P_{ij})_{i,j\in\mathcal{S}}$ with invariant measure $\eta = (\eta_i)_{i\in\mathcal{S}}$. If it reaches an empty node, then it immediately switches to some other node, still chosen according to $\mathbf{P}$. The switch-over time to go from node i to node j, for $i, j \in \mathcal{S}$, is exponentially distributed with mean τ_{ij}. All stochastic input sequences (inter arrival times, services, switch-over times) are supposed to be mutually independent. When the joint number of clients and the position of the server at time 0 are respectively given by $x = (x_1, \dots, x_N)$ and s, $Q(t, x, s) = \big(q_1(t, x, s), \dots, q_N(t, x, s)\big)$ and $S(t, x, s)$ represent the joint number of clients at each node and the position of the server at time t. As a rule, we shall write $S(t, x, s) = i$ if the server is serving some customer at node i and $S(t, x, s) = ij$ if it is in transit between nodes i and j. Set $\mathcal{S}_0 \stackrel{\text{def}}{=} \mathcal{S} \cup \mathcal{S}^2$, the state space of the server. Then

$$X_{x,s} \stackrel{\text{def}}{=} \Big\{\big(Q(t, x, s), S(t, x, s)\big),\ t \geq 0\Big\}$$

is a Markov process with generator R such that

$$Rf(x, s) = \sum_{(y,s')\in\mathbb{Z}_+^N\times\mathcal{S}_0} q(x, s; y, s')\Big(f(y, s') - f(x, s)\Big), \quad \forall (x, s) \in \mathbb{Z}_+^N \times \mathcal{S}_0,$$

where $f \in \mathcal{B}(\mathbb{Z}_+^N \times \mathcal{S}_0)$ and

$$q(x,s;y,s') = \begin{cases} \lambda_i, & \text{if } y = x + e_i, s' = s, \ \forall i \in \mathcal{S}, \\ \mu_i P_{ij}, & \text{if } x_i > 0, s = i, y = x - e_i, s' = ij, \ \forall i,j \in \mathcal{S}, \\ \dfrac{1}{\tau_{ji}}, & \text{if } x_i > 0, s = ji, y = x, s' = i, \ \forall i,j \in \mathcal{S}, \\ \dfrac{1}{\tau_{ji}} P_{il}, & \text{if } x_i = 0, s = ji, y = x, s' = il, \ \forall i,j,l \in \mathcal{S}, \\ 0, & \text{otherwise.} \end{cases}$$

Whenever no confusion arises, the initial state (x,s) will be dropped.

Let us introduce now a definition and a notation which will be of constant use in the sequel:

Definition 1.1 *For every $x = (x_1, \dots, x_N) \in \mathbb{R}_+^N$, denote by $\Lambda(x)$ the set of indices i such that $x_i > 0$. If Λ is a subset of $\mathcal{S}$, the subset of $\mathbb{R}_+^N$*

$$\{x \in \mathbb{R}_+^N / x_i > 0, \forall i \in \Lambda, x_i = 0, \forall i \in \Lambda^c\}$$

is called face Λ.

- For any set A, A^c will denote its complementary.
- For any space E, $\mathcal{B}(E)$, $\mathcal{M}(E)$, $\mathcal{P}(E)$, represent respectively the sets of bounded functions on E, of positive measures on E and of probability measures on E.
- $D\left([0,T], \mathbb{R}^N\right)$ is the space of right continuous functions $f : [0,T] \to \mathbb{R}^N$ with left limits, endowed with the Skorokhod metric denoted by d_d.

2 Previous work

A huge literature has been devoted to the study of polling systems because of their wide range of application. In [2, 8, 13], the necessary and sufficient conditions of ergodicity has been established for systems with one or several servers under a rich variety of service policies. However, the problem of determining the invariant measure for such systems is still open. Even, for limited policies, the mean waiting time can be computed only under symmetry assumptions [3]. The reader is referred to [17] for an overview about polling systems. In the present paper, a sample path large deviation principle or a sample path LDP for the rescaled process $n^{-1}Q_{nt}$ is established. This could be a preliminary step in order to obtain large deviations estimates for the stationary distribution. In view of future applications, some particular attention is devoted to the computation of the rate function governing the sample path LDP. All this program falls into the framework of LDP for Markov processes with discontinuous statistics i.e those for which the coefficients of their generator are not spatially continuous.

It seems that one of the first paper dealing with such processes is [12], where large deviations problems for Jackson networks were investigated using partial differential equations techniques. Quite recently, the LDP for a large class of Markov processes with discontinuous statistics has been proved in [10]. Roughly speaking, the authors of [10] express the logarithm of large deviation probabilities as the minimal cost of some stochastic optimal-control problem, and the limit of the optimal cost is shown to exist by means of a sub-additivity argument. However, the rate function is not explicit. Note that in [11], an explicit upper bound of large deviations involving Legendre transforms is proved. While for Jackson networks and some processor sharing models this bound is tight [1], in general the problem of the lower bound remains open. It is worth emphasizing that in our case, Q_t is not a Markov process so that the polling model does not satisfy the assumptions of [10]. Besides, the rate function can be explicitly described in terms of entropy functions.

Until now, the identification of the rate function has been carried out in some particular cases and usually for low dimensional systems. In [6], using the contraction principle, the exponential decay of the stationary distribution of the waiting time is computed for a two dimensional tandem networks taking advantage that it can be expressed simply as a continuous function of the input processes. It should be noted that in this setting, a sample path LDP for processes with independent increments over infinite intervals of time is needed [7]. General results were obtained in [9, 15] where the LDP has been established for random walks whose generator has a discontinuity along an hyperplane. These results are applied in [15] to compute the exponential decay of the stationary distribution of ergodic random walks in $\mathbb{Z}_+^2$. Nevertheless, in such examples, there are at most two boundaries with codimension one or two where discontinuity arise. Ultimately, the identification of the rate function governing the LDP for Jackson networks has been carried out in [1, 14].

3 Local bounds, empirical generator and entropy

Following [10], in order to get a sample path LDP for polling systems, the main step is to prove the forthcoming large deviations local bounds:

Theorem 3.1 [Local bounds] *Take $x \in \mathbb{R}_+^N$ and $D \in \mathbb{R}^N$ such that $D_i = 0$,*

$\forall i \in \Lambda^c(x)$. *Then, for any* τ *satisfying* $x_i + D_i\tau > 0$, $\forall i \in \Lambda(x)$,

$$\lim_{\delta\to 0}\lim_{\epsilon\to 0}\liminf_{n\to\infty}\frac{1}{n}\inf_{|y-nx|<\epsilon n}\log \mathbb{P}\left[\sup_{t\in[0,n\tau]}|Q(t,y)-nx-Dt|<\delta n\right]$$

$$= \lim_{\delta\to 0}\lim_{\epsilon\to 0}\limsup_{n\to\infty}\frac{1}{n}\sup_{|y-nx|<\epsilon n}\log \mathbb{P}\left[\sup_{t\in[0,n\tau]}|Q(t,y)-nx-Dt|<\delta n\right]$$

$$\stackrel{\text{def}}{=} -\tau L(\Lambda(x),D). \tag{1}$$

Note that for $D_i = 0$, $\forall i \in \Lambda^c(x)$, the conditions $x_i + D_i\tau > 0$, $\forall i \in \Lambda(x)$ is equivalent to $x + Dt$ lies in the face $\Lambda(x)$, for all $t \in [0,\tau]$.

In establishing theorem 3.1, one must know in some sense how the different transition rates have to be modified in order that Q_t follows a given drift D. This means that rather than studying Q_t itself, we focus on the so called empirical generator

$$G_t \stackrel{\text{def}}{=} \left(\frac{N_t}{t}\hat{L}_{N_t}, L_t, \frac{Q_t - Q_0}{t}\right) \in \mathcal{G},$$

where

- N_t is the number of jumps of the server until time t;
- S_n is the embedded process of the server just before it jumps. Note that it is not a Markov chain;
- $\hat{L}_n \stackrel{\text{def}}{=} \frac{1}{n}\sum_{i=0}^{n}\delta_{S_i} \in \mathcal{P}(\mathcal{S}_0)$ is the empirical measure of the process S_n;
- $L_t \stackrel{\text{def}}{=} \frac{1}{t}\int_0^t \delta_{S_u}du \in \mathcal{P}(\mathcal{S}_0)$ is the empirical measure of the process S_t.

The set of empirical generators is denoted by $\mathcal{G}$. On the long run, exactly like the pair empirical measure for Markov chains, the empirical generator tends to be *balanced*: this is (2), which means that the server exits a node i as often as it goes toward i. The set of balanced generators will be denoted by $\mathcal{G}_s$. When the transitions are restrained to lie in the support of $\mathbf{P}$ (i.e. $a_{ij} = 0$ if $P_{ij} = 0$), the set is denoted by $\mathcal{G}_s(\mathbf{P})$.

Definition 3.2 (Generators) *The set* $\mathcal{G}_s^\Lambda$ *of balanced generators is defined by the triples* $(A,\pi,D) \in \mathcal{M}(\mathcal{S}_0)\times\mathcal{P}(\mathcal{S}_0)\times\mathbb{R}^\Lambda$ *that verify*

$$\bar{a}_i \stackrel{\text{def}}{=} \sum_{j\in\mathcal{S}} a_{ji} = \sum_{j\in\mathcal{S}} a_{ij}, \quad \forall i \in \mathcal{S}, \tag{2}$$

$$a_i \le \bar{a}_i, \quad \forall i \in \Lambda^c, \tag{3}$$

$$a_i = \bar{a}_i, \quad \forall i \in \Lambda, \tag{4}$$

$$a_i + D_i \ge 0, \quad \forall i \in \Lambda. \tag{5}$$

We denote by $\mathbb{R}^\Lambda$, *for the sake of simplicity, the subspace of* $D \in \mathbb{R}^N$ *with* $D_i = 0$ *for* $i \in \Lambda^c$.

The three elements of a generator (A, π, D) are, roughly speaking, the stationary measure π of the position of the server, the mean number A of transitions and the mean drift of the queues length D. Note that, in general, for each generator (A, π, D) corresponds a polling system, described by its intensities $(\tilde{\lambda}_i, \tilde{\mu}_i, \tilde{\tau}_{ij}, \tilde{P}_{ij})$. The correspondence is given by

$$\tilde{\lambda}_i = a_i + D_i, \tag{6}$$

$$\tilde{\mu}_i = \frac{a_i}{\pi_i}, \tag{7}$$

$$\tilde{\tau}_{ij} = \frac{\pi_{ij}}{a_{ij}}, \tag{8}$$

$$\tilde{P}_{ij} = A_{ij} \stackrel{\text{def}}{=} \frac{a_{ij}}{\overline{a}_i}. \tag{9}$$

Theorem 3.3 [Generator's local bounds]
Let Λ be a face and $G = (\pi, A, D) \in \mathcal{G}_s^\Lambda$. Then,

$$\begin{aligned} & \lim_{\delta \to 0} \liminf_{t \to \infty} \frac{1}{t} \log \mathbb{P}\left[G_t \in B(G, \delta), \sup_{s \in [0,t]} |Q^\Lambda(s) - sD| < t\delta\right] \\ = & \lim_{\delta \to 0} \limsup_{t \to \infty} \frac{1}{t} \log \mathbb{P}\left[G_t \in B(G, \delta), \sup_{s \in [0,t]} |Q^\Lambda(s) - sD| < t\delta\right] \\ = & -H(G \| R). \end{aligned}$$

In theorem 3.3, we prove large deviations bounds with a rate function $H(.\|R)$ defined in definition 3.4, for a uniform version of G_t^Λ, where G_t^Λ is the empirical generator associated to a *localized* polling system X^Λ. The transition mechanism describing the evolution of X^Λ is identical to X's except that the components indexed by Λ for X^Λ can be negative. In order to prove this result, we use a change of measure (chosen in a restricted class) which gives rise to a new polling system, for which the fluid limits can be completely characterized, and this is a key ingredient in the proof of the lower bound, in the sense that it allows to invert the equations (6)–(9). Let us recall that, if $\{x_n, \ n \geq 0\}$ is a sequence of $\mathbb{Z}_+^N$ such that $|x_n| \to \infty$, then every limiting point in distribution of the sequence of processes $|x_n|^{-1} X_{x_n,s}$ is called a fluid limit. This ensures that for D as in theorem 3.1, after a suitable change of measure, X lies in any neighborhood of D. This way of reasoning is quite classical and was used among others in [14] for identifying the rate function governing the LDP for Jackson networks. However, owing to the presence of S_t, our approach has much more in common with the method used in [4] to prove the weak Sanov LDP for jump Markov processes in continuous time. Moreover, the function $H(.\|R)$ governing the large deviations bounds for G_t^Λ appears as an entropy and is then easily seen to possess good properties.

Definition 3.4 (relative entropy) *Let $R = (\lambda_i, \mu_i, \tau_{ij}, P_{ij})$ denotes the generator of the polling system, $G = (A, \pi, D) \in \mathcal{G}_s^\Lambda$ be a generator and $(\tilde{\lambda}_i, \tilde{\mu}_i, \tilde{\tau}_{ij}, \tilde{P}_{ij})$ its representation as a polling system. The relative entropy of G with respect to R*

is[1]

$$H(G\|R) \stackrel{\text{def}}{=} \sum_{i\in\mathcal{S}} \left(I_p(\tilde{\lambda}_i\|\lambda_i) + \pi_i I_p(\tilde{\mu}_i\|\mu_i) \right) + \sum_{i,j\in\mathcal{S}} \pi_{ij} I_p(\tilde{\tau}_{ij}^{-1}\|\tau_{ij}^{-1}) + H_d(A\|P),$$

where $H_d(A\|P) \stackrel{\text{def}}{=} \sum_{i,j\in\mathcal{S}} a_{ij} \log\left(\frac{A_{ij}}{P_{ij}}\right)$ *and* $I_p(\nu\|\lambda) \stackrel{\text{def}}{=} \nu \log\frac{\nu}{\lambda} - \nu + \lambda.$

This entropy has an easy interpretation in terms of information theory. The relative entropy can be defined as the *mean information gain.* $H(.\|R)$ is decomposed in definition 3.4 as the sum of the information gains, first $I_p(\tilde{\lambda}_i\|\lambda_i)$ for the arrivals, second $I_p(\tilde{\mu}_i\|\mu_i)$ for the service times, multiplied by the time π_i spent at each queue, third $\pi_{ij} I_p(\tilde{\tau}_{ij}^{-1}\|\tau_{ij}^{-1})$ for the transit times and fourth $H_d(A\|P)$ for the routing. This natural interpretation leads to natural properties :

Proposition 3.5 *The relative entropy* $H(.\|R)$ *is positive, finite, continuous and strictly convex on* $\mathcal{G}_s^{\Lambda}(\mathbf{P})$*; it is infinite otherwise; it is null if, and only if,* $G = R$*. It has compact level sets.*

Definition 3.6 *The rate function* $L(\Lambda, D)$ *is defined by*

$$L(\Lambda, D) \stackrel{\text{def}}{=} \inf_{G\in f_{\Lambda}^{-1}(D)} H(G\|R), \quad \forall D \in \mathbb{R}^{\Lambda}, \tag{10}$$

where $f_{\Lambda} : \mathcal{G}_s^{\Lambda} \mapsto \mathbb{R}^{\Lambda}$ *is the projection* $f_{\Lambda}(G) = D$*. For all* $x \in \mathbb{R}_+^N$*,* $L(x, D)$ *is defined to be* $L(\Lambda(x), D)$*.*

Note that $L(\Lambda, D)$ is a rate function derived by the contraction of $H(.\|R)$, which is a good rate function (see proposition 3.5). Even though the corresponding LDP are not proved, since f_{Λ} is linear, L inherits lots of good properties of H (see [4]).

Proposition 3.7 *The rate function* $L(\Lambda, D)$ *is positive, finite, continuous and strictly convex with respect to* $D \in \mathbb{R}^{\Lambda}$*; it is null if, and only if,* D *is the drift of the localized polling system; it has compact level sets. Moreover, the infimum* $G \in \mathcal{G}_s^{\Lambda}$ *such that* $H(G\|R) = L(\Lambda, D)$ *is reached at a unique point* $G(D) \in \mathcal{G}_s^{\Lambda}(\mathbf{P})$*, and* $G(D)$ *is a continuous function of* D*.*

At this stage, as an application of a kind of contraction principle, one can get theorem 3.1. Since $L(\Lambda, D)$ is the rate function derived from $H(.\|R)$ by a contraction principle, many properties can be derived without much effort and it turns out that $L(\Lambda, D)$ is solution of a convex program : (10). In general, for Markov processes in $\mathbb{Z}_+^N$, the fluid limits cannot be characterized, even in the case of maximal spatial homogeneity. So, it seems that the sole traditional change of measure would be ineffective for the identification of the rate function in this general setting.

[1]Note that the relative entropy $H(G\|R)$ is independent of Λ.

4 Sample path LDP

The rate function $I_T(.)$ for the sample path LDP is expressed as

$$I_T(\varphi) = \begin{cases} \int_0^T L\left(\varphi(t), \dot{\varphi}(t)\right) dt, & \text{if } \varphi \text{ is absolutely continuous,} \\ +\infty, & \text{otherwise.} \end{cases} \tag{11}$$

Remark 1 *In theorem 4.1, it is proved that the rate function $I_T(.)$ has the expected form given in (11). It is worth noting that in [10], the rate function governing the LDP was defined as the lower semi-continuous regularization of a functional defined on the set of piecewise linear functions. Nevertheless, it was conjectured that the rate function takes a form like in (11).*

Set

$$Q^n_{x,s} \stackrel{\text{def}}{=} \left\{ \frac{1}{n} Q(nt, [nx], s),\ t \geq 0 \right\},$$

and

$$\Phi_x(K) = \{\varphi \in D([0,T], \mathbb{R}^N_+) : I_T(\varphi) \leq K,\ \varphi(0) = x\}.$$

Using the Markov property and the linear bounds, one deduces easily the large deviations bounds for the probability that the process evolves around some piecewise linear function, which are expressed using $I_T(.)$. Then, using various properties of the rate function $I_T(.)$ and the exponential tightness of $\{Q^n_{x,s},\ n \geq 1\}$, one can get the sample path LDP:

Theorem 4.1 **[Sample path LDP]** *The sequence $\{Q^n_{x,s},\ n \geq 1\}$ satisfies a LDP in $D\big([0,T], \mathbb{R}^N_+\big)$ with good rate function $I_T(.)$: for every $T > 0$, $x \in \mathbb{R}^N_+$ and s,*

1. *For any compact set $C \subset \mathbb{R}^N_+$, $\bigcup_{x\in C} \Phi_x(K)$ is compact in $\mathcal{C}([0,T], \mathbb{R}^N_+)$,*
2. *for each closed set F of $D\big([0,T], \mathbb{R}^N_+\big)$,*

$$\limsup_{n\to\infty} \frac{1}{n} \log \mathbb{P}\left[Q^n_{x,s} \in F\right] \leq -\inf\{I_T(\phi),\ \phi \in F, \phi(0) = x\};$$

3. *for each open set O of $D\big([0,T], \mathbb{R}^N_+\big)$,*

$$\liminf_{n\to\infty} \frac{1}{n} \log \mathbb{P}\left[Q^n_{x,s} \in O\right] \geq -\inf\{I_T(\phi),\ \phi \in O, \phi(0) = x\}.$$

References

[1] Atar, R., and Dupuis, P. Large deviations and queueing networks: methods for rate function identification. *Stochastic Process. Appl. 84*, 2 (1999), 255–296.

[2] Borovkov, A. A., and Schassberger, R. Ergodicity of a polling network. *Stochastic Process. Appl. 50*, 2 (1994), 253–262.

[3] Boxma, O., and Weststrate, J. Waiting time in polling systems with markovian server routing. In *Messung, Modellierung und Bewertung von Rechensysteme* (1989), Springer-Verlag, pp. 89–104.

[4] de La Fortelle, A. Large deviation principle for markov chains in continuous time. Tech. Rep. 3877, INRIA, 2000.

[5] Delcoigne, F., and de La Fortelle, A. Large deviations for polling systems. Tech. Rep. 3892, INRIA, 2000.

[6] Dobrushin, R. L., and Pechersky, E. A. Large deviations for tandem queueing systems. *J. Appl. Math. Stochastic Anal. 7*, 3 (1994), 301–330.

[7] Dobrushin, R. L., and Pechersky, E. A. Large deviations for random processes with independent increments on infinite intervals. In *Probability theory and mathematical statistics (St. Petersburg, 1993)*. Gordon and Breach, Amsterdam, 1996, pp. 41–74.

[8] Down, D. On the stability of polling models with multiple servers. *J. Appl. Probab. 35*, 4 (1998), 925–935.

[9] Dupuis, P., and Ellis, R. S. Large deviations for Markov processes with discontinuous statistics. II. Random walks. *Probab. Theory Related Fields 91*, 2 (1992), 153–194.

[10] Dupuis, P., and Ellis, R. S. The large deviation principle for a general class of queueing systems. I. *Trans. Amer. Math. Soc. 347*, 8 (1995), 2689–2751.

[11] Dupuis, P., Ellis, R. S., and Weiss, A. Large deviations for Markov processes with discontinuous statistics. I. General upper bounds. *Ann. Probab. 19*, 3 (1991), 1280–1297.

[12] Dupuis, P., Ishii, H., and Soner, H. M. A viscosity solution approach to the asymptotic analysis of queueing systems. *Ann. Probab. 18*, 1 (1990), 226–255.

[13] Fricker, C., and Jaibi, M. Stability of multi-server polling models. Tech. Rep. 3347, INRIA, 1998.

[14] Ignatiouk-Robert, I. Large deviations of jackson networks. Tech. Rep. 14/99, Université de Cergy-Pontoise, 1999.

[15] Ignatyuk, I. A., Malyshev, V. A., and Shcherbakov, V. V. The influence of boundaries in problems on large deviations. *Uspekhi Mat. Nauk 49*, 2(296) (1994), 43–102.

[16] Nagot, I. *Grandes déviations pour des processus d'apprentissage lent à statistique discontinue sur une surface.* Université Paris XI, Orsay, 1995. Ph.D. Thesis.

[17] TAKAGI, H. Queuing analysis of polling models. *ACM Comput. Surveys 20*, 1 (1988), 5–28.

Trends in Mathematics, © 2000 Birkhäuser Verlag Basel/Switzerland

A nonlinear integral operator encountered in the bandwidth sharing of a star-shaped network

GUY FAYOLLE, JEAN-MARC LASGOUTTES[1] *INRIA – Domaine de Voluceau, BP 105, Rocquencourt 78153 Le Chesnay cedex, France.*
[Guy.Fayolle,Jean-Marc.Lasgouttes]@inria.fr

Abstract. *We consider a symmetrical star-shaped network, in which bandwidth is shared among the active connections according to the "min" policy. Starting from a* chaos propagation *hypothesis, valid when the system is large enough, one can write equilibrium equations for an arbitrary link of the network. This paper describes an approach based on functional analysis of nonlinear integral operators, which allows to characterize quantitatively the behaviour of the network under heavy load conditions.*

1 Model description

Consider a network comprising N links, where several data sources establish connections along routes going through these links. The main concern here is about the policies that can be used to share the bandwidth of the links between active connections, and the effect of these policies on the dynamics of the network.

In this paper, the network is star-shaped (see Figure 1) and all routes are of length 2, which is a reasonable model for a router. Each star branch contains two links ("in" and "out") and each route is isomorphic to a pair of links.

Let $r = (i, j)$ denote a route on links i and j, and $\mathcal{R}$ be the set of all possible routes (with cardinal $N^2/2$). Connections are created on r according to a Poisson process with intensity $2\lambda/N$, so that the total arrival intensity on each link is λ. A connection lasts until it has transmitted over the network its data, the volume of which follows an exponential law with mean v. Each link i, $1 \leq i \leq N$, has a bandwidth equal to 1 and its load is $\rho \stackrel{\text{def}}{=} \lambda v$.

The state of the system at time $t \in \mathbb{R}$ is given by the number of active connections on each route $\left(c_r^{(N)}(t),\, r \in \mathcal{R}\right) \stackrel{\text{def}}{=} X^{(N)}(t)$. The vector $X^{(N)}(t)$ is in general Markovian and

$$X_i^{(N)}(t) \stackrel{\text{def}}{=} \sum_{r \ni i} c_r^{(N)}(t)$$

is the total number of active connections on link i, $\forall 1 \leq i \leq N$.

It is now necessary to describe how bandwidth allocation is achieved. The "max-min fairness" policy, popular in telecommunication models, being too difficult to be studied rigorously, this article focuses on the "min" policy, proposed by L. Massoullié and J. Roberts [5], in which a connection on (i, j) gets bandwidth

$$\min\left(\frac{1}{X_i^{(N)}(t)}, \frac{1}{X_j^{(N)}(t)}\right). \tag{1.1}$$

[1]This work has been partly supported by a grant from *France Télécom R&D*.

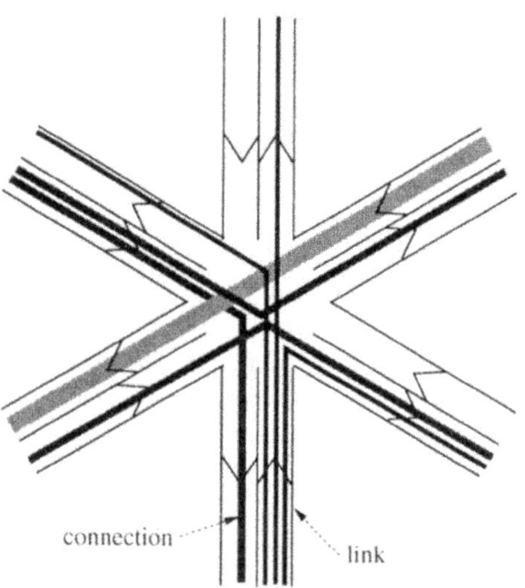

Figure 1: The star-shaped network

This allocation clearly satisfies the capacity constraints of the system, and can be shown to be sub-optimal with respect to max-min fairness. Its invariant measure equations are however too complicated to be solved explicitly. An efficient method in such situations is to study the model in so-called *thermodynamical limit*, using *mean field* analysis.

In order to study the stationary behaviour of the network as $N \to \infty$, the idea is to assume the conditions of *chaos propagation*, under which any finite number of links tend to become mutually independent. In this paper, this hypothesis will be considered as a heuristic, to be proved in further studies. Some rigorous studies of this type can be found in Vvedenskaya *et al.* [8] and Delcoigne et Fayolle [2].

From now on, $\rho < 1$ and the system is assumed to be in stationary state $X = (c_r, r \in \mathcal{R})$. For all $k \geq 0$, the empirical measure of the number of links with k connections is

$$\alpha_k^{(N)} \stackrel{\text{def}}{=} \frac{1}{N} \sum_{1 \leq i \leq N} \mathbb{1}_{\{X_i^{(N)}=k\}}.$$

Symmetry considerations imply that, for all $i \leq N$, $\mathbb{P}(X_i^{(N)} = k) = \mathbb{E}\alpha_k^{(N)}$. Besides, from the chaos propagation hypothesis, a law of large numbers is assumed to hold for $\alpha_k^{(N)}$:

$$\alpha_k \stackrel{\text{def}}{=} \lim_{N \to \infty} \alpha_k^{(N)} = \lim_{N \to \infty} \mathbb{P}(X_i^{(N)} = k) \stackrel{\text{def}}{=} \mathbb{P}(X = k).$$

The $\alpha_k^{(N)}$'s, traditionally named *mean field*, drives the dynamics of the system. The following notation will also prove useful:

$$\bar{\alpha}^{(N)} \stackrel{\text{def}}{=} \mathbb{E}X_i^{(N)} = \sum_{k>0} k\alpha_k^{(N)}, \qquad \bar{\alpha} \stackrel{\text{def}}{=} \sum_{k>0} k\alpha_k.$$

A heuristic computation (detailed in [3]), yields the following equations:

$$\begin{aligned} u_k &= \sum_{\ell>0} (k \wedge \ell)\alpha_l, \\ \alpha_{k+1}u_{k+1} &= \rho\bar{\alpha}\alpha_k, \qquad \forall k \geq 0. \end{aligned} \tag{1.2}$$

While (1.2) resembles a "birth and death process" equation, it is in fact highly non-linear, due to the form of u_k and of $\bar{\alpha}$.

The purpose of this paper is to show how the asymptotic behaviour of the system (as $\rho \to 1$) can be derived from the analytical study of the generating functions built from (1.2). This work is a part of the wider study [3], which also gives ergodicity conditions for any topology under the min and max-min policies, shows how equations like (1.2) are derived (also in the case where routes are longer than 2) and presents comprehensive numerical results.

The main byproduct of Theorem 3.3, is the following asymptotic expansions, valid in a neighborhood of $\rho = 1$.

$$\begin{aligned} \bar{\alpha} &\approx \frac{1}{(1-\rho)^2 A}, \\ \lim_{k\to\infty} \rho^{-k}\alpha_k &\approx (1-\rho)B\exp\Big[\frac{1}{(1-\rho)A}\Big], \end{aligned}$$

where A and B are non-negative constants. Moreover, if $c(z,1)$ and $v(z,1)$ are the solutions of the system of differential equations (3.5), then A can be written as follows:

$$A = \int_0^\infty c(z,1)dz = \lim_{z\to\infty} zv'(z,1) \approx 1.30.$$

Since this system is numerically highly unstable, it has proven difficult (with the "Livermore stiff ODE" solver from MAPLE) to get a better estimate for A.

2 An integral equation for the generating function

Let $\mathcal{C}(r)$ (resp. $\mathcal{D}(r)$), be the circle (resp. the open disk) of radius r in the complex plane.

Let $\alpha : z \to \alpha(z)$ be the generating function, *a priori* defined for z in the closed unit disk

$$\alpha(z) \stackrel{\text{def}}{=} \sum_{k\geq 0} \alpha_k z^k.$$

Denoting α' the derivative of α, (1.2) can be rewritten as

$$\alpha_{k+1}u_{k+1} = \rho\alpha'(1)\alpha_k, \quad \forall k \geq 0. \tag{2.1}$$

Lemma 2.1.

(a) If (2.1) has a probabilistic solution, then, necessarily $\rho < 1$, and

$$\lim_{k\to\infty} \alpha_k \rho^{-k} = K(\rho),$$

where $K(\rho)$ is a positive constant, bounded $\forall \rho < 1$.

(b) The function α satisfies the nonlinear integral equation

$$\alpha'(1)(1-\rho z)\alpha(z) = \frac{1}{2i\pi}\int_{C(r)} \alpha(\omega)\,\alpha\Big(\frac{z}{\omega}\Big)\frac{d\omega}{(1-\omega)^2}, \tag{2.2}$$

where $|z| < \rho^{-1}$ et r is an arbitrary positive number, with $1 < r < \rho^{-1}$.

Proof. When (1.2) has a probabilistic solution, necessarily $\bar{\alpha} = \lim_{m\to\infty} u_m$. Therefore, $\forall \epsilon > 0$, there exists a number $M(\epsilon) > 0$, such that

$$\rho \le \frac{\alpha_{m+1}}{\alpha_m} \le \frac{\rho}{1-\epsilon}, \quad \forall m \ge M(\epsilon),$$

which implies that $\rho < 1$. Moreover, under the same existence hypothesis, one can write

$$\alpha_k \rho^{-k} = \alpha_0 \prod_{l\ge 0}^{k} \frac{\bar{\alpha}}{\bar{\alpha} - D_{\ell+1}}, \tag{2.3}$$

where $D_\ell = \sum_{m\ge\ell}(m-\ell)\alpha_m$.

When $\ell \to \infty$, the convergence of the product in (2.3) is equivalent to the convergence of the series of general term D_ℓ, which holds since

$$D_\ell \le \alpha_0 \sum_{m\ge\ell}(m-\ell)(\rho+\epsilon)^m, \ \forall \ell \ge M(\epsilon),$$

so that $\sum_\ell D_\ell$ behaves like

$$\sum_{\ell\ge 0}\sum_{n\ge 0} n(\rho+\epsilon)^{\ell+n} = \frac{\rho+\epsilon}{(1-\rho-\epsilon)^3}.$$

Point (a) of the lemma is proven. Point (b) is an application of (a) and of an integral representation used by Hadamard, recalled below (see e.g. [7]).

Let a and b be functions

$$a(z) = \sum_{n\ge 0} a_n z^n, \qquad b(z) = \sum_{n\ge 0} b_n z^n,$$

analytic in the respective disks $\mathcal{D}(R)$ and $\mathcal{D}(R')$. Then the function

$$c(z) = \sum_n a_n b_n z^n$$

has a radius of convergence greater than RR' and has the integral form

$$c(z) = \frac{1}{2i\pi} \int_{\mathcal{L}} a(\omega)\, b\Big(\frac{z}{\omega}\Big) \frac{d\omega}{\omega},$$

where $\mathcal{L}$ is a closed contour containing the origin, and on which $|\omega| < R, \left|\frac{z}{\omega}\right| < R'$.

From (a), α has a radius of convergence at least equal to ρ^{-1}: this property, used in Hadamard's formula, leads directly to (2.2). □

The following Lemma provides a finer description of $\alpha(z)$.

Lemma 2.2. *The function α is meromorphic and can be written as*

$$\alpha(z) = \sum_{i=1}^{\infty} \frac{a_i}{\rho^{-i} - z}. \tag{2.4}$$

Moreover, for any sequence of circles $C(R_n)$, such that

$$(1+\epsilon)\rho^{-n} \le R_n \le (1-\epsilon)\rho^{-(n+1)}, \quad 0 < \epsilon < \frac{1-\rho}{1+\rho},$$

one has $|\alpha(R_n)| = o(1)$ as $n \to \infty$.

Proof. The integral equation (2.2) allows for the analytic continuation of α in the whole complex plane. Indeed, from point (a) of Lemma 2.1, α is holomorphic in $\mathcal{D}(\rho^{-1})$ and its first singularity is a simple pole at the point $z = \rho^{-1}$. An application of Cauchy's theorem to the integral in (2.2) leads to

$$\frac{1}{2i\pi} \int_{C(R_1)} = \frac{1}{2i\pi} \int_{C(r)} + \text{Residue}(\rho^{-1}),$$

where R_1 is defined above. Since $\alpha(\omega)$ and $\alpha(z/\omega)$ are analytic in the regions

$$|\omega| < \rho^{-1}, \quad \left|\frac{z}{\omega}\right| < \rho^{-1},$$

the integral is a function of z analytic in the ring-shaped area $\rho^{-1} < |z| < \rho^{-2}$. The same holds for α, thanks to the left-hand side of (2.2). Equation (2.4) follows by recurrence.

The second part of the lemma is obtained by application of Cauchy's theorem to the integral in (2.2) along the circle $C(R_n)$:

$$\alpha'(1)(1-\rho z)\alpha(z) - \sum_{i=1}^{n} \frac{a_i \alpha(z\rho^i)}{(\rho^{-i}-1)^2} = \int_{C(R_n)} \alpha(\omega)\, \alpha\Big(\frac{z}{\omega}\Big) \frac{d\omega}{(1-\omega)^2}.$$

One can make the analytic continuation of the above equality, the left-hand side of which is analytic in $|z| \le R_n$, letting z reach the circle $C(R_n)$ along a simple curve which avoids the polar singularities $\rho^{-i}, i = 1, \ldots, n$. Bounding the modulus of the integral yields the inequality

$$(\rho R_n - 1)\mathcal{M}(R_n) \leq A_n \sup_{i \leq n-1} (\mathcal{M}(R_i)) + D\mathcal{M}(R_n) \int_{C(R_n)} \frac{|d\omega|}{|1-\omega|^2},$$

where D is a positive constant and A_n is bounded $\forall n \leq \infty$. By induction, $\lim_{n\to\infty} \mathcal{M}(R_n) = 0$ follows easily, and the proof is completed. □

Combining (2.4) with (2.2), a straightforward computation of residues gives

$$\alpha'(1)(1-\rho z)\sum_{i=1}^{\infty} \frac{a_i}{\rho^{-i} - z} = \sum_{i,j=1}^{\infty} \frac{a_i a_j \rho^{-j}}{(\rho^{-(i+j)} - z)(1-\rho^{-j})^2}. \tag{2.5}$$

Let $a : z \to a(z)$ be the generating function

$$a(z) \stackrel{\text{def}}{=} \sum_{k \geq 0} a_{k+1} z^k,$$

defined for z in a bounded domain of the complex plane, including the origin. Using point (a) of Lemma 2.1 and (2.4), the following relations hold

$$\begin{cases} a_1 = \rho K(\rho) \\ \alpha_{k-1} = \rho^k a(\rho^k), \quad \forall k \geq 1, \end{cases} \tag{2.6}$$

Since $\alpha_0 < 1 - \rho$, the function a is thus analytic in the disk $\mathcal{D}(\rho)$. Identifying the coefficients of the power series in z in (2.5), one gets

$$\alpha'(1)(\rho^{k-1} - 1)a_k = \sum_{j=1}^{k-1} \frac{a_j a_{k-j} \rho^{j+k-1}}{(1-\rho^j)^2}, \quad \forall k \geq 1. \tag{2.7}$$

It follows easily by recurrence that the a_i's are of alternate signs, with $a_1 > 0$.

Let

$$f(t) \stackrel{\text{def}}{=} \sum_{j=1}^{\infty} \left(\frac{\rho^j}{1-\rho^j}\right)^2 t^j = t \sum_{j=1}^{\infty} \frac{j\rho^{j+1}}{1 - t\rho^{j+1}}, \quad \rho < 1, \ |t| < \rho^{-2}. \tag{2.8}$$

Hadamard's formula, when applied in (2.7), implies the integro-differential equation

$$\alpha'(1)[a(\rho z) - a(z)] = \frac{a(\rho z)}{2i\pi} \int_{C(r)} a(\omega) f\left(\frac{z}{\omega}\right) d\omega, \tag{2.9}$$

valid in the domain $\{r \leq \rho, \ |z| < \rho^{-2}\}$.

Taking the second form for f in (2.8), which in fact converges in the domain $\{\rho < 1, \ \Re(z) \leq 0\}$, and applying Cauchy's theorem to the integral in (2.9), one obtains the functional equation

$$\alpha'(1)\big[a(\rho z) - a(z)\big] = a(\rho z) b(z), \quad |z| \leq \rho, \ \rho < 1, \tag{2.10}$$

where

$$b(z) \stackrel{\text{def}}{=} z \sum_{j=1}^{\infty} j \rho^{j+1} a(\rho^{j+1} z).$$

From (2.10), let's make now the analytic continuation of a in the nested disks $\mathcal{D}(\rho^{-n}), n \geq 1$. It appears that a has no singularity at finite distance, and consequently is an *integral* function. From the general theory [7], it follows that a is completely characterized by its zeros and its *order* at infinity.

Let z_0 be an arbitrary zero of a. From (2.10) again, $a(\rho^{-i} z_0) = 0$ and the zeros of a form families of points in geometric progression with parameter ρ^{-1}. It suffices to determine the zeros of smallest modulus, but, alas they do not have any explicit form and numerical schemes are highly unstable. However, from (2.6), $b(1) = \alpha'(1)$, so that (2.10) implies $a(1) = 0$ together with

$$a(\rho^{-i}) = 0, \quad \forall i \geq 0. \tag{2.11}$$

3 On the asymptotic behaviour around $\rho = 1$

In order to assess the practical value of the "min" policy, it is important to evaluate the system behaviour in heavy traffic conditions. The numerical calculations in [3] show that the distribution of any X_i is *modal*, which is not common in known models.

In this section, it will be convenient to consider ρ not only as a parameter but as a plain variable. Therefore, in all quantities of interest, ρ will appear as an explicit variable, e.g. $a(z,\rho)$, $f(z,\rho)$ or $a_k(\rho)$.

The fundamental ideas of the analysis will be given after the next lemma, which proposes a scaling—likely to be the only interesting one—for the function $a(z,\rho)$.

Let

$$\begin{cases} \xi(\rho) \stackrel{\text{def}}{=} \dfrac{-a_1(\rho)}{(\log\rho)^3 \alpha'(1,\rho)}, \\ c_k(\rho) \stackrel{\text{def}}{=} \dfrac{-a_k(\rho)}{(\log\rho)^3 \alpha'(1,\rho)\xi(\rho)^k}, \quad \forall k \geq 1, \\ c(z,\rho) \stackrel{\text{def}}{=} \displaystyle\sum_{i=0}^{\infty} c_{k+1}(\rho) z^k. \end{cases} \tag{3.1}$$

The reader will easily convince himself that the factor $(1-\rho)^3$ arises rather naturally; however the factor $-\log^3 \rho$ has been chosen here, since it provides more compact formulas in the forthcoming results.

Lemma 3.1. *Let*

$$v(z,\rho) \stackrel{\text{def}}{=} \frac{(\log\rho)^2}{2i\pi} \int_{\mathcal{C}(1)} c(\omega,\rho) f\left(\frac{z}{\omega},\rho\right) d\omega, \quad |z| < \rho^{-2}.$$

The following functional relations hold:

$$\begin{cases} a(z,\rho) = a_1(\rho)\, c(z\xi(\rho),\rho), \\ c(0,\rho) = 1, \\ c(z,\rho) = c(\rho z,\rho)\big[1 + \log\rho\, v(z,\rho)\big], \\ \qquad\quad = c(\rho z,\rho)\big[1 + z(\log\rho)^3 \displaystyle\sum_{i=1}^{\infty} i\rho^{i+1} c(\rho^{i+1}z,\rho)\big]. \end{cases} \tag{3.2}$$

The coefficients $c_k(\rho)$, $k \geq 1$, are of alternate signs and the function $c(z,\rho)$ has the following properties.

(a) there exists only one solution $c(z,\rho)$ of (3.2), which is integral with respect to z and bi-analytic in (z,ρ) in the region $0 < \rho < 1$.

(b) Define

$$g(t) \stackrel{\text{def}}{=} \sum_{j=1}^{\infty} \frac{t^j}{j^2} = \int_0^t \frac{-\log(1-u)du}{u}, \quad \forall |t| \leq 1,$$

denoted by some authors as dilog$(1-t)$. *Then $c(z,1)$ exists and satisfies the integro-differential equation*

$$z\frac{\partial c(z,1)}{\partial z} = -\frac{c(z,1)}{2i\pi} \int_{C(r)} c(\omega,1) g\Big(\frac{z}{\omega}\Big) d\omega, \quad |r| = 1,\ \forall |z| \leq 1, \tag{3.3}$$

which rewrites in the form

$$z\frac{\partial c(z,1)}{\partial z} = -c(z,1) \int_0^z c(\omega,1) \log\Big(\frac{z}{\omega}\Big)\, d\omega, \tag{3.4}$$

which is equivalent to the non-linear differential system

$$\begin{cases} z\dfrac{\partial c(z,1)}{\partial z} + c(z,1)v(z,1) = 0, \\ z\dfrac{\partial^2 v(z,1)}{\partial z^2} + \dfrac{\partial v(z,1)}{\partial z} = c(z,1), \end{cases} \tag{3.5}$$

with initial conditions

$$v(0,1) = 0, \qquad \frac{\partial v(z,1)}{\partial z}_{|z=0} = 1, \quad c(0,1) = 1.$$

(c) Moreover, $c(z,1)$ is analytic in the open complex plane, except at a negative real point q, and $c(z,1) \neq 0$, $\forall z \neq q \cup \infty$.

Proof. The first three equations in (3.2) follow directly from the definition (2.7) and (2.10) of the coefficients $a_k(\rho)$, the fourth one coming from the analytic continuation of (2.9). Existence and uniqueness are simple consequences of the convolution equation (2.7).

The properties relative to the morphology of $c(z,\rho)$, $\rho \leq 1$, are more intricate. First, the reader will notice that (3.3) can be obtained rigorously from (2.7) or (2.8), but not from (3.2)! Then, there is a *phase transition* when $\rho = 1$. We will return to this topic in Section 4. □

It is interesting to note that the function $w(y) \stackrel{\text{def}}{=} v(e^y) + 1$ satisfies the so-called *Blasius* [1] third-order differential equation

$$w'''(y) + w(y)w''(y) = 0,$$

which arises in hydrodynamics to describe the stationary evolution of a laminar boundary layer along a flat plate! The explicit solution of this equation is still unknown, albeit it has been studied by many authors over the last decades (see e.g. [6]).

Starting from Lemma 3.1, it is now possible to sketch the main ideas of the proposed method. The Gordian knot amounts to the evaluation of $\xi(\rho)$, defined in (3.1). This can be done via the *anchoring equation*

$$c(\xi(\rho), \rho) = 0, \tag{3.6}$$

which follows from 2.11 and from the first equation of (3.2). From the structure of the third equation of (3.2), it appears that the smallest positive solution of $c(u, \rho) = 0$ satisfies

$$\frac{-1}{u \log \rho} = (\log \rho)^2 \sum_{i \geq 1} i\rho^{i+1} c(u\rho^{i+1}, \rho),$$

the right-hand side of which is an analytic function, bounded in any compact set $\forall \rho \leq 1$: as $\rho \to 1$, necessarily $u \to \infty$ and all positive zeros of the anchoring equation are sent to infinity. The key is to find the asymptotic behaviour in z of the various functions, in the cone $0 \leq z \leq \mathcal{U}(\rho)$, which contains $\xi(\rho)$: in this cone, $c(z, \rho)$ is close to $c(z, 1)$—in some sense—and

$$v(z, \rho) \approx w(z, \rho) \stackrel{\text{def}}{=} \log^2 \rho \sum_{i \geq 1} i\rho^{i+1} c(z\rho^{i+1}, 1). \tag{3.7}$$

Since $w(z, 1)$ has a logarithmic behaviour, $\xi(\rho)$ can be obtained by direct inversion. The sketch of the proof is outlined below.

We will need the Mellin transform (see e.g. [4]) of $c(z, 1)$, defined as

$$c^*(s) \stackrel{\text{def}}{=} \int_0^\infty x^{s-1} c(x) dx.$$

The behaviour of $c(z, 1)$ in the region $\Re(z) > 0$, given in Section 4, implies the existence of $c^*(s, 1), \forall s, \Re(s) > 0$ and of all the moments of $c(z, 1)$ on the positive real axis.

Lemma 3.2. *The function $w(z, \rho)$ defined by (3.7) admits, $\forall z, \Re(z) > 0$, the asymptotic expansion*

$$\begin{aligned} w(z, \rho) = c^*(1, 1) \log(\rho z) - \frac{\partial c^*(1, 1)}{\partial s} \\ + (\log \rho)^2 \left[(\log z) \Phi(z, \rho) + \Psi(z, \rho) \right] + \mathcal{O}(z^{-d}), \end{aligned} \tag{3.8}$$

where d is an arbitrary positive number, Φ and Ψ are fluctuating functions of small amplitude for $\rho \neq 1$, which vanish for $\rho = 1$.

Proof. From the Mellin transform inversion formulas,

$$w(z,\rho) = \frac{1}{2i\pi}\int_{\sigma-i\infty}^{\sigma+i\infty} z^{-s}c^*(s+1,1)\Big[\frac{\log\rho}{1-\rho^s}\Big]^2 ds, \quad \forall\sigma\in(-1,0). \tag{3.9}$$

Let $s_n \stackrel{\text{def}}{=} 2in\pi\log^{-1}\rho$, $\forall n\in\mathbb{Z}$. Cauchy's theorem can be applied to (3.9), by integrating along the vertical line $\Re(s) = d > 0$, so that

$$w(z,\rho) = \sum_{n\in\mathbb{Z}} z^{-s_n}\Big[c^*(s_n+1,1)\log(\rho z) - \frac{\partial c^*(s_n+1,1)}{\partial s}\Big] + \mathcal{O}(z^{-d}). \tag{3.10}$$

The series above is equal to the sum of the residues, taken on the vertical line $\Re(s) = 0$, and is uniformly bounded, $\forall\rho\le 1$. Indeed, an integration by parts gives the inequality

$$\big|c^*(s_n+1,1)\big| = \Big|\int_0^\infty y^{s_n}c(y,1)dy\Big| \le \Big|\frac{\Gamma(s_n+1)}{\Gamma(s_n+k+1)}\Big|\int_0^\infty y^k|c^{(k)}(y,1)|dy,$$

where $c^{(k)}$ is the k-th derivative of $c(z,1)$ and $\Gamma(x)$ is the usual Eulerian function. Since

$$\Big|\frac{\Gamma(s_n+1)}{\Gamma(s_n+k+1)}\Big| < \frac{\log^k\rho}{k!},$$

the proof of (3.8) and of the lemma is concluded. □

Theorem 3.3. *For some real number $d > 1$, the following expansions hold.*

$$\rho\xi(\rho) = \exp\Big[-\frac{1}{\log\rho\, c^*(1,1)} - \frac{\partial\log c^*(1,1)}{\partial s}\Big](1+\mathcal{O}((\log\rho)^d)), \tag{3.11}$$

$$\alpha'(1,\rho) = \frac{1}{(\log\rho)^2c^*(1,1)} + \mathcal{O}((\log\rho)^d). \tag{3.12}$$

Proof. From (3.6), $\xi(\rho)$ is solution of the equation in x

$$1+\log\rho\, v(x,\rho) = 0.$$

A deep analysis, which is not included here, shows that this equation can be replaced by the locally equivalent equation

$$1+\log\rho\, w(x,\rho) + \mathcal{O}\big((\log\rho)^p\big) = 0, \tag{3.13}$$

where p is a positive number, $p > 1$. Then, from Lemma 3.2,

$$-\frac{1}{\log\rho} = \log(\rho x)c^*(1,1) - \frac{\partial c^*(1,1)}{\partial s} + \log(\rho x)\Phi(x,\rho) + \Psi(x,\rho) + \mathcal{O}(x^{-d}), \tag{3.14}$$

which implies (3.11).

For the computation of $\alpha'(1,\rho)$, one uses the simple relation

$$\rho\alpha'(1,\rho) = \frac{\alpha_1}{\alpha_0} - \alpha_1,$$

obtained by derivation of (2.2) at $z = 0$, and (3.12) follows. □

4 Remarks and complements

It is important to note that one of the main technical difficulties of the problem, besides its strongly non-linear feature, comes from the *phase transition* which appears for $\rho = 1$. Actually, $c(z, \rho) > 0$ for $0 \leq z < \xi(\rho)$ and $c(\xi(\rho), \rho) = 0$. Then, for $z \gg \xi(\rho)$, $c(z, \rho)$ has wild unbounded oscillations. In particular, this implies that $\int_0^\infty c(x, \rho)dx$ does not exist.

On the other hand, when $\rho = 1$, the $c(z, 1)$ is no more an integral function: it has a singularity (which seems to be a pole of order 3) located on the negative real axis; it does not vanish for $z \geq 0$, and the quantity $\int_0^\infty c(x, 1)dx = A$ is finite. In the half plane $\Re(z) > 0$, the following expansions hold:

$$\begin{aligned} c(z,1) &= \exp\left[-\frac{c^*(1,1)}{2}\log^2 z + B\log z + \frac{D\log z}{z^2} + o\left(\frac{\log z}{z^2}\right)\right], \\ v(z,1) &= c^*(1,1)\log z + B + \frac{D\log z}{z^2} + o\left(\frac{\log z}{z^3}\right), \end{aligned}$$

where B and D are some constants.

Finally, iterating (3.2), one could improve some of the estimates given in the previous section, rewriting $c(z, \rho)$ as

$$c(z,\rho) = c(z\rho^{I+1},\rho)\prod_{i=0}^{I}[1 + \log\rho\, v(z\rho^i,\rho)],$$

where I is an arbitrary positive integer. The above product is uniformly convergent, $\forall I \leq \infty$, for all z in a compact set of the complex plane, since it behaves like the series

$$z(\log\rho)^3 \sum_{k\geq 0}\sum_{i\geq 1} i\rho^{i+k+1} c(z\rho^{i+k+1},\rho).$$

This series has its modulus bounded by $|z\rho^2 c(\rho z, \rho)|$, so that, from the maximum modulus principle, it converges uniformly. Then, one can choose I to ensure

$$0 < z\rho^I \leq 1, \text{ i.e. } I \leq -\frac{\log z}{\log\rho},$$

and make use of the properties of the Γ function to estimate the product.

References

[1] Blasius, H. Grenzschichten in flüssigkeiten mit kleiner reibung. *Z. Math Phys. 56* (1908), 1–37. English translation in NACA TM 1256.

[2] Delcoigne, F., and Fayolle, G. Thermodynamical limit and propagation of chaos in polling systems. *Markov Processes and Related Fields 5*, 1 (1999), 89–124.

[3] Deneux, T., Fayolle, G., de La Fortelle, A., and Lasgouttes, J.-M. Allocation de bande passante dans les grands réseaux : autour de la politique “min”. Tech. rep., INRIA, June 2000. *to appear.*

[4] Doetsch, G. *Handbuch der Laplace Transformation*, vol. 1-3. Birkhaüser-Verlag. Basel, 1955.

[5] Massoulié, L., and Roberts, J. Private communication, 1998.

[6] Schlichting, H. *Boundary layer theory.* McGraw-Hill Book Co., Inc., New York, 1960. Translated by J. Kestin. 4th ed. McGraw-Hill Series in Mechanical Engineering.

[7] Titchmarsh, E. C. *The Theory of Functions*, 2nd ed. Oxford University Press, 1939.

[8] Vvedenskaya, N. D., Dobrushin, R. L., and Karpelevich, F. I. A queueing system with a choice of the shorter of two queues—an asymptotic approach. *Problems Inform. Transmission 32* (1996), 15–27.

Part V

Other Topics

Trends in Mathematics, © 2000 Birkhäuser Verlag Basel/Switzerland

A new proof of Yaglom's exponential limit law

JOCHEN GEIGER[1] *Universität Frankfurt, Fachbereich Mathematik, Postfach 11 19 32, D-60054 Frankfurt, Germany*

Abstract. *Let $(Z_n)_{n\geq 0}$ be a critical Galton-Watson branching process with finite variance. We give a new proof of a classical result by Yaglom that Z_n conditioned on $Z_n > 0$ has an exponential limit law.*

Introduction

Consider a Galton-Watson branching process $(Z_n)_{n\geq 0}$ starting with a single founding ancestor at generation 0 where each particle independently has probability p_k of producing k offspring. Let $\mu := \sum_k kp_k$ be the mean number of children per particle. The most basic and well-known fact about Galton-Watson processes is that the population gets eventually extinct if and only if $\mu \leq 1$ and $p_1 \neq 1$. In the critical case $\mu = 1$ the asymptotic behavior of the nth generation size Z_n, if conditioned on non-extinction at generation n, is described by the following limit theorem due to Kesten, Ney and Spitzer [5]. The result was originally proved by Yaglom [12] under a third moment assumption.

Theorem 1 *Suppose that $\mu = 1$ and let $0 < \sigma^2 := \sum_k k(k-1)p_k < \infty$. Then*

$$\lim_{n\to\infty} \mathrm{P}(n^{-1}Z_n \geq x \,|\, Z_n > 0) \;=\; \exp(-2x/\sigma^2), \;\; x \geq 0.$$

Classical proofs of Yaglom's exponential limit law are by means of generating functions (see e.g. [1]). The purpose of this note is to give a proof of Theorem 1 which explains the exponential limit law. Note that if X_1 and X_2 are independent copies of an exponential random variable X, and U a uniform random variable on $[0,1]$ independent of the X_i, then

$$X \stackrel{d}{=} U(X_1 + X_2). \tag{1}$$

(See this by considering the ratio of the first and the second point of a homogeneous Poisson process on $\mathbb{R}^+$.) In the context of Theorem 1 the random variables UX_1 and UX_2 will be identified as the rescaled descendancies of the children of the most recent common ancestor (MRCA) of the particles at generation n. Given non-extinction at generation n there are asymptotically no more than two children of the MRCA with a descendant at n. These children found independent Galton-Watson processes conditioned on non-extinction at n. The scaling factor U reflects the fact that the conditional distribution of the generation of the MRCA is asymptotically uniform.

The harder part of our proof is to establish convergence in the first place which we do using the so-called contraction method [7, 9, 11]. There we will make use of the fact that property (1) characterizes the exponential distribution.

For a proof of Yaglom's limit law via some other characterization of the exponential distribution see [8].

[1] Research supported by Deutsche Forschungsgemeinschaft

Identification of the limiting distribution

Consider the random family tree T produced by the Galton-Watson branching process. We think of T as a rooted planar tree with the distinguishable offspring of each vertex ordered from left to right. On the event $Z_n > 0$ decompose the nth generation size as

$$Z_n = \sum_{j=1}^{N_n} Y_{n,j}, \tag{2}$$

where $Y_{n,j}$ is the number of descendants at generation n of the jth child of the MRCA having non-empty descendancy at n, and N_n is the number of those children. To avoid separating the case where generation n consists of a single particle we define $N_n = Y_{n,1} = 1$, if $Z_n = 1$. Let H_n be the generation of the MRCA and set

$$V_n := n - H_n - 1,$$

that is $V_n + 1$ is the distance to the MRCA. It is easy to see that given $N_n = k$ and $V_n = m$, $0 \le m \le n-1$, the $Y_{n,j}$ are independent and identically distributed with

$$\mathcal{L}(Y_{n,j} \mid N_n = k, V_n = m) = \mathcal{L}(Z_m \mid Z_m > 0), \ 1 \le j \le k. \tag{3}$$

Denote by Z_n^* a random variable with distribution

$$\mathcal{L}(a_n^{-1} Z_n \mid Z_n > 0),$$

where a_n, $n \ge 0$, is a sequence of positive real numbers to be specified later. Furthermore, let (V_n^*, N_n^*) have distribution

$$\mathcal{L}((n^{-1}V_n, N_n) \mid Z_n > 0).$$

Then (2) and (3) imply the following identity in law.

$$Z_n^* \stackrel{d}{=} \frac{a_{nV_n^*}}{a_n} \sum_{j=1}^{N_n^*} Z_{nV_n^*,j}^*, \ n \ge 1, \tag{4}$$

where for any $i \ge 0$ the $Z_{i,j}^*$, $j \ge 1$, are independent copies of Z_i^*, independent of (V_n^*, N_n^*), and we have set $a_{-1}Z_{-1,1} := 1$.

A natural normalization is to choose

$$a_n = \mathrm{E}(Z_n \mid Z_n > 0), \ n \ge 0.$$

The growth of the a_n is described by Kolmogorov's asymptotic (see e.g. Theorem I.9.1 in [1]),

$$\lim_{n\to\infty} n^{-1} a_n = \lim_{n\to\infty} (n\,\mathrm{P}(Z_n > 0))^{-1} = \sigma^2/2. \tag{5}$$

Since a_n grows linearly, the first factor on the right-hand side of (4) is asymptotically equivalent to V_n^* which has a uniform weak limit (Proposition 3.2 in [3]),

$$V_n^* \stackrel{d}{\to} U \text{ as } n \to \infty. \tag{6}$$

Finally, for the asymptotic behavior of N_n^* note that taking expectations on either side of (4) yields

$$1 = \mathrm{E}\Big(\frac{a_{nV_n^*}}{a_n} N_n^*\Big). \tag{7}$$

Since $1 = 2\,\mathrm{E}U$ and $N_n^* \geq 2$ unless $nV_n^* = -1$, relation (7) shows that the MRCA asymptotically has exactly two children with a descendant at generation n,

$$N_n^* \stackrel{d}{\to} 2 \text{ as } n \to \infty. \tag{8}$$

Passing to the limit $n \to \infty$ in (4), we see from (5),(6) and (8), that if Z_n^* has a weak limit X, say, then X satisfies (1).

Proof of convergence

We now come to the question of convergence. The idea behind the contraction method is to view (1) as a fixed point equation and to use the recursive structure (4) of the law of Z_n^* to prove convergence towards this fixed point in some appropriate metric.

Recall that the L_2-Wasserstein (or Mallows) distance between two probability measures α and β on $\mathbb{R}$ with finite second moment is defined as

$$d_2(\alpha, \beta) := \inf_{X \stackrel{d}{=} \alpha,\ Y \stackrel{d}{=} \beta} \sqrt{\mathrm{E}(X-Y)^2}, \tag{9}$$

where the infimum is over all pairs of random variables X and Y, where X has law α and Y has law β. We record the following basic properties of d_2 (see Section 8 of [2] for a detailed discussion of such metrics).

$$\text{The infimum in (9) is attained.} \tag{10}$$

$$d_2(\alpha_n, \alpha) \to 0 \iff \alpha_n \to \alpha \text{ weakly and } \textstyle\int x^2 \alpha_n(dx) \to \int x^2 \alpha(dx). \tag{11}$$

We will show that the d_2-distance between $\mathcal{L}(Z_n^*)$ and the exponential distribution with mean 1 goes to 0 as $n \to \infty$ (note that Z_n^* has finite second moment due to the assumption $\sigma^2 < \infty$).

Let $\widetilde{Z}_n$ denote a random variable with distribution

$$\mathcal{L}(a_n^{-1} Z_n \,|\, Z_n > 0, N_n = 2).$$

By (2) and (3) we have

$$\widetilde{Z}_n \stackrel{d}{=} \frac{a_{n\widetilde{V}_n}}{a_n} \left(Z^*_{n\widetilde{V}_n,1} + Z^*_{n\widetilde{V}_n,2}\right),\ n \geq 1, \tag{12}$$

where the $Z_{i,j}^*$ are as in (4), independent of $\widetilde{V}_n$ having distribution

$$\mathcal{L}(n^{-1} V_n \,|\, Z_n > 0, N_n = 2).$$

Using the representation of $\mathcal{L}(\widetilde{Z}_n)$ in (12) and the fact that $\mathrm{E}Z_n^2 = 1 + \sigma^2 n$, it is easily shown that

$$\lim_{n\to\infty} \mathrm{E}\widetilde{Z}_n^2 = 2 = \lim_{n\to\infty} a_n^{-2}\,\mathrm{E}(Z_n^2 \mid Z_n > 0).$$

Consequently, by (8) and (11),

$$d_2(\mathcal{L}(\widetilde{Z}_n), \mathcal{L}(Z_n^*)) \to 0 \text{ as } n \to \infty. \tag{13}$$

We now proceed in much the same way as in the proof of Theorem 3.1 in [10]. Fix independent random variables X_1, X_2 and U, where the X_i have exponential distribution with mean 1 and U is uniform on $[0,1]$. By (10), we can choose versions $\widetilde{V}_n$ and $Z_{i,j}^*$, such that

$$\begin{aligned} \mathrm{E}(\widetilde{V}_n - U)^2 &= d_2(\mathcal{L}(\widetilde{V}_n), \mathcal{L}(U))^2; &(14)\\ \mathrm{E}(Z_{i,j}^* - X_j)^2 &= d_2(\mathcal{L}(Z_i^*), \mathcal{L}(X))^2,\ i \ge 0,\ j = 1,2. &(15)\end{aligned}$$

Write $b_n := d_2(\mathcal{L}(Z_n^*), \mathcal{L}(X))^2$. Using first (13) and then (1), (12) and the definition of d_2 in (9), we have

$$\begin{aligned} b_n &= d_2(\mathcal{L}(\widetilde{Z}_n), \mathcal{L}(X))^2 + o(1)\\ &\le \mathrm{E}\Big(\frac{a_{n\widetilde{V}_n}}{a_n}(Z^*_{n\widetilde{V}_n,1} + Z^*_{n\widetilde{V}_n,2}) - U(X_1+X_2)\Big)^2 + o(1)\\ &= 2\,\mathrm{E}\Big(\frac{a_{n\widetilde{V}_n}}{a_n} Z^*_{n\widetilde{V}_n,1} - UX_1\Big)^2 +\\ &\qquad 2\,\mathrm{E}\Big(\frac{a_{n\widetilde{V}_n}}{a_n} Z^*_{n\widetilde{V}_n,1} - UX_1\Big)\Big(\frac{a_{n\widetilde{V}_n}}{a_n} Z^*_{n\widetilde{V}_n,2} - UX_2\Big) + o(1). \end{aligned} \tag{16}$$

Since $\mathrm{E}X_j = \mathrm{E}Z^*_{n\widetilde{V}_n,j} = 1$, the second term on the right-hand side of (16) equals

$$2\,\mathrm{E}\Big(\frac{a_{n\widetilde{V}_n}}{a_n} - U\Big)^2,$$

which vanishes as $n \to \infty$ in view of (5),(6) and (14). Using first this argument again and then relation (15), we obtain

$$\begin{aligned} b_n &\le 2\,\mathrm{E}\Big(\frac{a_{n\widetilde{V}_n}}{a_n} Z^*_{n\widetilde{V}_n,1} - UX_1\Big)^2 + o(1)\\ &= 2\,\mathrm{E}\Big(\frac{a_{n\widetilde{V}_n}}{a_n}\Big(Z^*_{n\widetilde{V}_n,1} - X_1\Big) + X_1\Big(\frac{a_{n\widetilde{V}_n}}{a_n} - U\Big)\Big)^2 + o(1)\\ &= 2\,\mathrm{E}\Big(\frac{a_{n\widetilde{V}_n}}{a_n}\Big(Z^*_{n\widetilde{V}_n,1} - X_1\Big)\Big)^2 + o(1)\\ &= 2\sum_{i=0}^{n-1} \mathrm{P}(n\widetilde{V}_n = i)\Big(\frac{a_i}{a_n}\Big)^2 \mathrm{E}(Z_{i,1}^* - X_1)^2 + o(1)\\ &= 2\sum_{i=0}^{n-1} \mathrm{P}(n\widetilde{V}_n = i)\Big(\frac{a_i}{a_n}\Big)^2 b_i + o(1). \end{aligned} \tag{17}$$

Combining (17) with (5) and (6) we deduce

$$b := \limsup_{n\to\infty} b_n \le 2b \limsup_{n\to\infty} \mathrm{E}\widetilde{V}_n^2 = 2b\,\mathrm{E}U^2 = 2b/3,$$

which shows that $b_n \to 0$ as $n \to \infty$ and completes our proof of Theorem 1.

We remark that our proof shows that property (1) characterizes the exponential distribution among distributions with finite second moment. In fact, if a random variable X with positive finite mean satisfies (1), then X is exponential ([6]; see [7] for a comprehensive treatment of fixed point equations of this type). We note that the proofs of (5) and (6) do not depend on Yaglom's exponential limit law. Elementary probabilistic proofs of these and other limit theorems can be found in [4].

References

[1] Athreya, K. B. and Ney, P. (1972) *Branching Processes*, Springer, New York.

[2] Bickel, P. J. and Freedman, D. A. (1981) Some asymptotic theory for the bootstrap, *Ann. Statist.* **9**, 1196–1217.

[3] Fleischmann, K. and Siegmund-Schultze, R. (1977) The structure of reduced critical Galton-Watson processes, *Math. Nachr.* **79**, 233–241.

[4] Geiger, J. (1999) Elementary new proofs of classical limit theorems for Galton-Watson processes, *J. Appl. Prob.* **36**, 301–309.

[5] Kesten, H., Ney, P. and Spitzer, F. (1966) The Galton-Watson process with mean one and finite variance, *Theory Prob. Appl.* **11**, 513–540.

[6] Kotz, S. and Steutel, F. W. (1988) Note on a characterization of exponential distributions, *Stat. Prob. Lett.* **6**, 201–203.

[7] Liu, Q. (1998) Fixed points of a generalized smoothing transformation and applications to the branching random walk, *Adv. Appl. Prob.* **30**, 85-112.

[8] Lyons, R., Pemantle, R. and Peres, Y. (1995) Conceptual proofs of $L \log L$ criteria for mean behavior of branching processes, *Ann. Prob.* **25**, 1125–1138.

[9] Rachev, S. T. and Rüschendorf, L. (1995) Probability metrics and recursive algorithms, *Adv. Appl. Prob.* **27**, 770–799.

[10] Rösler, U. (1991) A limit theorem for "Quicksort", *Inform. Theor. Appl.* **25**, 85–100.

[11] Rösler, U. (1992) A fixed point theorem for distributions, *Stoch. Proc. Appl.* **37**, 195–214.

[12] Yaglom, A. M. (1947) Certain limit theorems of the theory of branching processes, *Dokl. Acad. Nauk. SSSR* **56**, 795–798.

Trends in Mathematics, © 2000 Birkhäuser Verlag Basel/Switzerland

The branching measure, Hausdorff and packing measures on the Galton-Watson tree

QUANSHENG LIU *IRMAR, Université de Rennes 1, Campus de Beaulieu, 35042 Rennes, France.* liu@univ-rennes1.fr

Abstract. *We present some recent results concerning the branching measure, the exact Hausdorff measure and the exact packing measure, defined on the boundary of the Galton-Watson tree. The results show that in good cases, these three measures coincide each other up to a constant, that the branching measure is homogeneous (it has the same local dimension at each point) if and only if a certain simple condition is satisfied, and that it is singular with respect to the equally splitting measure. Similar results on marked trees are also presented, and are applied to the study of flows in networks and to the search of exact gauges for statistically self-similar fractals in* $\mathbb{R}^n$.

1 Introduction and notations

We study natural measures defined on the boundary of the genealogical tree of a supercritical Galton-Watson process.

First, under simple conditions, we calculate the exact Hausdorff measure, show that it coincides with the branching measure up to a multiplicative constant, and obtain a similar result for the packing measure.

Next, we investigate carefully the local dimensions of the branching measure, and we find its various critical exponents: its maximum and minimum local dimensions, its exact local dimension at typical points, etc. In particular we obtain a criterion which enables us to know when exactly the branching measure is homogeneous: it has the same local dimension at any point.

Then, we compare the branching measure with the equally splitting measure; in particular, we find that they are mutually singular.

The study of the three measures enables us to obtain new geometrical descriptions of the exponents of the tail probability of the limit variable of the natural martingale, considered by T.E. Harris [6] from the very beginning of the development of the theory of branching processes.

Finally, we give an extension of certain results to the marked trees, which enables us to determine the exact gauges (in the sense of Hausdorff or of packing) for a large class of statistically self-similar random fractals in $\mathbb{R}^n$.

Let us introduce some notations that we shall use in all the following.

Set $\mathbb{N}^* = \{1, 2...\}, \mathbb{N} = \{0\} \cup \mathbb{N}^*$ and let $\mathbf{U} = \{\emptyset\} \cup \bigcup_{n=1}^{\infty} (\mathbb{N}*)^n$ be the set of all finite sequences $u = u_1 \cdots u_n = (u_1, ..., u_n)$, containing the null sequence $\emptyset$. If $u = u_1...u_k = (u_1, ..., u_n)$ $(u_i \in \mathbb{N})$ is a finite sequence, we write $|u| = n$ and $u|k = u_1...u_k, k \leq n$; by convention $|\emptyset| = 0$ and $u|0 = \emptyset$. If $v = v_1...v_l$ is another sequence, we write uv for the juxtaposition $u_1...u_n v_1...v_l$; by convention $u\emptyset = \emptyset u = u$ for all u. We write $u < u'$ or $u' > u$ if $u' = uv$ for some sequence v. These notations are extended in an obvious way to infinite sequences.

Let $(\Omega, \mathcal{F}, P)$ be a probability space, and $\{N_u\}_{u \in \mathbb{U}}$ be a family of independent random variables, defined on $(\Omega, \mathcal{F}, P)$, each distributed according to the law $\{p_n\}$ on $\mathbb{N}$. Let $\mathbf{T} = \mathbf{T}(\omega)$ be the Galton-Watson tree with defining elements $\{N_u\}$ [29]: we have $\emptyset \in \mathbf{T}$ and, if $u \in \mathbf{T}$ and $i \in \mathbb{N}^*$, then the juxtaposition $ui \in \mathbf{T}$ if and only if $1 \leq i \leq N_u$. Let $\partial\mathbf{T} = \{u_1 u_2 ... : \forall n \geq 0, u_1...u_n \in \mathbf{T}\}$ be the boundary of $\mathbf{T}$ equipped with the ultra-metric

$$d(u,v) = e^{-|u \wedge v|},$$

where $u \wedge v$ is the common sequence of u and v: i.e. $u \wedge v = u|n = v|n$ with $n = \max\{k \in \mathbb{N} : u|k = v|k\}$; by convention $d(u,v) = 0$ if $u = v$.

We are interested to the super-critical case. For simplicity, we write $N = N_\emptyset$, $m = EN$, and we suppose that $p_0 = 0$, that N is not almost surely (a.s.) constant and that $EN \log N < \infty$. Therefore the limit

$$Z = \lim_{n \to \infty} \text{ card } \{u \in \mathbf{T} : |u| = n\}/m^n$$

exists a.s. with $EZ = 1$ and $P(Z > 0) = 1$. Let μ $(= \mu_\omega)$ be the unique Borel measure on $\partial\mathbf{T}$ such that, for all balls $B_u = \{v \in \partial\mathbf{T} : u < v\}$, $u \in \mathbf{T}$,

$$\mu(B_u) = m^{-|u|} Z_u, \text{ where } Z_u = \lim_{n \to \infty} \frac{\text{card}\{v \in \mathbf{T} : u < v, |v| = n\}}{m^{n-|u|}};$$

in other words,

$$\mu(B_u) = Z \lim_{n \to \infty} \frac{\text{card}\{v \in \mathbf{T} : u < v, |v| = n\}}{\text{card}\{v \in \mathbf{T} : |v| = n\}}.$$

It will be useful to notice that B_u is a ball of diameter $|B_u| = e^{-|u|}$. The measure μ is of mass Z; as in [21] and [22], we call it the **branching measure** on $\partial\mathbf{T}$ (or simply on $\mathbf{T}$) because it describes the asymptotic proportion of the number of descendants of each individual $u \in \mathbf{T}$ in the total population size of n-th generation (as $n \to \infty$), and it plays an essential role in the study of geometric measures on the Galton-Watson tree $\mathbf{T}$. This measure is studied by many authors, see for example [11, 30, 9, 27, 24].

2 The exact Hausdorff measure

If $A \subset \partial\mathbf{T}$ and if $g : \mathbb{R}_+ \to \mathbb{R}_+$ is a gauge function (non-decreasing and right continuous with $g(0) = 0$), we write

$$g\text{-}H(A) = \lim_{\delta \to 0} \inf\{\sum_i g(|U_i|) : A \subset \bigcup_i U_i, |U_i| \leq \delta\}$$

for the Hausdorff measure of A under the gauge g. As $(\partial\mathbf{T}, d)$ is an ultra-metric space, the value of $g\text{-}H(A)$ is invariant if we replace the U_i by balls $B_u, u \in \mathbf{T}$ [15, Lemma 0].

Hawkes [9] showed that if N has a geometric law on $\mathbb{N}^*$, then ϕ-$H\ (\partial\mathbf{T}) = Z$ a.s., where $\phi(t) = t^\alpha \log\log\frac{1}{t}$ and $\alpha = \log m$. In solving a conjecture of Hawkes, we have extended in [15] his result to the general case. To introduce our main result, let us define

$$\overline{m} := \ \operatorname{ess\,sup} N, \ \beta_+ = 1 - \log m/\log\overline{m}, \tag{1.1}$$

$$r_+ = \sup\left\{t \geq 0 : \quad E\exp(tZ^{1/\beta_+}) < \infty\right\} = \liminf_{x\to\infty} -x^{-1/\beta_+}\log P\{Z > x\} \tag{1.2}$$

(cf. [16] for the equality (1.2)). Then $1 < m < \overline{m} \leq \infty$, $0 < \beta_+ < 1$ if $\overline{m} < \infty$ and $\beta_+ = 1$ if $\overline{m} = \infty$, and $0 \leq r_+ \leq \infty$. The following theorem gives the exact gauge of $\partial\mathbf{T}$ in the sense of Hausdorff.

Theorem 1.1 [15]. *Let β_+ and r_+ be defined in (1.1) and (1.2), and write $\phi_+(t) = t^\alpha(\log\log\frac{1}{t})^{\beta_+}$. Then*

$$\phi_+\text{-}H\ (\partial\mathbf{T}) = (r_+)^{\beta_+}Z \qquad \text{a.s.}$$

Moreover, $0 < r_+ < \infty$ if either of the following two conditions is satisfied: (a) $\overline{m} < \infty$; (b) $E\exp(tN) < \infty$ for some but not all $t > 0$.

(By convention $0^a = 0, \infty^a = \infty$ and $\infty a = \infty$ if $0 < a < \infty$.) The result shows that when the right tail probability $P(Z > x)$ decreases exponentially, there is a gauge ϕ_+ such that $0 < \phi_+$ $H(\partial\mathbf{T}) < \infty$. When $EN^p = \infty$ for some $p > 1$, it is evident that $r_+ = 0$, so that ϕ_+-$H(\partial\mathbf{T}) = 0$, which means that the function ϕ_+ is too small to be a good gauge. In fact, in this case there exists an exponent $0 < \gamma < \infty$ such that, for $\psi_b(t) = t^\alpha(\log\frac{1}{t})^b$, we have ψ_b-$H(\partial\mathbf{T}) = 0$ if $b < \gamma$ and ϕ_b-$H(\partial\mathbf{T}) = \infty$ if $b > \gamma$ [15].

More precisely, we can prove that the exact Hausdorff measure coincides with the branching measure up to a multiplicative constant:

Theorem 1.2 [20]. *Let ϕ_+ be the function defined in Theorem 1.1. If either (a) or (b) of Theorem 1.1 is satisfied, then a.s. for all Borel set $A \subset \partial\mathbf{T}$,*

$$\phi_+\text{-}H(A) = r_+^{\beta_+}\mu(A).$$

3 The exact packing measure

In replying to a question of S.J.Taylor, we have shown in [20] a result similar to Theorem 1.1 for the packing measure (cf. [33]).

If $A \subset \partial\mathbf{T}$ and if g is a gauge, the *g-packing pre-measure* of A is by definition

$$g\text{-}P\,(A) = \lim_{\delta\to 0+}\ \sup\{\sum_u g(|B_u|) : B_u \text{ are disjoint, } B_u \cap A \neq \emptyset \text{ and } |B_u| \leq \delta\ \}.$$

In general, g-$P(.)$ is finite-additive but not σ-additive. We define by

$$g\text{-}P^*(A) = \inf\{\sum_u g\text{-}P(B_u) : \ A \subset \cup_u B_u\}$$

the *spherical g-packing measure* of A. $g\text{-}P^*(.)$ is an outer measure. Its restriction to the class of Borel sets is a measure; when $g\text{-}P(\partial\mathbf{T}) < \infty$, this measure coincides with the g-premeasure on the class of finite unions of balls B_u [20].

When $p_1 = 0$, we write

$$\underline{m} := \operatorname{ess\,inf} N, \quad \beta_- = 1 - \log m / \log \underline{m}, \tag{2.1}$$

$$r_- := \sup\left\{t \geq 0 : \quad E\exp(tZ^{1/\beta_-}) < \infty\right\} = \liminf_{x\to 0} -x^{-1/\beta_-} \log P\{Z < x\}; \tag{2.2}$$

thus $1 < \underline{m} < m < \infty$, $-\infty < \beta_- < 0$ and $0 < r_- < \infty$.

The following result is similar to Theorem 1.1:

Theorem 2.1 [20]. *Suppose that* $p_1 = 0$. *Let* $\beta_- \in (-\infty, 0)$ *and* $r_- \in (0, \infty)$ *be defined in (2.1) and (2.2), and set* $\phi_-(t) = t^\alpha(\log\log\frac{1}{t})^{\beta_-}$. *Then*

$$\phi_-\text{-}P^*(\partial\mathbf{T}) = (r_-)^{\beta_-} Z \quad \text{a.s.}$$

The result shows that when $p_1 = 0$, i.e. when the left tail probability $P(Z \leq x)$ decreases exponentially as $x \to 0$, then there exists a gauge ϕ_- such that $0 < \phi_-\text{-}P^*(\partial\mathbf{T}) < \infty$ a.s. The case where $p_1 > 0$ is also discussed in [20], but the situation is not yet quite clear.

Just as in the case of Hausdorff measure, the exact packing measure also coincides with the branching measure up to a constant:

Theorem 2.2 [20]. *Under the conditions of Theorem 2.1, a.s. for all Borel sets* $A \subset \partial\mathbf{T}$,

$$\phi_-\text{-}P^*(A) = (r_-)^{\beta_-}\mu(A).$$

4 Local dimensions of the branching measure

Let $\underline{d}(\mu, u)$ and $\overline{d}(\mu, u)$ be the lower and upper local dimensions of μ at $u \in \partial\mathbf{T}$:

$$\underline{d}(\mu, u) = \liminf_{n\to\infty} \frac{-\log\mu(B_{u|n})}{n}, \quad \overline{d}(\mu, u) = \limsup_{n\to\infty} \frac{-\log\mu(B_{u|n})}{n}.$$

It is well-known that (cf. [9] and [27]) a.s.

$$\underline{d}(\mu, u) = \overline{d}(\mu, u) = \alpha \text{ for } \mu\text{-almost all } u \in \partial\mathbf{T}. \tag{3.1}$$

A natural question is to know when (3.1) holds for *all* $u \in \partial\mathbf{T}$. This question is solved in [22]; the answer will be given below. In the proof, more precise informations are given about the uniform bounds of $\underline{d}(\mu, u)$ and $\overline{d}(\mu, u)$. To find these bounds, we are led to a study of asymptotic properties of

$$m_n = \min_{u\in\mathbf{T}:|u|=n} \mu(B_u) = \min_{u\in\partial\mathbf{T}} \mu(B_{u|n}) \text{ and } M_n = \max_{u\in\mathbf{T}:|u|=n} \mu(B_u) = \max_{u\in\partial\mathbf{T}} \mu(B_{u|n}),$$

which is interesting by its own. To prove the asymptotic properties, we first establish an interesting result concerning the convergence of iterations of a probability generating function.

Our results show that the branching measure behaves like the occupation measure of a stable subordinator or a Brownian motion if and only if the law of reproduction satisfies a certain condition.

Another interesting question is to calculate the exact local dimension of μ for μ-almost all $u \in \partial\mathbf{T}$ (the typical u). This question is also solved in [22].

Let us quote some results of [22]. We shall use the convention that $1/\infty = 0$, $1/0 = \infty$ and $\log \infty = \infty$, and the following notations:

$$p_- = \frac{\log(1/p_1)}{\log m} \quad \text{and} \quad p_+ = \sup\{a \geq 1 : EN^a < \infty\}.$$

Then it is evident that $p_- = \infty$ if and only if $p_1 = 0$, and that $p_+ = \infty$ if and only if $EN^a < \infty$ for all $a > 1$. By classic results on the limit variable Z (cf. e.g. [1]), it is easy to verify that $p_- = \lim_{x\to 0} \frac{\log P(Z\leq x)}{\log x}$ and $p_+ = \liminf_{x\to\infty} \frac{-\log P(Z>x)}{\log x}$.

Theorem 3.1 [22]. *A.s.* $\lim_{n\to\infty} \frac{-\log m_n}{n} = (1+\frac{1}{p_-})\alpha$ *and* $\liminf_{n\to\infty} \frac{-\log M_n}{n} = (1-\frac{1}{p_+})\alpha$. *If additionally* $p_+ = \lim_{x\to\infty} \frac{-\log P(Z>x)}{\log x}$, *then* a.s. $\lim_{n\to\infty} \frac{-\log M_n}{n} = (1-\frac{1}{p_+})\alpha$.

The proof of the theorem uses the "first moment method" and the following result about the iteration of a probability generating function.

Theorem 3.2 [22]. *Let* $f(t) = \sum_{n=0}^{\infty} p_n t^n$ *be a probability generating function, and write* $f_1(t) = f(t)$ *and* $f_{n+1}(t) = f(f_n(t))$ *for* $n \geq 1$. *Let* ρ *and* c *be two numbers in* $(0,1]$. *Under the only condition* $p_0 = 0$ *and* $m := \sum_{n=0}^{\infty} np_n < \infty$, *we have the following assertions:*

(i) if $\rho > 1/m$, *then there exist two constants* $0 < \lambda < 1$ *and* $0 < K < \infty$ *such that, for all* n *large enough,* $f_n(1-c\rho^n) \leq K\lambda^n$;

(ii) if $\rho = 1/m$, *then* $\liminf_n f_n(1-c\rho^n) \geq e^{-c}$;

(iii) if $\rho < 1/m$, *then* $\lim_n f_n(1-c\rho^n) = 1$.

If $\rho < 1$, the assertions are also valid for $c > 1$, so that for all $0 < c < \infty$. If $\rho = 1$, we need the hypothesis $c \leq 1$ to ensure that $1 - c\rho^n \geq 0$.

Part (a)(i) of the following theorem gives a necessary and sufficient condition under which a.s. there is no exceptional point in (3.1).

Theorem 3.3 [22]. *Suppose that* $EN^{1+\epsilon} < \infty$ *for some* $\epsilon > 0$ *and that* $p_+ = \lim_{x\to\infty} \frac{-\log P(Z>x)}{\log x}$.

(a) We have the following assertions:

(i) a.s. $\underline{d}(\mu, u) = \overline{d}(\mu, u) = \alpha$ *for all* $u \in \partial\mathbf{T}$ *if and only if* $p_+ = p_- = \infty$;

(ii) a.s. $\underline{d}(\mu, u) = \alpha$ *for all* $u \in \partial\mathbf{T}$ *if and only if* $p_+ = \infty$.

(b) More precisely, we have:

(i) if $p_+ = p_- = \infty$, *then* a.s. $\underline{d}(\mu, u) = \overline{d}(\mu, u) = \alpha$ *for all* $u \in \partial\mathbf{T}$;

(ii) if $p_+ = \infty$ *and* $p_- < \infty$, *then* a.s. $\underline{d}(\mu, u) = \alpha$ *for all* $u \in \partial\mathbf{T}$, *and* $\overline{d}(\mu, u) > \alpha$ *for some* $u \in \partial\mathbf{T}$;

(iii) if $p_+ < \infty$, *then* a.s. $\underline{d}(\mu, u) < \alpha$ *for some* $u \in \partial\mathbf{T}$.

(c) Moreover, a.s. $\sup_{u\in\partial\mathbf{T}} \underline{d}(\mu, u) = \alpha$ *and* $\inf_{u\in\partial\mathbf{T}} \underline{d}(\mu, u) = (1 - 1/p_+)\alpha$.

Part (b)(i), the conclusions for $\underline{d}(\mu, u)$ in Parts (b)(ii) and (c), and therefore the "if" parts of (a)(i) and (a)(ii), all hold without the conditions of the theorem.

In the proof, in addition to the use of Theorem 3.1, to prove that a certain set is not empty, we construct a non-homogeneous Galton-Watson process by choosing "good" generations and "good" individuals of the initial process, and we prove that the new process is not extinct with positive probability.

Part (b)(ii) shows that, when $p_+ = \infty$ and $p_- < \infty$, the branching measure and the occupation measure of a stable process [10] have the same property that a.s. the lower local dimension is constant but the upper local dimension is not so. Parts (b)(i) and (b)(iii) show that in the other cases, a new phenomenon occurs for the branching measure compared with the stable occupation measure.

The following result shows that Theorem 3.1 can be improved in the case where Z has an exponential right tail probability.

Theorem 3.4 [22]. *(a) Suppose that* $p_1 = 0$. *Let* β_- *and* r_- *be defined in (2.1) and (2.2), and let* $C_- := (\alpha/r_-)^{\beta_-}$ $(\in (0,\infty))$. *Then* $\liminf_{n\to\infty} \frac{m^n m_n}{n^{\beta_-}} = C_-$ *a.s.; if additionally* $r_- = \lim_{x\to\infty} \frac{-\log P\{Z<x\}}{x^{1/\beta_-}}, =$

then $\lim_{n\to\infty} \frac{m^n m_n}{n^{\beta_-}} = C_-$ *a.s.*

(b) Let β_+ *and* r_+ *be defined in (1.1) and (1.2), and let* $C_+ := (\alpha/r_+)^{\beta_+} \in [0,\infty]$. *Then* $\limsup_{n\to\infty} \frac{m^n M_n}{n^{\beta_+}} = C_+$ *a.s.; if additionally* $r_+ = \lim_{x\to\infty} \frac{-\log P\{Z>x\}}{x^{1/\beta_+}}$, *then* $\lim_{n\to\infty} \frac{m^n M_n}{n^{\beta_+}} = C_+$ *a.s.*

In Part (b), the first conclusion was first established in [25]; in the case where N has a geometric law (so that $C_+ = 1$), the second conclusion was proved by Hawkes [9, Theorem 3]. Notice that the equalities about the limits can be written in the form $\lim_{n\to\infty} \min_{u\in\partial\mathbf{T}} \frac{\mu(B_{u|n})}{\phi_+(|B_{u|n}|)} = C_-$ and $\lim_{n\to\infty} \max_{u\in\partial\mathbf{T}} \frac{\mu(B_{u|n})}{\phi_+(|B_{u|n}|)} = C_+$, where $\phi_+(t) = t^\alpha(\log\frac{1}{t})^{\beta_+}$; in this form the results are similar to asymptotic laws for a stable subordinator or for a Brownian motion: cf. [9, Theorem 1], [10, (3.1)], [13, Théorème 52,2, p.172], [8, Theorem 2] and [31, Lemma 2.3 and Corollary 5.2].

The following theorem gives the exact uniform bounds of local dimensions of μ.

Theorem 3.5 [22]. *(a) If* $r_+ = \lim_{x\to\infty} \frac{-\log P\{Z>x\}}{x^{1/\beta_+}}$, *then*

$$\sup_{u\in\partial\mathbf{T}} \limsup_{n\to\infty} \frac{m^n \mu(B_{u|n})}{n^{\beta_+}} = C_+ \quad \text{a.s.}$$

(b) Suppose that $p_1 = 0$, *that* $EN^p < \infty$ *for all* $p > 1$, *and that*

$r_- = \lim_{x\to\infty} \frac{-\log P\{Z<x\}}{x^{1/\beta_-}}$, *then*

$$\inf_{u\in\partial\mathbf{T}} \liminf_{n\to\infty} \frac{m^n \mu(B_{u|n})}{n^{\beta_-}} = C_- \quad \text{a.s.}$$

Our last result of this section concerns the exact local dimension of μ at typical points; it gives a precise estimation of large values of $\mu(B_{u|n})$ for μ-almost all u:

Theorem 3.6 [22]. *For P-almost all $\omega \in \Omega$ and μ_ω-almost all $u \in \partial\mathbf{T}$,*

$$\limsup_{n\to\infty} \frac{m^n \mu(B_{u|n})}{(\log n)^{\beta_+}} = \frac{1}{r_+^{\beta_+}}.$$

When N has a geometric law, the result was proved by Hawkes [9].

In view of the above results, it is then natural to calculate the Hausdorff dimensions of some sets of exceptional points; this has been done very recently in [32] and [26].

5 The branching measure and the equally splitting measure

There is another natural measure on $\partial\mathbf{T}$, called the **equally splitting measure**. It is the unique Borel measure ν $(= \nu_\omega)$ on $\partial\mathbf{T}$ such that $\nu(\partial\mathbf{T}) = 1$ and, for all $u \in \mathbf{T}$ with $|u| \geq 1$,

$$\nu(B_u) = \prod_{k=0}^{|u|-1} \frac{1}{N_{u|k}}.$$

In other words, ν is the probability measure on $\partial\mathbf{T}$ whose mass at each individual $u \in \mathbf{T}$ is equally split to its descendants ui in the sense that for all $1 \leq i \leq N_u$, $\nu(B_{ui}) = \nu(B_u)/N_u$.

O'Brien [30] proved that the branching measure μ has no atom a.s.; A. Joffe [11] asked under which conditions the measure μ could be absolutely continuous with respect to ν. In replying to the question of Joffe, we proved with A. Rouault the following result:

Theorem 4.1 [24]. *A.s. the measures μ and ν have no atom, and are mutually singular.*

Under classic moment conditions, we can compare precisely the two measures by calculating their Hölder exponents at typical points:

Theorem 4.2 [24]. Suppose that $EN\log N < \infty$. Then:

(a) For P-almost all $\omega \in \Omega$ and μ_ω- almost all $u \in \partial\mathbf{T}$, as $n \to \infty$,

$$\frac{\log \mu_\omega(B_{u|n})}{n} \to -\log m \text{ and } \frac{\log \nu_\omega(B_{u|n})}{n} \to -\frac{EN\log N}{m}.$$

(b) For P-almost all $\omega \in \Omega$ and ν_ω-almost all $u \in \partial\mathbf{T}$,

$$\frac{\log \mu_\omega(B_{u|n})}{n} \to -\log m \text{ and } \frac{\log \nu_\omega(B_{u|n})}{n} \to -E\log N.$$

Of course, the first formula in (a) is nothing but (3.1).

In fact, in [24], Theorems 4.1 and 4.2 are extended to marked trees; such an extension enables us to generalize a result of Kahane and Peyrière [12] concerning the dimension of the Mandelbrot measure for multiplicative cascades [28].

6 Marked trees

Just as in the preceding section, the results of sections 1 and 2 can also be extended to marked trees: this time each vertex $u \in \mathbf{T}$ is marked with a sequence of positive weights $A_{ui}, 1 \leq i \leq N_u$, where $(N_u, A_{u1}, A_{u2}, ...)_{u\in\mathbb{U}}$ is a family of independent random variables with values in $N \times \mathbb{R}_+ \times \mathbb{R}_+ \times ...$, each distributed as $(N, A_1, A_2, ...)$. We can also think that each edge linking the vertices ui and u is marked by the weight $A_{ui}, 1 \leq i \leq N_u$; or, equivalently, each vertex $u = u_1...u_n$ is equipped with the weight

$$X_{u_1\ldots u_n} = A_{u_1}A_{u_1u_2}\ldots A_{u_1u_2\ldots u_n} \quad (n \geq 1). \tag{5.1}$$

Suppose that for some $\alpha \in (0, \infty)$,

$$E\sum_{i=1}^{N} A_i^\alpha = 1, E(\sum_{i=1}^{N} A_i^\alpha)\log^+(\sum_{i=1}^{N} A_i^\alpha) < \infty \text{ and } -\infty < E\sum_{i=1}^{N} A_i \log A_i < 0, \tag{5.2}$$

then $Y_n = \sum_{u\in\mathbf{T},|u|=n} X_u^\alpha$ $(n \geq 1)$ is a positive martingale which converges a.s. to a random variable, also denoted by Z, satisfying $EZ = 1$. Similarly, for all $u \in \mathbb{U}$, the limit

$$Z_u = \lim_{n\to\infty} \sum_{v=v_1\ldots v_n\in\mathbf{T}_u} A_{uv_1}^\alpha \ldots A_{uv_1\ldots v_n}^\alpha$$

exists a.s. It is easy to verify that a.s. for all $u \in \mathbb{U}$, $X_uZ_u = \sum_{i=1}^{N_u} X_{ui}Z_{ui}$. Therefore a.s. there exists a unique Borel measure, also denoted by $\mu = \mu_\omega$, such that

$$\mu(B_u) = X_uZ_u \text{ for all } u \in \mathbf{T}, \tag{5.3}$$

or, equivalently,

$$\mu(B_u) = Z \lim_{n\to\infty} \frac{\sum_{|v|=n,v\in\mathbf{T},u<v} X_v}{\sum_{|v|=n,v\in\mathbf{T}} X_v}.$$

In [14], [19] and [23], we have used this measure to study flows in networks and exact gauges of random fractals in $\mathbb{R}^n$. As in [21], we call μ the **generalized Mandelbrot measure** because, when $N = r$ is a constant ≥ 2 and when the A_i are i.i.d., it coincides with the Mandelbrot measure for multiplicative cascades on r-ary trees [28, 12]. If $EN < \infty$ and $A_i = 1/EN$, it reduces to the branching measure; so we can also call it the **generalized branching measure**.

We can equip $\partial\mathbf{T}$ with the ultra-metric $d_c(u,v) = c^{-|u\wedge v|}$, with $c > 1$ given, and study the Hausdorff and packing measures of the support of μ: this is what we do in [12], [24] and [21].

We can also equip $\partial\mathbf{T}$ with a distance associated with the weights $X = (X_u)_u$, and study the Hausdorff and packing measures of $\partial\mathbf{T}$: this is what we are going to explain. This study is interesting due to its applications to the study of flows in networks and to that of random fractals in $\mathbb{R}^n$, as will be seen below. For simplicity, we suppose that

$$P(N \geq 1, 0 < A_i \leq 1 \ \forall i \in \{1, ..., N\}) = 1 \ \text{ and } P(A_1 = ... = A_N = 1) < 1. \quad (5.4)$$

Thus $Z > 0$ a.s. and

$$d_X(u,v) = X_{u\wedge v} \quad (5.5)$$

is an ultra-metric on $\partial\mathbf{T}$ (by convention $d_X(u,v) = 0$ if $u = v$). In the following statements, we use this distance, and we write

$$r_b = \sup\{r \geq 0 : E\exp(rZ^b) < \infty\}, \quad -\infty < b < \infty, \quad (5.6)$$

which is equivalent to $r_b = \liminf_{x\to\infty} \frac{-\log P(Z>x)}{x^b}$ if $b \geq 0$, and $r_b = \liminf_{x\to 0} \frac{-\log P(Z<x)}{x^b}$ if $b < 0$. Of course $0 \leq r_b \leq \infty$. Set

$$\phi_b(t) = t^\alpha (\log\log \frac{1}{t})^b, \quad -\infty < b < \infty. \quad (5.7)$$

As in sections 1 and 2, we denote by ϕ_b-H the Hausdorff measure and by ϕ_b-P^* the spherical packing measure, with the gauge ϕ_b, and we admit the' convention that $\infty a = \infty$ if $a > 0$.

Theorem 5.1 [23]. *Assume (5.2) and (5.4). Then for all $b > 0$, a.s.*

$$\phi_b\text{-}H(\partial\mathbf{T}) = r_{1/b}^b Z. \quad (5.8)$$

If additionally $r_{1/b} < \infty$, then a.s. for all Borel sets $A \subset \partial\mathbf{T}$,

$$\phi_b\text{-}H(A) = r_{1/b}^b \ \mu(A). \quad (5.9)$$

According to the theorem, ϕ_b-$H(\partial\mathbf{T})$ and $r_{1/b}$ are simultaneously zero, positive and finite, or infinite. It is therefore important to find the critical value of $b > 0$ such that $0 < r_{1/b} < \infty$; the following result gives this value under some conditions.

Theorem 5.2 [23]. *Suppose that $\|N\|_\infty < \infty$ and that $\|\sum_{i=1}^N T_i^x\|_\infty \leq 1$ for some $x > 0$. Let $\gamma > \alpha$ be the least solution of the equation $\|\sum_{i=1}^N T_i^\gamma\|_\infty = 1$, and set $\beta = 1 - \alpha/\gamma$. Then*

$$\phi_b\text{-}H(\partial\mathbf{T}) = \begin{cases} 0 & \text{if } 0 < b < \beta, \\ r_{1/\beta}^\beta Z \ \text{ with } r_{1/\beta} > 0 & \text{if } b = \beta. \end{cases} \quad (5.10)$$

We have also $r_{1/\beta} < \infty$ *if additionally there exist some constants* $0 < \delta < 1$, $a \geq 0$ *and* $c > 0$ *such that, for all* $x > 0$ *small enough,*

$$P\{\sum_{i=1}^{N} T_i^{\gamma} > 1 - x \text{ and } T_i \leq \delta \text{ for all } 1 \leq i \leq N\} \geq cx^a. \tag{5.11}$$

There is an interesting interpretation of the above results in terms of flows in the network composed with the tree $\mathbf{T}$ and a capacity $C_u > 0$ assigned to each vertex $u \in \mathbf{T}$. A positive flow in the network is a function $f : \mathbf{T} \to \mathbb{R}_+$ such that $f(\emptyset) > 0$ and that, for all $u \in \mathbf{T}$, $f(u) = \sum_{i=1}^{N_u} f(ui)$ and $f(u) \leq C_u$. If g is a gauge, then g-$H(\partial\mathbf{T}) > 0$ if and only if a positive flow is possible in the tree $\mathbf{T}$ with capacities $C_u = g(X_u), u \in \mathbf{T}$. (Using the famous "max-flow min-cut" theorem [4], Falconer [2] proved that the positivity of the liminf of "cut-set sums" of $g(X_u)$ is equivalent to the existence of a positive flow in the tree $\mathbf{T}$ with capacities $C_u = g(X_u), u \in \mathbf{T}$; we have observed in [14] that the liminf above is just g-$H(\partial\mathbf{T})$. A simple and direct proof of the criterion is given in [22].) According to Theorem 5.2, under simple conditions, the gauge ϕ_β is the minimal function ϕ for which a positive flow is possible in the tree $\mathbf{T}$ with capacities $C_u = \phi(X_u), u \in \mathbf{T}$:

in other words, a flow is possible in the tree $\mathbf{T}$ with capacities $C_u = \phi_\beta(X_u), u \in \mathbf{T}$, and is impossible if the capacities are $C_u = \phi(X_u)$ with ϕ less then ϕ_β in the sense that $\lim_{t\to 0} \phi(t)/\phi_\beta(t) = 0$.

The following result for the packing measure is similar to that for the Hausdorff measure.

Theorem 5.3 [19]. *Suppose that the conditions (5.2) and (5.4) are satisfied, that* $N \geq 2$ *a.s. and that for some* $b < 0$, $0 < r_{1/b} < \infty$ *(b is necessarily unique). Then a.s.*

$$\phi_b - P^*(\partial\mathbf{T}) = r_{1/b}^b Z \quad \text{and} \quad \phi_b - P^*(A) = r_{1/b}^b\, \mu(A) \tag{5.12}$$

for all Borel set $A \subset \partial\mathbf{T}$,

In some cases (for example, the case where the sequence $\{A_i\}$ is i.i.d. and independent of N, and the case where $A_1 = A_2 = \ldots$ and independent of N) and under some conditions, we can calculate explicitly the critical value of $b < 0$ such that $0 < r_{1/b} < \infty$ [17, 18].

The results can be easily applied to the determination of exact gauges for some statistically self-similar random fractals in $\mathbb{R}^n$, and thus enable us improve and generalize the results of Graf, Mauldin and Williams [5] and of Falconer [2, 3].. In fact, a Cantor-like fractal, constructed from an initial compact $J_\emptyset \subset \mathbb{R}^n$ and by a self-similar procedure, is naturally associated with a tree; thus a compact J_u of generation n gives birth to N_u new compacts $J_{ui} \subset J_u$ $(1 \leq i \leq N_u)$ of generation $n+1$. We are interested to the determination of exact gauges for the limit set $K = \cap_{n\geq 1} \cup_{u\in\mathbf{T},|u|=n} J_u$ in $\mathbb{R}^n$. Each vertex u is equipped with the weight $X_u = |J_u|$ = diameter of J_u. Writing $A_{ui} = |J_{ui}|/|J_u|$, we can apply the preceding theorems to find exact gauges of $\partial\mathbf{T}$ equipped with the metric $d(u,v) = X_{u\wedge v} = |J_{u\wedge v}|$. Notice that the exact gauges of $\partial\mathbf{T}$ depend only on the algebraic aspect (the diameters of generators) of the fractal K, which are therefore easier to calculate than those of $K \subset \mathbb{R}^n$. However, under simple geometric conditions (which are often satisfied), we can prove that the exact gauges of the boundary $\partial\mathbf{T}$ of the tree $\mathbf{T}$ are equal to those of the fractal K in $\mathbb{R}^n$. (cf. [14, 19, 23].)

Acknowledgment. The author thanks a referee for valuable comments and remarks.

References

[1] Bingham, N.H.(1988): On the limit of a supercritical branching process. J. Appl. Prob. 25A (1988) 215-228.

[2] Falconer, K.J. (1986) : Random fractals. Math.Proc.Camb.Phil.Soc., 100, 559-582.

[3] Falconer, K.J. (1987) : Cut set sums and tree processes. Proc. Amer. Math. Soc. (2), 101, 337-346.

[4] Ford, L.R. and D.R. Fulkerson (1962): Flows in networks, Princeton Univ. Press

[5] Graf, S. Mauldin, R.D. and Williams, S.C. (1988) : The exact Hausdorff dimension in random recursive constructions. Mem. Amer. Math. Soc. 71, 381.

[6] Harris, T.E. (1948) : Branching processes, Ann. Math. Stat. 19, 474-494.

[7] Harris, T.E. (1963) : Branching processes, Springer - Verlag.

[8] Hawkes, J. (1971): A lower Lipschitz condition for the stable subordinator. Z. Wahr. verw. Geb. 17, 23-32.

[9] Hawkes, J. (1981): Trees generated by a simple branching process. J. London Math. Soc. **24** 373-384.

[10] Hu, X. and Taylor, S.J. (1997): The multifractal structure of stable occupation measure. Stoch. Proc. and their Appl. 66, 283-299.

[11] Joffe, A. (1978): Remarks on the structure of trees with applications to supercritical Galton-Watson processes. *In Advances in Prob.* **5**, ed. A.Joffe and P.Ney, Dekker, New- York p.263 – 268.

[12] Kahane, J.P. and Peyrière (1976) : Sur certaines martingales de Benoit Mandelbrot . Adv. Math., 22, 131-145.

[13] Lévy, P. (1954): Théorie de l'addition des variables aléatoires. 2ème éd., Gautier-Villars, Paris.

[14] Liu, Q. (1993): Sur quelques problèmes à propos des processus de branchement, des flots dans les réseaux et des mesures de Hausdorff associées. Thèse, Université Paris 6, 1993.

[15] Liu, Q. (1996a): The exact Hausdorff dimension of a branching set. Prob. Th. Rel. Fields, 104, 1996, 515-538. [24 pages],

[16] Liu,Q. (1996b): The growth of an entire characteristic function and the tail probability of the limit of a tree martingale. In *Trees*, Progress in Probability, vol.40, pp 51-80, 1996, Birkhäuser: Verlag Basel, Eds.B.Chauvin, S.Cohen, A.Rouault.

[17] Liu, Q. (1999a): Asymptotic properties of supercritical age-dependent branching processes. Stoch. Proc. Appl., 82 (1999) 61-87.

[18] Liu, Q. (1999b): An extension of a functional equation of Mandelbrot and Poincaré. Submitted.

[19] Liu, Q. (1999c): Exact packing of random fractals. Preprint.

[20] Liu, Q. (2000a): Exact packing measure on a Galton-Watson tree. Stoch. Proc. Appl. 85, 2000, 19-28.

[21] Liu, Q. (2000b): On generalized multiplicative cascades. Stoch. Proc. Appl., in press.

[22] Liu, Q. (2000c): Local dimensions of the branching measure on a Galton-Watson tree. To appear in Ann. Inst. Henri Poincaré.

[23] Liu, Q. (2000d): Flows in trees and exact Hausdorff measures in random constructions. Preprint.

[24] Liu, Q. and Rouault, A. (1996): On two measures defined on the boundary of a branching tree. In "Classical and modern branching processes", ed. K.B. Athreya and P. Jagers. IMA Volumes in Mathematics and its applications, vol. 84, 1996, pp.187-202. Springer-Verlag.

[25] Liu, Q. and Shieh, N.R. (1999): A uniform limit theorem for the branching measure on a Galton-Watson tree. Asian J. Math. Vol. 3, No. 2, 1999, 381-386.

[26] Liu, Q. and Wen, Z.-Y. (2000): L'analyse multifractale de la mesure de branchement. In preparation.

[27] Lyons, R., Pemantle R. and Peres Y. (1995): Ergodic theory on Galton-Watson trees, Speed of random walk and dimension of harmonic measure. Ergodic Theory Dyn. Systems 15, 593-619.

[28] Mandelbrot, B. (1974): Multiplications aléatoires et distributions invariantes par moyenne pondérée aléatoire. *C.R.Acad. Sci. Paris Ser.I* **278**, 289-292 and 355-358.

[29] Neveu, J. (1986): Arbre et processus de Galton -Watson. Ann. Inst. Henri Poincaré **22** 199-207.

[30] O'Brien, G.L. (1980): A limit theorem for sample maxima and heavy branches in Galton- Watson trees. *J. Appl. Prob.* **17**, 539 – 545.

[31] Shieh, N.R. and Taylor, S.J.(1998): Logarithmic multifractal spectrum of stable occupation measure. Stoch. Proc. and their Appl. 75, 249-261.

[32] Shieh, N.R. and Taylor, S.J.(1999): Multifractal spectra of branching measure on a Galton-Watson tree. Preprint.

[33] Taylor, S.J. and Tricot, C. (1985): Packing measure, and its evaluation for a brownian path. Trans. Amer. Math. Soc. 288, 679-699.

Trends in Mathematics, © 2000 Birkhäuser Verlag Basel/Switzerland

Likelihood ratio processes and asymptotic statistics for systems of interacting diffusions with branching and immigration

EVA LÖCHERBACH *Laboratoire de Probabilités et Modèles Aléatoires, Université Pierre et Marie Curie et C.N.R.S. 7599, 4 Place Jussieu, 75252 Paris CEDEX 05, France.* locherba@ccr.jussieu.fr

Abstract. *We consider statistical models for finite systems of branching diffusions with immigrations. We give necessary and sufficient conditions for local absolute continuity of laws for such branching particle systems on a suitable path space and derive an explicit version of the likelihood ratio process. For ergodic parametric submodels, under assumptions which combine smoothness properties of the parametrization at a fixed parameter point ϑ and integrability of certain information processes with respect to the invariant measure of the process under ϑ or with respect to an associated Campbell measure, one deduces local asymptotic normality at ϑ (LAN(ϑ)). Moreover, for null recurrent models, local asymptotic mixed normality (LAMN) at ϑ holds in situations where the right limit theorems for integrable additive functionals of the process are hand. These limit theorems follow from dividing the trajectory into independent pieces between successive returns to the void configuration and from limit theorems to stable processes.*

Key words. *branching diffusions, particle systems, likelihood ratio processes, local asymptotic normality.*

AMS subject classification. *60J60, 60J80, 62F99*

1 Basic assumptions on the model

The theory of spatially branching particle systems originates in questions from population biology and has been widely developed, see for instance Wakolbinger (1995) and the references therein for a study of infinite systems of spatially branching diffusions and their long time behaviour. In this work we are concerned with statistical models for branching particle systems and restrict our attention to processes with finite particle configurations where particles are moving in $I\!R^d$ in accordance with the following model assumptions:

1. Within a system of l particles

$$X^l = \begin{pmatrix} X^{1,l} \\ \vdots \\ X^{l,l} \end{pmatrix},$$

the i-th particle moves in $I\!R^d$ according to

$$dX_t^{i,l} = b(X_t^{i,l}, X_t^l)dt + \sigma(X_t^{i,l}, X_t^l)dW_t^i, \; 1 \le i \le l, \tag{1}$$

with independent m-dimensional Brownian motions $W^1, \ldots, W^l$ and coefficients $b(.,.)$ and $\sigma(.,.)$ which are Lipschitz continuous and symmetric with respect to the second component. We note $a := \sigma\sigma^T$ and suppose that $a(x^i, x)$ is invertible for all $x = (x^1, \dots, x^l)$ and all $1 \le i \le l$.

2. A particle located at position $x^i \in \mathbb{R}^d$ at time $t > 0$ which belongs to a configuration $x = (x^1, \dots, x^l)$ of l particles dies with probability

$$\kappa(x^i, x)\, h + o(h) \text{ as } h \to 0$$

in the small time interval $(t, t+h]$; here $\kappa(.,.)$ is a nonnegative function which is continuous with respect to both components and symmetric with respect to the second component x.

At its death time the particle in position x^i gives rise to a random number of offspring according to a position and configuration dependent reproduction law

$$F((x^i, x), dn) \in \mathcal{M}^1(\mathbb{N}_0),$$

such that $F(x^i, x, \{1\}) = 0$, $F(x^i, x, \{0\}) > 0$; here x is the configuration of coexisting particles. The newborn particles start their motion at their parent's position in space.

3. Additionally, there is immigration of new particles at a configuration dependent rate $c(x)$. At an immigration time, only one new particle immigrates, and it chooses its position in space randomly, according to a probability law $\nu(x, dy)$ on $\mathbb{R}^d$, depending on the configuration x of already existing particles. The immigration of particles into occupied positions in space is not allowed. We assume that $c(\Delta) > 0$, where Δ is the void configuration of particles – as a consequence, Δ will not be a trap for the particle process.

Example 1. l-particle motions with mean field interaction have been considered – in a different context – in Sznitman (1991) and Méléard (1996) and can also be treated within our frame. Take Lipschitz functions $\tilde{b} : \mathbb{R}^d \times \mathbb{R}^d \to \mathbb{R}^d$ and $\tilde{\sigma} : \mathbb{R}^d \times \mathbb{R}^d \to \mathbb{R}^{d \times m}$; then (1) takes the form

$$dX_t^{i,l} = \frac{1}{l} \sum_{j=1}^{l} \tilde{b}(X_t^{i,l}, X_t^{j,l})dt + \frac{1}{l} \sum_{j=1}^{l} \tilde{\sigma}(X_t^{i,l}, X_t^{j,l})dW_t^i,\ 1 \le i \le l.$$

Example 2. Of interest in physics are interacting l-particle systems with interaction potential

$$\Psi(x^1, \dots, x^l) = \gamma_l \sum_{1 \le i \ne j \le l} V(x^i - x^j),$$

$d = 1$, where $V : \mathbb{R} \to \mathbb{R}_+$ is a symmetric smooth pair potential function having compact support, with $V(0) > 0$ (cf. Spohn (1987)). In this case we take

$$b(x^i, x) := \tilde{b}(x^i) - D_i \Psi(x)$$

for some Lipschitz continuous function $\tilde{b} : \mathbb{R} \to \mathbb{R}$, $x = (x^1, \dots, x^l)$.

Example 3. Similarly, we can model branching mechanisms where the branching rate $\kappa(x^i, x)$ of a particle in position x^i depends on the number of neighbours of

x^i in some given (finite) neighbourhood (branching with finite interaction range). Also, systems where the branching or the immigration activity depend just on the total population size can be treated in our context.

Throughout this note we assume the following:

Assumption 1. $F((x^i, x), dn) \leq G(dn)$ in the sense of convolution of probability measures, where $G(dn) \in \mathcal{M}^1(\mathbb{N}_0)$ is a fixed reproduction law which does not depend on space and configuration, having finite mean offspring number. The immigration rates $c(.)$ are bounded away from infinity, and the branching rates $\kappa(.,.)$ are both bounded away from zero and infinity.

We introduce a canonical path space $(\Omega, \mathcal{A}, \mathbb{F})$ for branching particle systems. The set of all ordered finite particle configurations is $S := \bigcup_{l=0}^{\infty} (\mathbb{R}^d)^l$, with $(\mathbb{R}^d)^0 := \{\Delta\}$, the void configuration. We write $l(x) := l$ if $x \in (\mathbb{R}^d)^l$ for the number of particles within a configuration $x \in S$. Then the canonical path space Ω will be a subset of the space $D(\mathbb{R}_+, S)$ of all càdlàg functions taking values in S consisting of those functions $\psi \in D(\mathbb{R}_+, S)$ which have a strictly increasing sequence of jump times t_n, $t_n \uparrow \infty$, such that in between successive jumps the function $\psi_{|[t_n, t_{n+1}[}$ is continuous taking values in some fixed $(\mathbb{R}^d)^l$ for some $l \geq 0$. At jump times $(t_n)_n$ either a new particle is added to the already existing configuration of particles (we call this event **immigration event**) or one of the existing particles is replaced by some offspring particles or just removed from the configuration in case of a pure death (we call this event **branching event**). Let ϕ be the canonical process on Ω, then $\mathcal{A} = \sigma(\phi_t : 0 \leq t < \infty)$ is the σ-field generated by ϕ and $\mathbb{F} := (\mathcal{F}_t)_{t \geq 0}$, where $\mathcal{F}_t := \bigcap_{T > t} \sigma(\phi_r : r \leq T)$. As a consequence of results obtained by Ikeda, Nasagawa and Watanabe (1968) on the construction of Markov processes by "piecing out", we obtain the following

Proposition 1. (Ikeda, Nasagawa and Watanabe (1968, theorem 2.2) *For all $x \in S$ there is a unique probability measure Q_x on $(\Omega, \mathcal{A}, \mathbb{F})$, such that ϕ under Q_x is strongly Markov and satisfies the model assumptions* **1.-3.** *stated above with $\phi_0 = x$ a.s.*

Remark. Note that assumption 1 ensures that no accumulation of jumps already at a finite time is possible (cf. Löcherbach (1999b, proposition 5.13)). See also Ikeda and Watanabe (1970) and Löcherbach (2000) for conditions for non-explosion of branching diffusions.

For statistical purposes, we are interested in the following question: If the underlying functions and probability measures b, κ, F, c and ν are unknown – is it then possible to draw any statistical inference on them based on observation of ϕ continuously in time? The first problem in this context which we address to is a Girsanov theorem: Suppose that another set of $(b', \kappa', F', c', \nu')$ is given, satisfying all conditions made in **1.-3.** and in assumption 1 above ($\sigma(.)$ is supposed to be fixed since we observe continuously in time) and write Q'_x for the associated probability measure on $(\Omega, \mathcal{A}, \mathbb{F})$.

Theorem 1.
a) We have $Q'_x \overset{\text{loc}}{\ll} Q_x$ *for all* $x \in S$ *if (i)-(iii) hold for all* $x = (x^1, \ldots, x^l) \in S$.

(i) $\kappa(x^i, x) = 0$ *implies* $\kappa'(x^i, x) = 0$ *for all* $1 \le i \le l, l > 0$, *and* $c(x) = 0$ *implies* $c'(x) = 0$.

(ii) $F'((x^i, x), dn) \ll F((x^i, x), dn)$ *for all* $1 \le i \le l, l > 0$.

(iii) $\nu'(x, dy) \ll \nu(x, dy)$.

We write $\Gamma^i(x) := a^{-1}(b' - b)(x^i, x)$ *for all* $1 \le i \le l, l > 0$, $\Gamma := (\Gamma^1, \ldots, \Gamma^l)$.
b) Under conditions (i)-(iii) of a), the likelihood ratio process of Q'_x *to* Q_x *relative to* $\mathbb{F}$ *is*

$$L_t = (L1)_t^{b'/b} (L2)_t^{\kappa'/\kappa} (L3)_t^{F'/F} (L4)_t^{(c',\nu')/(c,\nu)}, \tag{2}$$

with factor processes

$$(L1)_t^{b'/b} = \exp\left(Y_t^{b'/b} - \frac{1}{2} < Y^{b'/b} >_t \right), \tag{3}$$

where $Y^{b'/b} \in \mathcal{M}_{loc}^{2,c}(Q_x, \mathbb{F})$ *with* $< Y^{b'/b} >_t = \int_0^t \left[\Gamma^T a \Gamma\right](\phi_s) ds$, *where* Γ *is chosen as in a) and*

$$a(x^1, \ldots, x^l) := \begin{pmatrix} a(x^1, x) & 0 & \cdots & 0 \\ \vdots & \cdots & \ddots & \vdots \\ 0 & \cdots & 0 & a(x^l, x) \end{pmatrix}.$$

$$(L2)_t^{\kappa'/\kappa} = \left(\prod_{j \ge 1, T_j^B \le t} \frac{\kappa'}{\kappa}(\zeta_j^B, \phi_{T_j^B -}) \right) e^{-\int_0^t \phi_s(\kappa' - \kappa) ds}, \tag{4}$$

where $(T_j^B)_j$ *is the sequence of jump times of* ϕ *which correspond to branching events,* ζ_j^B *the position of the branching particle at time* T_j^B, *and where we interpret* ϕ_s *as measure on* $\mathbb{R}^d$ *by defining* $\phi_s(f) := \sum_{i=1}^{l(\phi_s)} f(\phi_s^i, \phi_s)$, $\Delta(f) := 0$ *for any measurable function* $f : \mathbb{R}^d \times S \to \mathbb{R}$.

$$(L3)_t^{F'/F} = \prod_{j \ge 1, T_j^B \le t} \frac{F'}{F}((\zeta_j^B, \phi_{T_j^B -}), \{l_j^B\}), \tag{5}$$

with l_j^B *the number of newborn particles at time* T_j^B, *and*

$$(L4)_t^{(c',\nu')/(c,\nu)} = \left(\prod_{j \ge 1, T_j^I \le t} \frac{c'}{c}(\phi_{T_j^I -}) \frac{d\nu'}{d\nu}(\phi_{T_j^I -}, \zeta_j^I) \right) e^{-\int_0^t (c' - c)(\phi_s) ds}, \tag{6}$$

with $(T_j^I)_j$ the sequence of jump times of ϕ which correspond to immigration events and ζ_j^I the position of the immigrating particle at time T_j^I.

The proof is based on a consideration of ϕ between successive jump events and using a Girsanov theorem for diffusions (cf. Jacod and Shiryaev (1987)) and for marked point processes (cf. Brémaud (1981) and Jacod (1975)) there. For details, we refer the reader to Löcherbach (1999b). See also Eisele (1981) for related results obtained under more restrictive conditions, in particular on the diffusion part of the branching process.

2 Local asymptotic normality and local asymptotic mixed normality for branching particle systems

We now turn to questions of parametric statistical inference of branching particle systems and suppose that the "characteristics" $(b^\vartheta, \kappa^\vartheta, F^\vartheta, c^\vartheta, \nu^\vartheta)$ depend on some unknown finite dimensional parameter $\vartheta \in \Theta \subset \mathbb{R}^k$ open. We write $Q_{x,\vartheta}$ for the associated probability measure on $(\Omega, \mathcal{A}, \mathbb{F})$. In this section our aim is to show that – when observing the process continuously in time up to time n and letting n tend to infinity – the sequence of models converges (in the sense of Le Cam) locally over shrinking neighbourhoods of a fixed parameter point ϑ to a limit experiment which is a Gaussian shift experiment in the ergodic case (the sequence of models is termed **local asymptotic normal at ϑ - LAN(ϑ)** in this case) or a mixed normal experiment in some null recurrent cases (we speak of **local asymptotic mixed normality - LAMN(ϑ)** in this situation). The notion LAN(ϑ) is due to Le Cam (cf. Le Cam (1960)), the notion LAMN(ϑ) to Jeganathan (1982). In both cases, thanks to the convolution theorem of Hájek (1970) and of Jeganathan (1982) we are able to determine the optimal behaviour of estimators. We start with a condition concerning the asymptotic behaviour of ϕ.

Assumption 2. We assume that ϕ under $Q_{x,\vartheta}$ is invariant in the sense of Harris with recurrent point Δ and invariant measure m^ϑ. If $m^\vartheta(S) = \infty$, we assume further that for $R := \inf\{T_n : n > 0, \phi_{T_n} = \Delta\}$ either
(i) $\mathcal{L}(R|Q_{\Delta,\vartheta})$ belongs to the domain of attraction of a positive stable law with index α, $0 < \alpha < 1$ (i.e. $x \mapsto Q_{\Delta,\vartheta}(R > x)$ is regularly varying at infinity with index $-\alpha$)
or
(ii) $\mathcal{L}(R|Q_{\Delta,\vartheta})$ is relatively stable (i.e. for X_n, $n \in \mathbb{N}$, i.i.d., distributed according to $\mathcal{L}(R|Q_{\Delta,\vartheta})$, there exists a sequence of normalizing constants $a_n \uparrow \infty$ such that $S_n/a_n \to 1$ in probability, $S_n := \sum_{i=1}^n X_i$).

Remarks.
1. The invariant measure is given by $m^\vartheta(A) = E_{\Delta,\vartheta}\left(\int_0^R 1_A(\phi_s ds)\right)$ up to multiplication by a constant, $A \in \mathcal{S}$.
2. In case $m^\vartheta(S) = \infty$, condition (i)-(ii) is necessary and sufficient for weak con-

vergence of additive functionals of ϕ

$$\left(\frac{1}{v(n)}\int_0^{t\cdot n} 1_A(\phi_s)ds\right)_{t\geq 0} \xrightarrow{\mathcal{L}} m^\vartheta(A)\cdot V,\ m^\vartheta(A)<\infty,$$

(weak convergence in $D(\mathbb{R}_+,\mathbb{R})$ as $n\to\infty$, under $Q_{x,\vartheta}$)
where $v(n)\uparrow\infty$ as $n\to\infty$, with V continuous, having non-decreasing paths with $V_0\equiv 0, V_t\uparrow\infty$ as $t\to\infty$. For details, we refer the reader to Bingham, Goldie and Teugels (1987), Greenwood and Resnick (1979) and Touati (1988).

Example 4.
1. Assumption 2 holds with $m^\vartheta(S)<\infty$, if the dominating reproduction measure $G(dn)$ of assumption 1 is subcritical.
2. Assumption 2 holds with $m^\vartheta(S)=\infty$, if $F^\vartheta,\kappa^\vartheta$ and c^ϑ are purely population size dependent such that the process $(l(\phi_t))_t$ is a birth-and-death process having strictly positive birth and death rates λ_k and μ_k everywhere and such that for all $k\geq k_0$ for some fixed k_0, $\lambda_k=Ck+D$ and $\mu_k=CK$ for $0<D<C$. In this case $\alpha=1-D/C$ (cf. Karlin and McGregor (1961)).

In order to obtain LAN or LAMN at ϑ, we have to assume that the parametrization of the model is sufficiently smooth at ϑ. For this purpose we introduce a Campbell measure $\bar{m}^\vartheta$ on $\mathcal{B}(\mathbb{R}^d\times S)$ which is associated to the invariant measure m^ϑ in the following way. For $B\in\mathcal{B}(\mathbb{R}^d)$ and $C\in\mathcal{B}(S)$ we define

$$\bar{m}^\vartheta(B\times C):=\int_S x(B)1_C(x)m^\vartheta(dx)$$

where for $x\in S$ $x(B):=\sum_{i=1}^{l(x)}1_B(x^i)$ is the number of particles of x in B.

Remark. If $d=1$, in purely position dependent situations (no interactions between coexisting particles), under regularity assumptions on drift and diffusion coefficient, Höpfner and Löcherbach (1999) show that $\bar{m}(\cdot\times S)$ as a measure on $\mathbb{R}$ has a Lebesgue density u satisfying the following equation

$$A^*u-\kappa(1-\varrho)u=-r$$

where A^* is the adjoint of the generator of the diffusion process driving each particle's motion, where $\varrho(x):=\sum_k F(x,\{k\})k$ is the mean number of offspring to be produced in position x and where we suppose the immigration measure to be of the form $c\cdot\nu(dy)=r(y)dy$ independently of the configuration of already existing particles.

We state the following smoothness condition.

Assumption 3.
1. The function $\xi\mapsto\kappa^\xi(.,.)$ is logarithmic differentiable in ϑ with derivative $\dot{\kappa}^\vartheta$: $\mathbb{R}^d\times S\to\mathbb{R}^k$ in the following sense.

(i) For all $\vartheta' \in \Theta$, for all $y \in \mathbb{R}^d$ and $z \in S$,

$$\left(\frac{\kappa^{\vartheta'}}{\kappa^{\vartheta}}(y,z) - 1 - (\vartheta' - \vartheta)^T \dot{\kappa}^{\vartheta}(y,z)\right)^2 \leq f_{\vartheta}(y,z,|\vartheta' - \vartheta|)\,|\vartheta' - \vartheta|^2$$

for a function $f_\vartheta : \mathbb{R}^d \times S \times \mathbb{R}_+ \to \mathbb{R}_+$ such that $f_\vartheta(y,z,.)$ is non-decreasing for every (y,z), $\lim_{c\downarrow 0} f_\vartheta(y,z,c) = 0$ for all (y,z) and $f_\vartheta(.,.,c)$ measurable. We suppose further that

$$\int_0^{\cdot} \phi_s\left(\kappa^{\vartheta} f_{\vartheta}(.,.,\delta(\vartheta))\right) ds \text{ is locally integrable w.r.t. } Q_{x,\vartheta}$$

for all $x \in S$, for some $\delta(\vartheta) > 0$.

(ii)

$$\int_0^{\cdot} \phi_s\left(\kappa^{\vartheta}\left[(\dot{\kappa}^{\vartheta})^T \dot{\kappa}^{\vartheta}\right](.,.)\right) ds$$

is locally integrable with respect to $Q_{x,\vartheta}$ for all $x \in S$.

(iii) We suppose that $f_\vartheta(.,.,\delta(\vartheta)) \in L^1(\kappa^\vartheta \bar{m}^\vartheta)$ and that $(\dot{\kappa}^\vartheta)^T \dot{\kappa}^\vartheta \in L^1(\kappa^\vartheta \bar{m}^\vartheta)$ component-wise.

2. We assume analogous differentiability conditions for the other "characteristics" $(b^\xi, F^\xi, c^\xi, \nu^\xi)$ of the process. For details, we refer to Löcherbach (2000).

We define a local scale given by

$$\delta_n(\vartheta)\begin{cases} = 1/\sqrt{n}, & \text{if } m^\vartheta(S) = 1, \\ \sim (\Gamma(1-\alpha) Q_{\Delta,\vartheta}(R > n))^{1/2}, & \text{if } Q_{\Delta,\vartheta}(R > \cdot) \in RV_{-\alpha},\ 0 < \alpha < 1, \\ \sim (\frac{1}{n}\int_0^n Q_{\Delta,\vartheta}(R > t)dt)^{1/2}, & \text{if } \mathcal{L}(R|Q_{\Delta,\vartheta}) \text{ is relatively stable,} \end{cases} \tag{7}$$

where $RV_{-\alpha}$ is the space of functions that vary regularly at infinity with index $-\alpha$. Then we arrive at the following result.

Theorem 2. *At a point ϑ such that assumptions 2 and 3 hold, for $\vartheta_n := \vartheta + \delta_n(\vartheta)h$, $h \in \mathbb{R}^k$ fixed, the log-likelihood ratio process $\log L^{\vartheta_n/\vartheta}$ of Q_{x,ϑ_n} to $Q_{x,\vartheta}$ admits a decomposition*

$$\log L_n^{\vartheta_n/\vartheta} = h^T M_1^{n,\vartheta} - \frac{1}{2} h^T < M^{n,\vartheta} >_1 h + R_n \tag{8}$$

where $R_n \to 0$ in $Q_{x,\vartheta}$-probability and $M^{n,\vartheta} \in \mathcal{M}^2_{loc}(Q_{x,\vartheta}, (\mathcal{F}_{nt})_{t\geq 0})$ with

$$(M^{n,\vartheta}, < M^{n,\vartheta} >) \xrightarrow{\mathcal{L}} (Y, < Y >) \tag{9}$$

(weak convergence in $D(\mathbb{R}_+, \mathbb{R}^k \times \mathbb{R}^{k\times k})$ as $n \to \infty$, under $Q_{x,\vartheta}$).
Here, the limit martingale Y is given as follows. Let B be a continuous k-dimensional

Gaussian martingale with covariance matrix $J = J(\vartheta)$ where $J(\vartheta)$ is the Fisher-information matrix of the experiment, given as

$$J(\vartheta) = (J1)(\vartheta) + (J2)(\vartheta) + (J3)(\vartheta) + (J4)(\vartheta) + (J5)(\vartheta)$$

with $(J1)(\vartheta) = \int_{\mathbb{R}^d \times S} [\dot{\kappa}^\vartheta (\dot{\kappa}^\vartheta)^T] \kappa^\vartheta (y,z) \bar{m}^\vartheta (dy, dz)$. The other Fisher information matrices are given explicitly in Löcherbach (2000, (3.44-45)) and correspond to the other "subexperiments" of the whole experiment, i.e. to the diffusion part, the reproduction part, the immigration rate part and the choice of the position in space of the immigrating particle. Then

$$Y = B \text{ in case of relative stability or ergodicity} \tag{10}$$

or

$$Y = B \circ W^\alpha, \; <Y>= J \cdot W^\alpha \text{ if } Q_{\Delta,\vartheta}(R > \cdot) \in RV_{-\alpha}, \; 0 < \alpha < 1. \tag{11}$$

Here, W^α is the Mittag-Leffler-process of index α, the process inverse to the stable subordinator S^α.

Proof: The proof of the decomposition of the log-likelihood ratio process imitates a well-known scheme, see for instance the proof of theorem 1 in Luschgy (1992). The joint convergence of $M^{n,\vartheta}$ together with its angle bracket follows from a martingale convergence theorem which has been obtained by Touati (1988) and which is based on a decomposition of the trajectory of ϕ into independent life-cycles and results concerning convergence to stable processes. We refer to Löcherbach (2000) for the details.

Remark. Note that the techniques leading to LAN or LAMN can also be used to characterize – in a non-parametric context – an optimal rate of convergence of estimators. This approach has been presented in the problem of estimating the position dependent branching rate in Höpfner, Hoffmann and Löcherbach (2000) where exactly the same rates arise as in estimation problems related to classical diffusion processes.

References

Bingham, N. H., Goldie, C. M., Teugels, J. L. (1987): Regular variation. Cambridge New York Sydney: Cambridge University Press.

Brémaud, P. (1981): Point Processes and queues, martingale dynamics. Berlin Heidelberg New York: Springer.

Eisele, Th. (1981): Diffusions avec branchement et interaction. Ann. Sci. Univ. Clermont-Ferrand II 69, Math. 19, 21-34.

Greenwood, P., Resnick, S. (1979): A bivariate stable characterization and domains of attraction. J. Multivar. Anal. 9, 206-221.

Hájek, J. (1970): A characterization of limiting distributions of regular estimates. Z. Wahrscheinlichkeitsth. Verw. Geb. 14, 323-330.

Höpfner, R., Hoffmann, M., Löcherbach, E. (2000): Nonparametric estimation of the death rate in branching diffusions. Preprint 577, Laboratoire de Probabilités, Université Paris VI.

Höpfner, R., Löcherbach, E. (1999): On invariant measure for branching diffusions. Unpublished note,
see `http://www.mathematik.uni-mainz.de/~hoepfner`.

Ikeda, N., Nagasawa, M., Watanabe, S. (1968). Branching Markov processes II. J. Math. Kyoto Univ. 8, 365-410

Ikeda, N., Watanabe, S. On uniqueness and non-uniqueness of solutions for a class of non-linear equations and explosion problem for branching processes. J. Fac. Sci. Tokyo, Sect. I A 17, 1970, 187-214.

Jacod, J. (1975): Multivariate point processes: predictable projection, Radon-Nikodym derivates, representation of martingales. Z. Wahrscheinlichkeitsth. Verw. Geb. 31, 235-253.

Jacod, J., Shiryaev, A. N. (1987): Limit Theorems for Stochastic Processes. Berlin Heidelberg New York: Springer.

Jeganathan, P. (1982): On the asymptotic theory of estimation when the limit of the log-likelihood ratio is mixed normal. Sankhya 44, 173-212.

Karlin, S., McGregor, J. (1961): Occupation time laws for birth and death processes. In: Neyman, J. (Ed.), Proc. Fourth Berkeley Symp. Math. Stat. Prob. 2, pp. 249-273. Berkeley, University of California Press.

Le Cam, L. (1960): Locally asymptotic normal families of distributions. Univ. Calif. Publ. Statistics 3, 37-98.

Löcherbach, E. (1999a): Statistical models and likelihood ratio processes for interacting particle systems with branching and immigration. PhD-Thesis, University of Paderborn.

Löcherbach, E. (1999b): Likelihood ratio processes for Markovian particle systems with killing and jumps. Preprint 551, Laboratoire de Probabilités, Université Paris VI.

Löcherbach, E. (2000): LAN and LAMN for systems of interacting diffusions with branching and immigration. Preprint 590, Laboratoire de Probabilités, Université Paris VI.

Luschgy, H. (1992): Local asymptotic mixed normality for semimartingale experiments. Probab. Theory Relat. Fields. 92, 151-176.

Méléard, S. (1996): Asymptotic behavior of some interacting particle systems; McKean-Vlasov and Boltzmann models. In: Graham, C. (ed.) et al., Probabilistic models for nonlinear partial differential equations, Proc. Montecatini Terme 1995, pp. 42-95. Lect. Notes in Mathematics 1627, Berlin: Springer.

Spohn, H. (1998). Dyson's model of interacting Brownian motions at arbitrary coupling strength. Markov Process. Relat. Fields 4, 649-661.

Sznitman, A. S. (1991). Topics in propagation of chaos. Ecole d'été de Probabilités de Saint Flour XIX 1989, Lecture Notes in Mathematics 1464, New York: Springer.

Touati, A. (1988): Théorèmes limites pour les processus de Markov recurrents. Unpublished manuscript.

Wakolbinger, A. (1995): Limits of spatial branching processes. Bernoulli 1, 171-189.

Trends in Mathematics, © 2000 Birkhäuser Verlag Basel/Switzerland

Probabilistic Analysis of a Schröder Walk Generation Algorithm

GUY LOUCHARD *Université Libre de Bruxelles, Département d'Informatique, CP 212, Boulevard du Triomphe, B-1050 Bruxelles, Belgium.* louchard@ulb.ac.be
OLIVIER ROQUES *LaBRI, Université Bordeaux 1, 351 cours de la Libération, 33405 Talence Cedex, France.* roques@labri.u-bordeaux.fr

Abstract. *Using some tools from Combinatorics, Probability Theory, and Singularity analysis, we present a complete asymptotic probabilistic analysis of the cost of a Schröder walk generation algorithm proposed by Penaud et al.([13]). Such a walk* $S(.)$ *is made of northeast, southeast and east steps, but each east step is made of two time units (if we consider recording the time* t *on the abscissa and the moves on the ordinates). The walk starts from the origin at time* 0*, cannot go under the time axis, and we add the constraint* $S(2n) = 0$*. Five different probability distributions will appear in the study: Gaussian, Exponential, Geometric, Rayleigh and a new probability distribution, that we can characterize by its density Laplace Transform and its moments.*

1 Introduction

In [11], we have analyzed some asymptotic properties of an underdiagonal walk generation algorithm (GA). This can be represented as a walk on a plane with northeast, southeast and east steps, under the condition that the walk cannot go under the x axis.

As announced in [11], we intend to pursue this approach on other walks GA. In this paper, we analyze an algorithm proposed in [13] to generate Schröder walks of length $2n$. Such a walk $S(.)$ is also made of northeast, southeast and east steps, but each east step is made of two time units (if we consider recording the time t on the abscissa and the moves on the ordinates). As in [11], the walk starts from the origin at time 0, cannot go under the time axis, but now, we add the constraint $S(2n) = 0$. We denote by R_n the number of S of length $2n$.

Let us say for completeness that $\{R_n\}_{n\geq 0}$ was discovered by Schröder ([14]) looking at the generalized bracketing problem. Several combinatorial objects are enumerated by R_n, such as : the number of dissections of a convex polygon, planted rooted trees with $n+1$ leaves whose nodes have a degree greater than one (sometimes called ordered hierarchies ([7]) in classification theory) and some string edit problems.

Another walk will be used in the sequel: $SL(.)$ is a left factor of a S i.e. a walk starting from the origin at time 0 and only constrained to remain above the time axis. We denote by F_n the number of SL of length n.

The GA proposed in [13] proceeds as follows. Using a slightly different form of the cyclic lemma ([4]) and the Catalan factorization ([8]), the authors establish a mapping between a $S(.)$ walk of length $2n$, and $(n+1)$ $SL(.)$ walks of length $2n+1$ which end either by a southeast step or by an east step. Hence, sampling uniformly in this subset of SL leads to a Schröder walk with the uniform distribution. The first part of the GA given in [13], that we will call GA.1, uses a rejecting method, which builds a walk step by step using some constant probability, such that the probability of a northeast step (resp. southeast step) is p (resp. p) and the probability of an east step is p^2, where p is the positive root of the polynomial $1 - 2p - p^2$. GA.1 is allowed to fail (and then has to restart) either if the walk crosses the time axis, or if the walk stops at time $2n+2$ or if it stops at time $2n+1$ by a forbidden step (actually a northeast step). When GA.1 succeeds, it outputs a valid $SL(.)$ walk of height h, defined as the difference between the number of northeast steps and the number of southeast steps . Then, in the last part of GA (GA.2), a procedure called *lightning procedure* in [13], based on Catalan factorization, maps $\frac{h+1}{2}$ northeast steps into $\frac{h+1}{2}$ southeast steps, in order to make the walk ending at height -1. Finally, another combinatorial mapping, based on the cyclic lemma, maps the walk obtained previously in a Schröder walk.

The generating functions (GF) of R_n and F_n are given in [13] , respectively as

$$R(z) = \frac{1 - z - \sqrt{1 - 6z + z^2}}{2z}, \tag{1}$$

$$F(z) = \frac{2}{1 - 2z - z^2 + \sqrt{1 - 6z^2 + z^4}}. \tag{2}$$

It is proved in [13] that the complexity of the proposed GA (in terms of number of used letters) is $\mathcal{O}(n)$. The purpose of this paper is to give a complete asymptotic ($n \to \infty$) precise probabilistic analysis of this GA (in terms of used letters and calls to a random generator).

In the field of analysis of algorithms, the moments of cost distribution are usually the first steps in the complete study: asymptotic distributions of costs and related random variables (RV) are more informative and shed more light on their stochastic behaviour.

Using some tools from Combinatorics, Probability Theory, and Singularity analysis, we obtain a surprising list of 5 different probability distributions : Gaussian, Exponential, Geometric, Rayleigh and a new probability distribution, that we can characterize by its density Laplace Transform and its moments.

Another useful GF: $G(w, z)$ is related to the number Nt of transformed steps during the lightning procedure: marking these steps by w, we obtain (the proof is given in Sec 3.4), setting
$f_1 := \sqrt{1 - 6z^2 + z^4}$,

$$G(w, z) = \frac{2z(1 - z^2 + f_1 + 2wz^2)}{(-1 + z^2 - f_1 + 2\sqrt{w}\, z)(-1 + z^2 - f_1 - 2\sqrt{w}\, z)}. \tag{3}$$

The paper is organized as follows: Sec.2 presents the probabilistic aspects and the main results, Sec.3 gives some asymptotic analysis of several parameters and RV. Sec.4 is devoted to analysis of the GA. Sec.5 develops some simulation results and Sec.6 concludes the paper. An appendix provides rather technical proofs.

The following probabilistic notations will be used in the sequel:

- $\stackrel{\mathcal{D}}{\sim}$: convergence in distribution, for $n \to \infty$.

- $\Longrightarrow_{n\to\infty}$: weak convergence in the space of all right– continuous functions having left limits in R^2 and endowed with the Skorohod metric d_0 (see Billingsley, [2],Chapter III).
- $\mathcal{N}(M,V)$:= the Normal (or Gaussian) RV with mean M and variance V.
- Brownian Motion (BM) := Markovian Gaussian process, with mean $E[BM(t)] = 0$, variance $\text{VAR}[BM(t)] = t$, and covariance $E[BM(s)\cdot BM(t)] = s (s \le t)$: see Ito, McKean, [10].
- GEOM(P) : geometric RV with probability distribution $P(1-P)^{k-1}$.
- Rayleigh : RV with density $e^{-x^2/2}xdx$.
- Exponential : RV with density e^{-x}.

The computer algebra system MAPLE has been quite useful for detailed computations and simulations.

2 Probabilistic aspects and main results

In this section we present the connections between S and some random walks (RW) and stochastic processes. We also summarize our main results.

The GA is based on the fact that we can associate to each S a RW such that all S with the same lenght are endowed with the same probability. This amounts to define three time-space steps (up, down and horizontal) with probabilities:

$$\begin{aligned} p &:= Pr[X_i = 1, \Delta t_i = 1] = Pr[X_i = -1, \Delta t_i = 1], \\ r = p^2 &:= Pr[X_i = 0, \Delta t_i = 2], \\ \text{and } p^2 + 2p &= 1, \text{ i.e. } p = \sqrt{2}-1, r = 3 - 2\sqrt{2}. \end{aligned} \tag{4}$$

$S(i)$ is actually an Excursion, starting from 0, returning to 0 at time $2n$ and staying strictly positive for $0 < i < 2n$. $SL(i)$ is called a Meander, starting from 0 , conditionned to stay strictly positive for $0 < i \le n$.

Let us mention a third RW, the Bridge $B(.)$ (with the same time-space steps) which has been used in [12], in connection with some string edit problems. There the only condition is that $B(n) = 0$. The probabilites (4) have also been derived in [12]. It is also proved in the same paper that the number Nh of horizontal steps in a Bridge such that $B(n) = \alpha\sqrt{n}$,
$\alpha = \mathcal{O}(1)$, is characterized by

$$\frac{Nh - \mu_h \cdot n}{\sigma_h \sqrt{n}} \overset{\mathcal{D}}{\sim} \mathcal{N}(0,1), \quad n \to \infty, \quad \text{with} \tag{5}$$

$$\mu_h = (2-\sqrt{2})/4, \quad \sigma_h^2 = \sqrt{2}/16. \tag{6}$$

Let us note that the asymptotic behaviour of Nh is *independent of* α. This is *not the case* for the number of up or down steps in the Bridge. Another result from

[12] is related to the unconditioned RW $Y(.)$, endowed with (4) :

$$\frac{2^{1/4}Y([nt])}{\sqrt{n}} \Longrightarrow BM(t), \quad n \to \infty, \quad t \subset [0,1]. \tag{7}$$

We now present the main results of this paper. The following RV notations will be used in the sequel:

- κ_1 := number of letters necessary to generate a Schröder walk, where we associate a letter to each time unit.
- κ_2 := number of calls to a random generator necessary to generate a Schröder walk.
- T := number of trials in the rejecting procedure.
- Nh := number of horizontal steps in an Excursion or a Meander of length ℓ.
- Nr := number of Excursions generated before getting a length $2n$ Meander.
- Nt := number of transformed steps in the ligthning procedure.
- h := normalized(by $1/\sqrt{2n}$) height of the length $2n$ Meander.

Our main results are summarized in Table 1. We give the asymptotic distribution of the principal RV we have considered in the paper.

RV	Asymptotic characteristics($n \to \infty$)
κ_1	$\Psi_1(\alpha) := E[\exp(-\alpha\kappa_1/2n)] \sim \frac{1}{2}\phi_1(\alpha)/[1 - 1/2\phi_1(\alpha)]$ (ϕ_1 is given in (27))
$\kappa_2 \stackrel{\mathcal{D}}{\equiv} \mu_s\kappa_1$	$\Psi_2(\alpha) := E[\exp(-\alpha\kappa_2/2n)] = \Psi_1(\mu_s\alpha)$, (μ_s is given in (16))
T	$T \stackrel{\mathcal{D}}{\sim} \text{GEOM}(1/2)$
Nh	$\frac{Nh-\ell\mu_h}{\sqrt{\ell}\sigma_h} \stackrel{\mathcal{D}}{\sim} \mathcal{N}(0,1), \ell \to \infty$ (μ_h, σ_h are given in (6))
Nr	$\frac{Nr\cdot 2^{1/4}}{\sqrt{\pi n}} \stackrel{\mathcal{D}}{\sim}$ Exponential RV
$Nt/\sqrt{2n} \stackrel{\mathcal{D}}{\equiv} h/2$	$Nt/\sqrt{2n} \stackrel{\mathcal{D}}{\sim}$ Rayleigh RV with density $exp(-x^2 2\sqrt{2})4x\sqrt{2}$

Table 1: Asymptotic characteristics ($n \to \infty$).

3 Some asymptotic analysis

In this section, we present a precise asymptotic analysis of several parameters and RV we need in the sequel. We analyze successively the Excursion, the Meander, $R(z)$ and $F(z)$, and the lightning procedure. A good introduction to the kind of techniques we need here can be found in Flajolet, Sedgewick [6].

3.1 Excursions

The hitting time to 0, T_0, starting from 1, has a probability generating function (PGF) $F_1(z)$, which satisfies the equation:

$$\begin{aligned} F_1 &= pz + pzF_1^2 + rz^2F_1, \quad \text{hence} \\ F_1(z) &= (1 - p^2z^2 - (1 - 6p^2z^2 + p^4z^4)^{1/2}/(2pz). \end{aligned}$$

It is simpler to substitute $z^2 = u$ in $F_1(z) \cdot z = \varphi_1(u)$, say , and we have $Pr_1[T_0 = 2n - 1] = [u^n]\varphi_1(u)$. Similarly, the PGF of $Pr_1[T_0 \geq 2n - 1]$ is given by $\varphi_2(u) = \frac{1-\varphi_1(u)}{1-u}$.

The dominant algebraic singularity of φ_2 is given by 1. So we set $u = 1 - \varepsilon$ and expand φ_2 into ε. This gives a Puiseux series:

$$\varphi_2(u) \sim 2^{1/4}/\sqrt{\varepsilon} - (\sqrt{2} - 1)/2 + \mathcal{O}(\sqrt{\varepsilon}), \quad \varepsilon \to 0.$$

By classical singularity analysis, we obtain immediately

$$Pr_1[T_0 \geq 2n - 1] = [u^n]\varphi_2(u) \sim \frac{2^{1/4}}{\sqrt{\pi n}}, \quad n \to \infty. \tag{8}$$

T_0 is of course directly related to the length of an Excursion.

Similarly, starting from $\varphi_1(u)$, we derive: $\varphi_1(u) \sim 1 - 2^{1/4}\sqrt{\varepsilon} + \mathcal{O}(\sqrt{\varepsilon})$ and

$$Pr_1[T_0 = 2n - 1] \sim \frac{2^{1/4}}{2\sqrt{\pi}n^{3/2}}, \quad n \to \infty, \tag{9}$$

which is of course compatible with (8).

A more interesting RV is the number Nh of horizontal steps in an Excursion of length $2n$ (we mark Nh with w). The PGF $F_2(w, z)$ satisfies

$$\begin{aligned} F_2 &= pz + pzF_2^2 + rz^2wF_2, \quad \text{hence} \\ F_2(w,z) &= [1 - p^2z^2w - (1 - 2p^2z^2w + p^4z^4w^2 - 4p^2z^2)^{1/2}]/(2pz). \end{aligned}$$

Substitute again $z^2 = u$ in $F_2(w, z) \cdot z = \varphi_3(w, u)$, say. The dominant singularity is now given by

$$r(w) = (8\sqrt{2} + 12 + 6w + 4w\sqrt{2}) - 4(17 + 12\sqrt{2} + 12w\sqrt{2} + 17w)^{1/2}/(2w^2) \tag{10}$$

and $r(1) = 1$. More precisely

$$r(w) = 1 + C_1(w - 1) + C_2(w - 1)^2 + C_3(w - 1)^3 + \mathcal{O}((w - 1)^4), \quad w \to 1,$$

$$
\begin{aligned}
C_1 &= \sqrt{2}/2 - 1, \\
C_2 &= 5/4 - 13\sqrt{2}/16, \\
C_3 &= 69\sqrt{2}/64 - 25/16.
\end{aligned} \tag{11}
$$

Setting now $u = r(w) - \varepsilon$, we derive

$$
\varphi_3(w,u) \sim G_1(w) + G_2(w)\sqrt{\varepsilon} + \mathcal{O}(\varepsilon), \quad \varepsilon \to 0,
$$

with

$$
\begin{aligned}
G_1(w) &:= [2\sqrt{2} - 2 + (2 - 3\sqrt{2}/2)(w-1) + \mathcal{O}((w-1)^2)]/(2p) \\
&\sim 1 + \mathcal{O}(w-1), \quad w \to 1, \\
G_2(w) &:= -[2(3\sqrt{2}-4)^{1/2} + (3\sqrt{2}-4)^{1/2}/4(w-1) + \mathcal{O}((w-1)^2)]/(2p) \\
&\sim -2^{1/4} + \mathcal{O}(w-1), \quad w \to 1,
\end{aligned} \tag{12}
$$

and

$$
[u^n]\varphi_3(w,u) \sim \frac{G_2(w)}{-2\sqrt{\pi n^3}} [1/r(w)]^n \sqrt{r(w)}, \quad n \to \infty.
$$

Proceeding now as in Bender's Theorems 1 and 3 in [1], we set $w = e^{is}$ and expand $ln(1/r(w))$. This leads to

$$
is(1 - \sqrt{2}/2) - \sqrt{2}/16 s^2 + \mathcal{O}(s^3), \quad s \to 0, \tag{13}
$$

and finally, we obtain, normalizing by (9):

$$
\frac{Nh - n(1-\sqrt{2}/2)}{\sqrt{n\sqrt{2}/8}} \overset{\mathcal{D}}{\sim} \mathcal{N}(0,1), \quad n \to \infty, \tag{14}
$$

which is identical to (5) , *even with different constaints*. The *total number* M of steps in an excursion of length ℓ is given by $M = \tilde{N} + Nh$, where $\tilde{N}$ is the number of up and down steps. Hence $\tilde{N} + 2Nh = \ell$ and , by (14),

$$
\frac{Nh - \ell\mu_h}{\sqrt{\ell}\sigma_h} \overset{\mathcal{D}}{\sim} \mathcal{N}(0,1), \quad \ell \to \infty, \tag{15}
$$

(μ_h, σ_h are given in (6)). Therefore,

$$
\frac{M - l\mu_s}{\sqrt{l}\sigma_h} \overset{\mathcal{D}}{\sim} \mathcal{N}(0,1), \quad \ell \to \infty, \quad \text{where } \mu_s = (1/2 + \sqrt{2}/4). \tag{16}
$$

It is finally clear from (8), that the number Nr of returns to 0 before getting a length $2n$ Meander (i.e. the number of Excursions) is such that

$$
\frac{Nr \cdot 2^{1/4}}{\sqrt{\pi n}} \overset{\mathcal{D}}{\sim} \text{Exponential}, \quad n \to \infty, \tag{17}
$$

with Probability Distribution Function (PDF) $1 - e^{-x}$. (Indeed, GEOM has an asymptotic distribution given by the Exponential RV). Actually, this corresponds to the local time of $Y(.)$ at the origin.

3.2 Meanders

Let us now turn to the length $2n$ Meander distribution. Reverting the time axis, this amounts to analyze the hitting time $T_0^{(h)}$ to 0 , starting from an height $h\sqrt{2n}$. The PGF is of course given by $[F_1(z)]^{h\sqrt{2n}}$ (this is equivalent to the ladder epochs: see Feller[5], Chap.XII.7).

Setting again $z^2 = u$ and $u = 1 - \varepsilon$, we derive the PGF of $\tilde{T_0}^{(h)} = T_0^{(h)} + h\sqrt{2n}$ as

$$(\varphi_1(u))^{h\sqrt{2n}} \sim (1 - 2^{1/4}\sqrt{\varepsilon})^{h\sqrt{2n}}, \quad \varepsilon \to 0.$$

Now we set $u = e^{-\theta/n}$. We obtain asymptotically

$$e^{-\sqrt{\theta}h2^{1/4}\sqrt{2}}, \tag{18}$$

which means that $\tilde{T_0}^{(h)}/(2n)$ has, asymptotically, the density(wrt t):

$$h2^{1/4}exp(-h^2\sqrt{2}/(2t))/\sqrt{2\pi t^3}.$$

(We note that this is also true for $T_0^{(h)}/(2n)$). Setting $t = 1$ and normalizing, this leads to the asymptotic (normalized by $1/\sqrt{2n}$) Meander (of duration $2n$) density, given by

$$e^{-h^2\sqrt{2}/2}h\sqrt{2}. \tag{19}$$

This is the well known Rayleigh density, which gives the classical BM Meander density. So (19) could have been derived from (7). The GF approach will however be necessary in the next analysis. The moments of h are given by

$$E[h^k] = 2^{k/4}\Gamma(k/2+1). \tag{20}$$

Note that the Rayleigh distribution appears in other asymptotic analysis: for instance the number of trees in a random mapping and the number of points at distance d to a cycle are also asymptotically Rayleigh: see Drmota and Soria [3].

Let us now consider the number Nh of horizontal steps in a Meander of length $2n$. Again we derive (The proof is given in Appendix A) the distribution (14). The number M of steps is again characterized by (16). In Appendix A we prove also the asymptotic independence of Nh and h.

3.3 R_n and F_n

The dominant algebraic singularity of $R(z)$ (see (1)) is given by $z_1^* = 3 - 2\sqrt{2} \equiv p^2$. Standard analysis gives, with $z = z_1^* - \varepsilon$,

$$\begin{aligned} R(z) &\sim \sqrt{2}+1-2^{1/4}(3+2\sqrt{2})\sqrt{\varepsilon}+(17/2+6\sqrt{2})\varepsilon \\ &+ 2^{1/4}(-195\sqrt{2}-276)\varepsilon^{3/2}/16+\mathcal{O}(\varepsilon^2), \quad \varepsilon \to 0 \end{aligned}$$

Hence

$$R_n \sim \frac{C_4}{n^{3/2}z_1^{*n}} + \frac{C_5}{n^{5/2}z_1^{*n}}, \quad n \to \infty, \tag{21}$$

with

$$C_4 = 2^{1/4}(\sqrt{2}+1)/(2\sqrt{\pi}), C_5 = -3\cdot 2^{1/4}(14+11\sqrt{2})/(64\sqrt{\pi}).$$

We use the fact that

$$[z^n](1-z)^{-\alpha} \sim \frac{n^{\alpha-1}}{\Gamma(\alpha)}(1+\frac{\alpha(\alpha-1)}{2n}+\mathcal{O}(\frac{1}{n^2})), \quad n\to\infty.$$

The dominant singularity of $F(z)$ is given by the root (with smallest module) of the denominator, i.e. $z_2^* = \sqrt{2}-1 \equiv p$, and, with $z = z_2^* - \varepsilon$,

$$F(z) \sim \frac{1}{[(4-2\sqrt{2})^{1/2}\sqrt{\varepsilon}]} - \sqrt{2}/(4-2\sqrt{2}) + \mathcal{O}(\sqrt{\varepsilon}), \quad \varepsilon\to 0.$$

Hence

$$F_n \sim \frac{C_4}{\sqrt{n}z_2^{*n}}, \quad n\to\infty. \tag{22}$$

To get more precise information on F_{2n} and F_{2n-1}, we extract from (2) their GF, given respectively by

$$\begin{aligned} F^+(z) &:= \frac{F(z)+F(-z)}{2} = \frac{2(1-z^2+f_1)}{(1-z^2+f_1)^2-4z^2}, \\ \text{with } f_1 &:= \sqrt{1-6z^2+z^4}, \\ F^-(z) &:= \frac{F(z)-F(-z)}{2} = \frac{4z}{(1-z^2+f_1)^2-4z^2}. \end{aligned}$$

Set again $z^2 = u$ and $u = z_1^* - \varepsilon$, this leads to

$$F^+ \sim 2^{3/4}/(4\sqrt{\varepsilon}) - 2^{1/4}/32\sqrt{\varepsilon} + \mathcal{O}(\varepsilon^{3/2}), \quad \varepsilon\to 0.$$

Hence

$$\begin{aligned} F_{2n} &\sim \frac{1}{z_1^{*n}}[2^{1/4}(\sqrt{2}+2)/(4\sqrt{\pi n}) - (5+\sqrt{2})2^{1/4}/(64\sqrt{\pi}n^{3/2}) \\ &+ \mathcal{O}(n^{-5/2})], \quad n\to\infty. \end{aligned} \tag{23}$$

Also

$$zF^- \sim 2^{1/4}(2-\sqrt{2})/(4\sqrt{\varepsilon}) + 2^{1/4}(3\sqrt{2}+1)/32\sqrt{\varepsilon} + \mathcal{O}(\varepsilon^{3/2}), \quad \varepsilon\to 0.$$

Hence

$$F_{2n-1} \sim \frac{1}{z_1^{*n}}[2^{3/4}/(4\sqrt{\pi n}) - 5\cdot 2^{1/4}/(64\sqrt{\pi}n^{3/2}) + \mathcal{O}(n^{-5/2})], \quad n\to\infty.$$

It is combinatorically proved in [13] that $(n+1)R_n = F_{2n}+F_{2n-1}$. This amounts to prove that $F^+(z) + [zF^-](z) = R(u) + uR'(u)$, which is easily checked. It is easy to check that the first two terms of $F_{2n}+F_{2n-1}$ coincide with the first two terms of $(n+1)R_n$.

3.4 Lightning Procedure

First of all let us prove (3), the GF of Nt. Let h be the height of the walk sampled by the first part GA.1 of the GA given in [13], which is always odd, because the length of the walk is $2n+1$. A northeast step $\{(i,j),(i+1,j+1)\}$ of the walk is said to be lighted iff there is no southeast step $\{(m,j+1),(m,j)\}$ in the walk, for $m>i+1$. The lightning procedure simply transforms each lighted step whose height is less or equal than $\frac{h+1}{2}$, into a southeast step. Note that there are exactly $\frac{h+1}{2}$ lighted steps of height less or equal than $\frac{h+1}{2}$ because lighted steps are precisely those which increase the height of the walk. Hence in order to establish (3), we consider the bivariate generating function $F(w,z)$ of $SL(.)$ walks, where z counts the length and w counts the height. ¿From the functional equation for $F(z)$, given in [13], we have

$$F(w,z)=1+z^2R(z^2)F(w,z)+wzF(w,z)+z^2F(w,z),$$

where $R(z^2)$ is the generating function of $S(.)$ walks according to their length. The walk given by GA.1 is either a walk of length $2n$ in which a southeast step is added, or a walk of length $2n-1$ in which a (two times units) east step is added. Hence, the bivariate GF for this walk is

$$T(w,z)=\frac{z}{w}\frac{F(w,z)+F(w,-z)}{2}+z^2\frac{F(w,z)-F(w,-z)}{2}.$$

Now, if we want w to count only the half-height $(\frac{h+1}{2})$, we obtain

$$\begin{aligned}G(w,z) &= \sqrt{w}\,T(\sqrt{w},z)\\ &= \frac{2z(1-z^2+f_1+2wz^2)}{(-1+z^2-f_1+2\sqrt{w}\,z)(-1+z^2-f_1-2\sqrt{w}\,z)},\end{aligned}$$

as announced before.
Substitute $z^2=u$ in $G(w,z)/z=\varphi_4(w,u)$, say. Let us first note that, for $w=1$, we recover the GF of $(n+1)R_n$. Also, if $w=1$, the dominant singularity of $\varphi_4(1,u)$ is given by the root(with smallest module)of the denominator, i.e. $u^*=p^2=3-2\sqrt{2}$. The corresponding singularity of $\varphi_4(w,u)$ is given by

$$r(w)=[1+4w+w^2-(1+8w+14w^2+8w^3+w^4)^{1/2}]/(2w)$$

and $r(1)=u^*$. More precisely,

$$r(w)=u^*+C_6(w-1)^2+C_7(w-1)^3+\mathcal{O}((w-1)^4),\quad w\to 1,\quad \text{with}$$

$$C_6=(1/2-3/8\sqrt{2}),C_7=(-1/2+3/8\sqrt{2}).$$

As usual, we set $u=r(w)-\varepsilon$. We obtain

$$\begin{aligned}\varphi_4 &\sim \varphi_5/(\varphi_6\varepsilon)+\mathcal{O}(1),\quad \varepsilon\to 0,\quad \text{with}\\ \varphi_5 &= (w-1)[4-2\sqrt{2}+(5-3\sqrt{2})(w-1)+\mathcal{O}((w-1)^2)],\quad w\to 1,\\ \varphi_6 &= 4\sqrt{2}+6\sqrt{2}(w-1)+\mathcal{O}((w-1)^2),\quad w\to 1,\end{aligned}$$

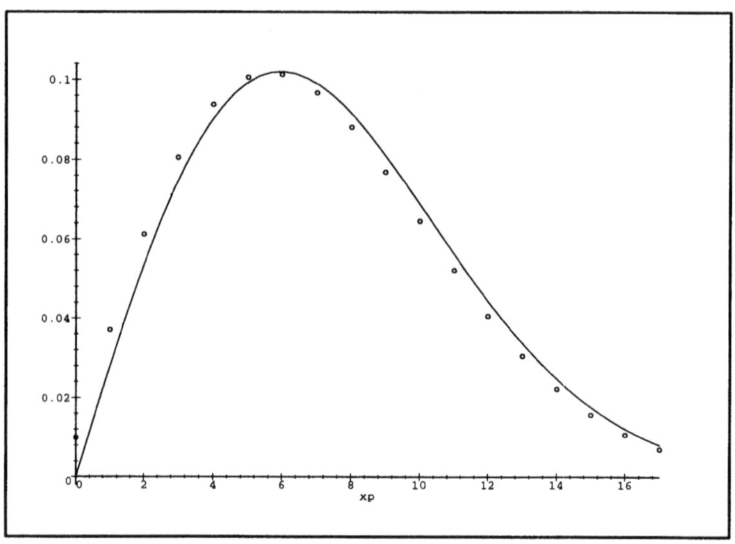

Figure 1: Observed and asymptotic normalized distribution of Nt

and

$$[u^n]\varphi_4(w,u) \sim \varphi_5/[r(w)\varphi_6\ r(w)^n], \quad n \to \infty. \tag{24}$$

Set now $w = e^{i\xi/\sqrt{2n}}$ in (24). Normalizing by $(n+1)R_n$, this leads to the asymptotic equivalent

$$i\sqrt{2\pi}\xi e^{-\xi^2\sqrt{2}/16}(\sqrt{2}/8)^{1/2}.$$

This corresponds exactly to the Rayleigh density (19), with a 1/2 factor, i.e. $Nt/\sqrt{2n} \overset{\mathcal{D}}{\sim} h/2$, as expected.

Let us check numerically our results. For $n = 100$, Fig.1 gives the asymptotic and observed normalized distribution of Nt (observed=circle, asymptotic=line). The observed one is extracted from $[u^{100}]\varphi_4$. The asymptotic one is given by $\exp(-x^2 2\sqrt{2})4x\sqrt{2}$.

There is an obvious bias: this can be explained as follows. Proceeding as in Hwang [9] ,Theorem 2, we analyze $\varphi_7(w)/\varphi_7(1)$, where φ_7 is extracted from (24): $\varphi_7(w) = \varphi_5/(s\ r(w)\varphi_6)$. Set now $w = e^{is}$ in $\ln[\varphi_7(w)/\varphi_7(1)]$ and expand into s. This gives the dominant term
$\frac{-\sqrt{2}}{4}is = -.3535539059381\ldots is = \beta \cdot is$, say and β is exactly the bias we have to reintroduce in Fig.1 ($n = 100$). This gives Fig.2, where the adjustment is now nearly perfect.

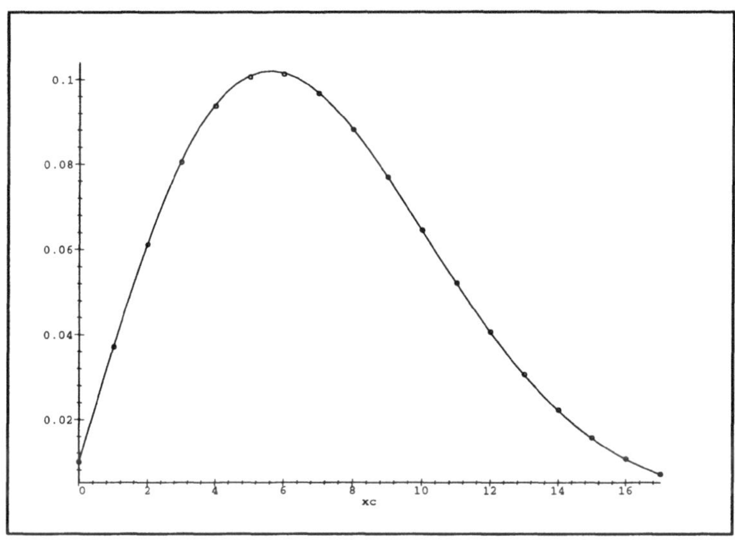

Figure 2: Observed and asymptotic normalized distribution of Nt, with bias

4 Asymptotic Analysis of the Generation Algorithm

In this section, we provide a complete asymptotic probabilistic analysis of the GA. We consider successively the rejecting procedure, the Meander generation, the Algorithm total cost and the cost moments.

4.1 Rejecting procedure

¿From the first part of the algorithm, we see that we accept the Meander with probability
$P = A_1/(A_1 + A_2 + A_3)$, where

$$\begin{aligned} A_1 &:= (n+1)R_n p^{2n+1}, \\ A_2 &:= F_{2n} p^{2n+2}, \\ A_3 &:= F_{2n} p^{2n+1}. \end{aligned} \tag{25}$$

Hence $P \sim \frac{1}{2}, n \to \infty$, by (21) and (22). An interesting alternative proof of this result is given in Appendix B, based on a renewal argument.

The number T of trials to obtain a suitable walk is such that, asymptotically,

$$T \overset{\mathcal{D}}{\sim} \text{GEOM(P)} \tag{26}$$

By (21) and (23), we obtain a more precise equivalent:
$P \sim 1/2 + (4 - 3\sqrt{2})/(32n) + \mathcal{O}(1/n^2)$.

4.2 Meander Generation

In terms of *the number* H_1 *of generated letters*, the RV $\tilde{H}_1 := H_1/(2n)$ has been analyzed in [11], $T2$: this is the cost of generating a Meander of length $2n$. We have obtained the Laplace transform

$$\phi_1(\alpha) = E[\exp(-\alpha\tilde{H}_1)] \sim e^{-\alpha}/[e^{-\alpha} + \sqrt{\pi\alpha}\ \mathrm{erf}(\sqrt{\alpha})], \quad n \to \infty. \tag{27}$$

(This was established for a Motzkin path, but the analysis is similar). This corresponds to a length 1 Brownian Meander generation. In terms of *the number* H_2 *of calls to the random generator* we must use the *total* number of steps M in an Excursion or a Meander. But from (16), we see that only $E(M)$ must be taken into account: the random (Gaussian) part is $\mathcal{O}(\sqrt{n})$ and is asymptotically negligible after normalizing by $1/(2n)$. Hence we obtain, with $\tilde{H}_2 := H_2/(2n)$: $\tilde{H}_2 \overset{\mathcal{D}}{\equiv} \mu_s\tilde{H}_1$ and

$$\phi_2(\alpha) = E[\exp(-\alpha\tilde{H}_2)] \sim e^{-\alpha\mu_s}/[e^{-\alpha\mu_s} + \sqrt{\pi\alpha\mu_s}\ \mathrm{erf}(\sqrt{\alpha\mu_s})], \quad n \to \infty. \tag{28}$$

4.3 Total Cost

The total cost is made of two parts: the first part $\kappa_.$ is related to the Meander generation : its asymptotic normalized density Laplace transform, namely $E[exp(-\alpha\tilde{\kappa}_.)]$, where $\tilde{\kappa}_. := \kappa_./(2n)$, is given by

$$\Psi_.(\alpha) = P\phi_.(\alpha)/[1 - (1-P)\phi_.(\alpha)], \tag{29}$$

where $\phi_.(\alpha) = \phi_1$ or ϕ_2, according to the cost elementary unit: letter or step.

The second part Nt is related to the ligthning procedure. The asymptotic normalized density of $Nt/\sqrt{2n}$ is given by $exp(-x^2 2\sqrt{2})4x\sqrt{2}$. As $Nt/\sqrt{2n} \overset{\mathcal{D}}{\equiv} h/2$, the moments of Nt are immediately derived from (20). The two costs are asymptotically independent.

4.4 Moments

The moments of Ψ_1 or Ψ_2 are easily computed. For instance, under Ψ_1 , with P:=1/2, $m_1 := E(\tilde{\kappa}_1) = 4, \mathrm{VAR}(\tilde{\kappa}_1) = 32/3$ and the third and fourth centered moments are given by $\mu_3(\tilde{\kappa}_1) = 352/5, \mu_4(\tilde{\kappa}_1) = 108544/105$. We just expand $e^{\alpha m_1}\Psi_.(\alpha)$ into α .

As a check, we extract from (21) the mean number of trials during the rejecting technique: this is given by

$$1/[(n+1)R_n p^{2n+1}] \sim \frac{\sqrt{n}}{C_4 p}, \quad n \to \infty. \tag{30}$$

The total mean number of letters used in a Meander generation is given by $E_1 + E_2 + E_3 + E_4$, with $E_1 := (2n+1)A_1, E_2 := (2n+2)A_2, E_3 := (2n+1)A_3$,

$$E_4 := p + \sum_{i=1}^{n-1}(2i+1)R_i p^{2i+1} \sim 2C_4 p \sum_{i=1}^{n-1} \frac{1}{\sqrt{i}} \sim 2C_4 p 2\sqrt{n}, \quad n \to \infty.$$

With (25), we obtain

$$E_6 := E_1 + E_2 + E_3 + E_4 \sim 2nC_4 2p/\sqrt{n} + 4C_4 p\sqrt{n} = 8C_4 p\sqrt{n}, \quad n \to \infty.$$

Multiplying by (30) leads to $8n = 2n.m_1$ as it should.

The moments of $\tilde{\kappa}_2$ are given by (as $\tilde{\kappa_2} \stackrel{\mathcal{D}}{\equiv} \mu_s \tilde{\kappa_1}$)
$m_1 = 2 + \sqrt{2}, \mathrm{VAR} = 4 + 8\sqrt{2}/3, \mu_3 = 22 + 77\sqrt{2}/5, \mu_4 = \frac{28832}{105} + \frac{6784}{35}\sqrt{2}$.

5 Simulations

We have done an extensive set of simulations, in order to check the quality of our asymptotic equivalents. For $n = 2000$, we have simulated $N = 1000$ generations. This has led to $NET = 134.553$ Excursions, $NM = 2083$ Meanders. To get reasonably long Excursions, we have observed in detail $NE = 11061$ Excursions of length ≥ 100.

The first picture Fig.3 gives (for $n = 30$) the RW $S(i)$. 4 Meanders were generated.

Fig.4 shows the Distribution Function (DF) of Nr : the number of returns to 0 (i.e. Excursions) before getting a length n=2000 Meander. According to (17), the limiting distribution is Exponential. The observations are based on NM generated Meanders.

The DF of the Meander height h is given in Fig.5, based on NM generated Meanders. According to (19), the limiting distribution is Rayleigh.

In Fig.6, we display the DF of Nh: the number of horizontal steps in Excursions, based on the NE Excursions of length ≥ 100. According to (15), the limiting distribution is Gaussian.

We have also checked in Fig.7 the same DF for the NM Meanders. Again we obtain a limiting Gaussian DF. Fig.8 shows the DF of T: the number of trials before getting a suitable Meander, based on N generations. According to (26), we obtain a GEOM(1/2) distribution. The fit is not as good as in our previous pictures. The observed frequency at 1 is .472. The corresponding standard deviation $\sigma = \sqrt{1/(4N)} = 0.016 \cdots$ So the observed frequency falls within 1.77σ of $1/2$, which is statistically acceptable.

In Fig.9, we give the observed and limiting Laplace transform $\Psi_1(\alpha)$ of the Meander generation cost (according to the number of generated letters). The fit with the expression given in (29) is quite good.

Fig.10 shows the same observation for the steps cost (number of calls to the random generator). Here we consider $\Psi_2(\alpha)$.

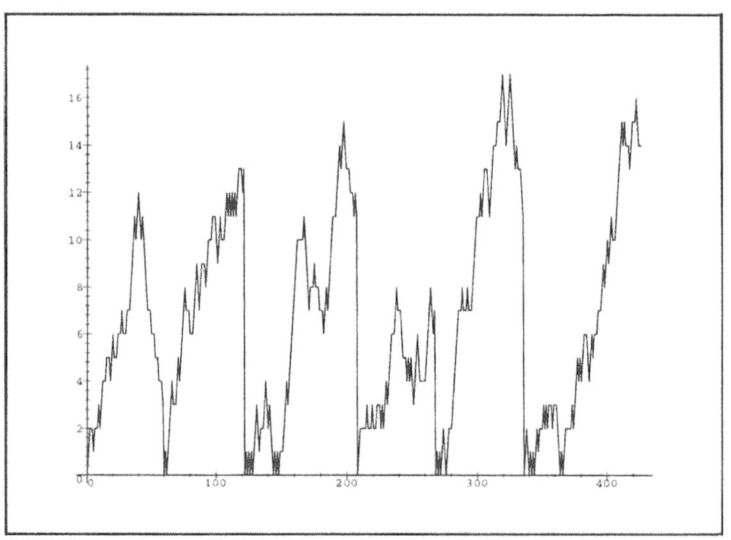

Figure 3: $S(i), n = 30$.

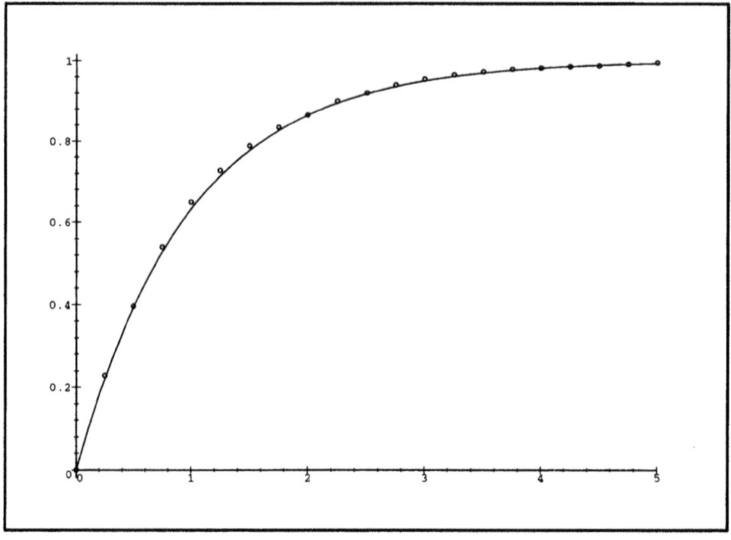

Figure 4: Nr : observed and limiting (exponential) DF.

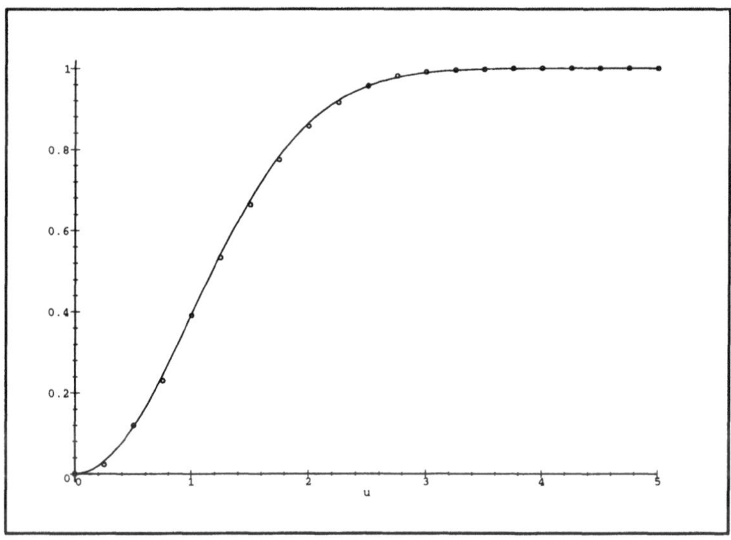

Figure 5: h : Observed and limiting (Rayleigh) DF.

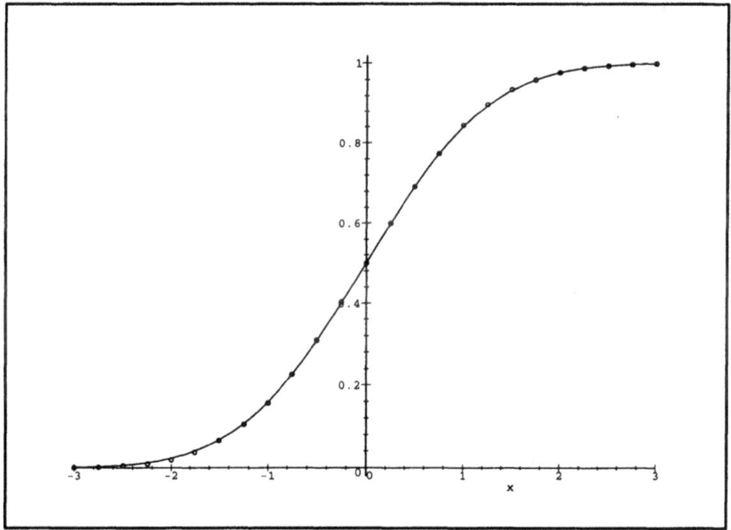

Figure 6: Nh in Excursions : observed and limiting (Gaussian) DF.

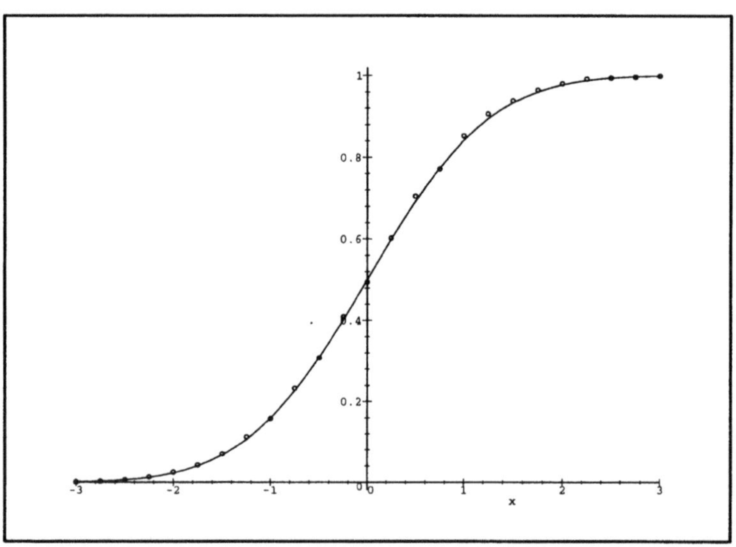

Figure 7: Nh in Meanders : observed and limiting (Gaussian) DF.

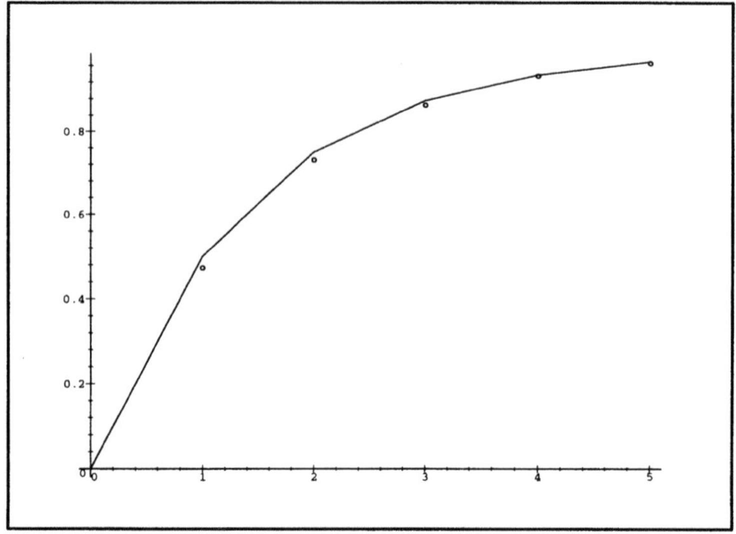

Figure 8: T : observed and limiting (GEOM(1/2)) DF.

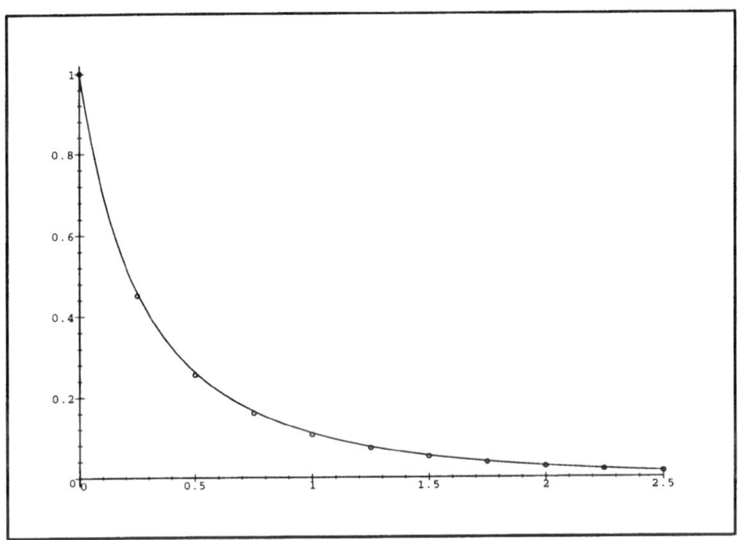

Figure 9: Cost(letters) : observed and limiting ($\Psi_1(\alpha)$) Laplace transform.

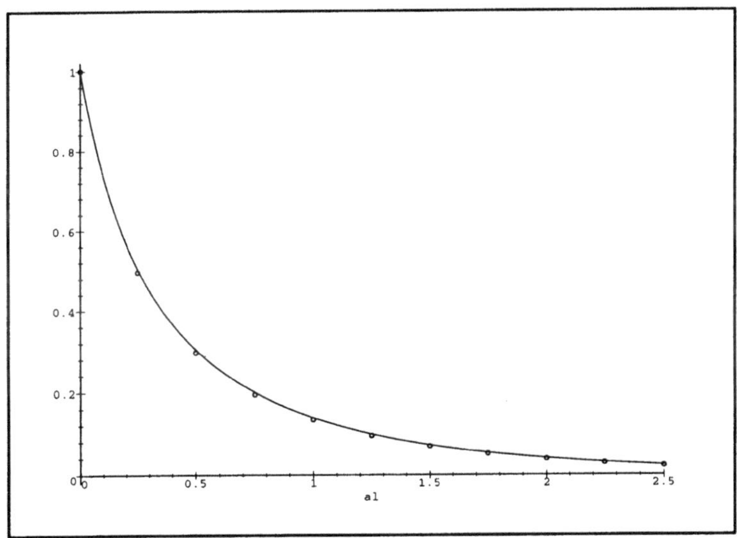

Figure 10: Cost(steps) : observed and limiting ($\Psi_2(\alpha)$) Laplace transform.

6 Conclusion

Using various tools from Combinatorics, Probability Theory, and Singularity analysis, we have achieved a complete asymptotic probabilistic analysis of the cost of a Schröder walk generation algorithm proposed by Penaud et al. ([13]). Five different probability distributions have been observed. We intend to pursue such approach in other generation algorithms found in the litterature.

References

[1] E. Bender. Central and local limit theorems applied to asymptotics= enumeration. *Journal of Combinatorial Theory,* Series A, 15:91–111, 1973.

[2] P. Billingsley. *Convergence of Probability Measures.* Wiley, 1968.

[3] M. Drmota and M. Soria. Marking in combinatorial constructions: Generating functions and limiting distributions. *Theoretical Computer Science*, 144(1–2):67–100,= 1995.

[4] A. Dvoretzky and T. Motzkin. A problem of arrangements. *Duke Mathematical Journal*, 14:305–313,= 1947.

[5] W. Feller. *Introduction to Probability Theory and its= Applications.* Wiley, 1970.

[6] P. Flajolet and R. Sedgewick. The average case analysis of algorithms: Multivariate asymptotics= and limit distributions. *INRIA T.R.*, 3162, 1997.

[7] P. Flajolet, P. Zimmerman, and B. V. Cutsem. A calculus for the random generation of labelled combinatorial structures. *Theoretical Computer Science*, 132:1–35,= 1994.

[8] D. Gouyou-Beauchamps and G. Viennot. Equivalence of the two-dimensional directed animal problem to a one-dimensional path problem. *Advances in Applied Mathematics*, 9:334–357,= 1988.

[9] H. Hwang. On convergence rates in the central limit theorems for= combinatorial structures. *European Journal of Combinatorics*, To= appear.

[10] K. Ito and H. McKean, Jr. *Diffusion Processes and their Sample Paths.* Springer, 1974.

[11] G. Louchard. Asymptotic properties of some underdiagonal walks generation algorithms. *Theoretical Computer Science*, 218:249–262,= 1999.

[12] G. Louchard and W. Szpankowski. A probabilistic analysis of a string edit problem and its= variations. *Combinatorics, Probability and Computing*, 4:143–166,= 1995.

[13] J. Penaud, E. Pergola, R. Pinzani, and O. Roques. Chemins de Schröder et hiérarchies aléatoires. *Theoretical Computer Science*, To= appear.

[14] E. Schröder. Vier combinatorische probleme. *Z. Math. Phys.*, 15:361–370, 1870.

A Appendix: Proof of (14) for Meanders.

Proof Set $\tau := T_0^{(h)}/(2n)$. We must compute

$$[u^\beta][\varphi_3(w,u)]^{h\sqrt{2n}}, \tag{31}$$

where β is such that $n \cdot \tau = \beta - h\sqrt{2n}/2$. Now we analyze

$$\varphi_3(w,u) := (1 - p^2uw)/(2p) - [G_3(w,u)]^{1/2}/(2p), \tag{32}$$

with

$$G_3(w,u) := 1 - 2p^2uw + p^4u^2w^2 - 4p^2u.$$

First, we set $w = e^{is}, u = e^{-\theta}$ in the first part of (32). This leads to $G_4(s) + \theta G_5(s) + \mathcal{O}(\theta^2)$, with $G_4(s) = 1 - isp/2, G_5(s) = p/2 + \mathcal{O}(s)$.
Wc should be tempted to proceed immediately in the same way for G_3. But this is not correct: as we shall see, we need a more precise analysis, based on the GF. So we set $u = 1 - \varepsilon$, this gives

$$\begin{aligned} G_3 &= [(-11 + 8\sqrt{2}) + (4\sqrt{2} - 6)w + (17 - 12\sqrt{2})w^2] \\ &+ [(12 - 8\sqrt{2}) + (6 - 4\sqrt{2})w + (-34 + 24\sqrt{2})w^2]\varepsilon + \mathcal{O}(\varepsilon^2), \quad \varepsilon \to 0. \end{aligned}$$

But, if $\varphi(u) = \sum p_k u^k$, then $[u^k][\varphi(u\alpha)] = p_k\alpha^k$. If we want to cancel the constant term $G_3(w,1)$ in G_3, we must choose α such that $G_3(w,\alpha) = 0$, namely $\alpha = r(w)$ as given by (10). This leads to

$$\tilde{G}_3 := G_3(w, r(w)u) = \varepsilon G_6(w) + \mathcal{O}(\varepsilon^2), \quad \varepsilon \to 0,$$

with

$$G_6 = -2p^4r(w)^2w^2 + 2p^2r(w)w + 4p^2r(w).$$

Now we can set $w = e^{is}, u = e^{-\theta}$. This leads to

$$\tilde{G}_3 = \theta[-16 + 12\sqrt{2} + is(24 - 17\sqrt{2}) + \mathcal{O}(s^2)] + \mathcal{O}(\theta^2), \quad \theta \to 0,$$

and

$$\sqrt{\tilde{G}_3}/(2p) = (2^{1/4} + \mathcal{O}(s))\sqrt{\theta} + \mathcal{O}(\theta^{3/2}), \quad \theta \to 0.$$

Now setting $u = r(w)u$ amounts to replace θ in the first part of (32) by $\theta - [is(\sqrt{2}/2 - 1) + \mathcal{O}(s^2)]$, by (13). Therefore, we obtain, for the probability distribution of τ, the asymptotic expression

$$\Big\{[G_4(s) + (\theta - is(\sqrt{2}/2 - 1) + \mathcal{O}(s^2))G_5(s) \\ -(2^{1/4} + \mathcal{O}(s))\sqrt{\theta} + \mathcal{O}(\theta^{3/2})]^{h\sqrt{2n}}\Big\}^{(-1)} e^{-\beta\, ln[r(w)]}, \quad \theta \to 0,$$

where $\varphi^{(-1)}(\theta)$ is the probability density corresponding to $\varphi(\theta)$. This leads to

$$\Big\{\exp[h\sqrt{2n}[-isp/2 - is(\sqrt{2}/2-1)p/2 + \mathcal{O}(s^2) - (2^{1/4}+\mathcal{O}(s))\sqrt{\theta} + \mathcal{O}(\theta)] \\ -(n\tau + h\sqrt{2n}/2)[is(\sqrt{2}/2-1) + \sqrt{2}/16s^2 + \mathcal{O}(s^3)]\Big\}^{(-1)}, \quad \theta \to 0.$$

First of all, the $\sqrt{n}s$ term disappears(as it should). Setting now $\theta = \theta/n$, we obtain, asymptotically, for the θ contribution:

$$e^{-\sqrt{\theta}h2^{1/4}\sqrt{2}},$$

which is identical to (18). We are left with

$$\exp[n\tau[is(1-\sqrt{2}/2) - \sqrt{2}/16s^2] + \mathcal{O}(ns^3) + \mathcal{O}(\sqrt{n}s^2)], \quad s \to 0,$$

which proves (14). ■

We observe that, asymptotically, for fixed τ, Nh is conditionally independent of h. This can be explained probabilistically as follows. Let us return to the original RW $Y(.)$. ¿From (5), in $Y(n)$,
$Nh \stackrel{\mathcal{D}}{\sim} n\mu_h + \sigma_h\sqrt{n}\xi_1$, where $\xi_1 = \mathcal{N}(0,1)$. The number of up and down steps $Nv = n - 2Nh \sim n(1-2\mu_h) - 2\sigma_h\sqrt{n}\xi_1$ and $Y(n) \sim \sum_1^{Nv} V_i$, with $V_i = \pm 1$, with probability 1/2. Hence $Y(n) \sim \sqrt{Nv}\xi_2$, with $\xi_2 = \mathcal{N}(0,1)$ (independently of ξ_1). So

$$\begin{aligned} E(e^{i[xY(n)+yNh]}) &\sim E(e^{-Nv\ x^2/2+iyNh}) \\ &\sim E\{\exp[-x^2/2(n\sqrt{2}/2 - 2\sigma_h\sqrt{n}\xi_1) \\ &\quad +iy(n\mu_h + \sigma_h\sqrt{n}\xi_1)]\} \\ &\sim \exp[-x^2/2n\sqrt{2}/2 + iyn\mu_h - \sigma_h^2 n/2[x^2/i+y]^2] \\ &= \exp(iyn\mu_h)\cdot\exp[-x^2/2n\sqrt{2}/2 - n\sigma_h^2/2[x^2/i+y]^2]. \end{aligned}$$

Now we replace x by $x/\sqrt{n}$ and y by $y/\sqrt{n}$ in the second term. We are left, asymptotically, with $\exp[-x^2/2\sqrt{2}/2 - \sigma_h^2 y^2/2]$, which shows the asymptotic independence of $Y(.)$ and Nh (incidentally, we recover (7)).

B Appendix: A Renewal Approach

If we look only at the time displacement D, we have $Pr[D=1] = 2p, Pr[D=2] = p^2$. The renewal theorem (for discrete RV) gives the following DF for the residual waiting time (see Feller [5],XI.4): $H(j) = \sum_0^{j-1}[1-F(k)]/m_1$, where F is the DF of D and m_1 is its mean. We obtain $H(0) = 0$, $H(1) = 1/m_1$, $H(2) = 1$. Now let Ph be the stationary distribution of an horizontal step covering time n ($n \to \infty$) and let $Pv := 1 - Ph$. We must have $H(1) = Ph.1 + Pv.2p$, hence
$Ph = (2-\sqrt{2})/4$, $Pv = (2+\sqrt{2})/4$. The asymptotic probability P of success is given by
$Ph + Pv\cdot p = 1/2$ as expected.

Trends in Mathematics, © 2000 Birkhäuser Verlag Basel/Switzerland

Gibbs Families

V. MALYSHEV *I.N.R.I.A. Rocquencourt, Domaine de Voluceau, 78153 Le Chesnay Cedex, France*

Abstract. *Gibbs random field is now one of the central objects in probability theory. We define a generalization of Gibbs distribution, when the space (lattice, graph) is not fixed but random. Moreover, randomness of the space is not given independently of the configuration but both depend on each other. We call such objects Gibbs families because they appear to parametrize sets of ordinary Gibbs distributions. Moreover, they are well suited to study local probability structures on graphs with random topology. First results of this theory are presented here.*

Gibbs random field [2, 3] is now one of the central objects in probability theory. Here we define a natural generalization of Gibbs distribution, when the space (lattice, graph) is not fixed but random. Moreover, randomness of the space is not given apriori and independently of the configuration but both depend on each other. We call the introduced objects Gibbs families because, as it will be shown, they parametrize sets of Gibbs distributions on fixed graphs.

We present here the foundations of such theory. Two central definitions are given. The first one is Gibbs family on the set of countable spingraphs (spingraph is a graph with a function on its vertices) with the origin (a specified vertex). To deal with countable graphs where no vertex is specified we define the notion of empirical distribution. The conditional measure on the configurations of a given graph G gives standard Gibbs field on G with the same potential. We discuss some examples of Gibbs families.

This paper is selfcontained, but it is a part of the series of papers by the author starting with [5] concerning new connections between computer science and mathematical physics.

1 Definitions

Spingraphs We consider non-directed connected (unless otherwise stated) graphs G (finite or countable) with the set $V = V(G)$ of vertices and the set $L = L(G)$ of links. The following properties are always assumed: between each pair of vertices there is 1 or 0 edges; each node (vertex) has finite degree (the number of edges incident to it).

A *subgraph* of G is a subset $V_1 \subset V$ of vertices together with some links connecting pairs of vertices from V_1 and inherited from L. A *regular subgraph* $G(V_1)$ of G is a subset $V_1 \subset V$ of vertices together with ALL links connecting pairs of vertices from V_1 and inherited from L.

The set V of nodes is a metric space with the following metrics: the distance $d(x, y)$ between vertices $x, y \in V$ is the minimum of the lengths of paths connecting these vertices. The lengths of all edges are assumed to be equal, say to some constant, assumed equal to 1.

Spingraph $\alpha = (G, s)$ is a graph G together with a function $s : V \to S$, where S is some set of spin values. Spingraphs are always considered up to isomorphisms. Isomorphism here is an isomorphism of graphs respecting spins. Denote $\mathcal{G}_N$ $(\mathcal{A}_N)$ - the set of equivalence classes (with respect to isomorphisms) of connected finite graphs (spingraphs) with N vertices.

Spingraphs with the origin Define the annular neighbourhood $\gamma(\alpha, v; a, b)$ to be the regular subgraph of α defined by the set $V(\gamma(\alpha, v; a, b)) = \{v' : a \leq d(v, v') \leq b\}$. $O_d(v) = \gamma(\alpha, v; 0, d)$ is the d-neighborhood of $v, O(v) = O_1(v)$. Let $R_0(G) = \max_{v \in V(G)} d(0, v)$ be the radius of graph G with respect to vertex 0. Thus $R_v(O(v)) = 1$. Radius of a graph is $R(G) = \min_{v \in V(G)} R_v(G)$.

We will also consider graphs with one specified vertex, the origin. In this case isomorphisms are also assumed to respect the origin. When we want to emphasize that the graphs are given together with an origin v we shall write it as $G^{(v)}, \mathcal{G}^{(v)}$ etc.

Let $\mathcal{G}_N^{(0)}$ $(\mathcal{A}_N^{(0)})$ - the set of equivalence classes of finite graphs (spingraphs) with respect to isomorphisms, with origin 0 and having radius N with respect to 0. Further on, spingraph will mean its equivalence class. Let $\mathcal{A}^{(0)} = \cup \mathcal{A}_N^{(0)}$ be the set of all finite spingraphs with the origin 0.

σ-algebra and free measure Let $\mathcal{A}$ be an arbitrary set of finite spingraphs $\alpha = (G, s)$. Let $\mathcal{G} = \mathcal{G}(\mathcal{A})$ be the set of all graphs G such that there exists s with $\alpha = (G, s) \in \mathcal{A}$. We assume always that if $\alpha = (G, s) \in \mathcal{A}$ then for any s' all (G, s') also belong to $\mathcal{A}$.

$\mathcal{A}$ is a topological space which is a discrete (finite or countable) union of topological spaces $T_G = S^{V(G)}, G \in \mathcal{G}(\mathcal{A})$. We consider the Borel σ-algebra on $\mathcal{A}$. It is generated by cylindrical subsets $A(G, B_v, v \in V(G))$, $G \in \mathcal{G}$, where B_v are some Borel subsets of S. $A(G, B_v, v \in V(G))$ is the set of all $\alpha = (G, s)$ such that G is fixed and functions (configurations) $s(v) : V(G) \to S$ are such that $s(v) \in B_v$ for all $v \in V(G)$.

Let some nonnegative measure λ_0 on S be given. The following nonnegative measure $\lambda_{\mathcal{A}}$ (not necessarily a probability measure) on $\mathcal{A}$

$$\lambda_{\mathcal{A}}(A(G, B_v, v \in V(G))) = \prod_{v \in V(G)} \lambda_0(B_v) \tag{1}$$

is called a free measure.

Regular potentials Potential is a function $\Phi : \cup_N \mathcal{A}_N \to R \cup \{+\infty\}$, that is a function on the set of finite spingraphs invariant with respect to isomorphisms of spingraphs. We say that Φ has a finite radius if $\Phi(\alpha) = 0$ for all α with radius greater than some $r_0 < \infty$. Unless otherwise stated we shall consider only radius 1 case, that is $r_0 = 1$. Fix potential Φ and measure λ_0 on S. The energy of spingraph α is defined as

$$H(\alpha) = \sum_{B \subset V(G)} \Phi(\Gamma(B)), \alpha = (G, s) \tag{2}$$

where the sum is over all subsets B of $V(G)$, $\Gamma(B)$ is the regular subgraph of α with $V(\Gamma(B)) = B$. Gibbs $\mathcal{A}$-family with potential Φ is the following probability measure $\mu_{\mathcal{A}}$ on $\mathcal{A}$

$$\frac{d\mu_{\mathcal{A}}}{d\lambda_{\mathcal{A}}}(\alpha) = Z^{-1}\exp(-\beta H(\alpha)), \alpha \in \mathcal{A} \tag{3}$$

assuming $Z \neq 0$ and the stability condition

$$Z = Z(\mathcal{A}) = \int_{\mathcal{A}} \exp(-\beta H(\alpha)) d\lambda_{\mathcal{A}} < \infty \tag{4}$$

Here $\beta \geq 0$ is the inverse temperature.

Markov property The set $\mathcal{A}_{\infty}^{(0)}$ of countable spingraphs with the origin is a topological space. The basis of its open subsets is defined as follows. Let C_N be an arbitrary open subset of the set $\mathcal{A}_N^{(0)}$ of all spingraphs with radius N with respect to the origin. Then the basis consists of sets of all countable spingraphs such that (for some N and some C_N) their N-neighborhood of the origin belongs to C_N. We again take the Borel σ-algebra as the basic σ-algebra.

For notational convenience we give definition only for potential of radius 1. For any spingraph α with the origin 0 we call $\gamma(\alpha, 0; N, N)$ its N-slice, $\gamma(\alpha, 0; 0, N)$ - the N-interior, $\gamma(\alpha, 0; N, \infty)$ - the N-exterior parts. Note that N-slice and the exterior part may be non-connected. There are no links connecting $(N-1)$-interior and $(N+1)$-exterior parts of the graph: this will give us an analog of the Markov property. For given N and given some finite graph γ (it may be non-connected) let $\mathcal{A}(\leq N, \gamma) \subset \mathcal{A}_N^{(0)}$ be the set of all finite connected spingraphs α having radius N with respect to the origin and such that γ is the N-slice of α.

Definition 1 *We call measure μ on $\mathcal{A}_{\infty}^{(0)}$ a Gibbs family with regular potential Φ if, for any γ and any N-exterior part with N-slice γ, the conditional distribution on the set of spingraphs α having fixed exterior part (that is on N-neighborhoods with N-slice γ) coincides a.s. with the Gibbs family with potential Φ on $\mathcal{A}(\leq N, \gamma)$.*

In particular, it depends only on the spingraph γ but not on the whole N-exterior part.

Boundary conditions Boundary condition is a sequence $\nu_N(\gamma)$ of measures on the set of finite (not necessary connected) spingraphs γ. Intuitively - on N-slices. Gibbs family on $\mathcal{A}_N^{(0)}$ with boundary conditions $\nu_{N+1}(\gamma)$ is defined as

$$\mu_N(\alpha) = Z^{-1}(N, \nu_{N+1}) \sum_{\xi \in A(\alpha,\gamma)} \int \exp(-\beta H(\xi)) d\nu_{N+1}(\gamma), \alpha \in \mathcal{A}_N^{(0)} \tag{5}$$

where $A(\alpha, \gamma)$ is the set of all spingraphs from $\mathcal{A}_{N+1}^{(0)}$ with $(N+1)$-slice γ and N-interior part α. Note that $A(\alpha, \gamma)$ is finite for any α, γ.

Non-regular potentials Gibbs family with non-regular potential Φ on $\mathcal{A}$ here is defined similarly but with the energy of α defined by the formula

$$H(\alpha) = \sum_{\gamma \subset \alpha} \Phi(\gamma), \alpha = (G, s) \tag{6}$$

where the sum is over all (not necessary regular) subspingraphs γ of α.

A particular case is the "chemical" potential. It is given by a function Φ equal to a constant z_0 for each vertex (independently of spins in this vertex), equal to a constant z_1 for each link (independently of spins in both vertices of the link) and equal 0 otherwise. Then

$$H_N(\alpha) = z_0 V(\alpha) + z_1 L(\alpha) \tag{7}$$

where $V(\alpha), L(\alpha)$ are the numbers of vertices and links in α.

2 Limiting correlation functions

Compactness Gibbs families on the set $\mathcal{A}_{\infty}^{(0)}$ of connected countable spingraphs with fixed vertex 0 can be obtained as weak limits of Gibbs families on finite graphs. There are 3 sources of non-existence of limiting Gibbs families: non-existence of finite Gibbs families (see examples below), non-compactness of S, the distribution for a finite Gibbs family can be concentratred on the graphs with large degrees of vertices (ultraviolet problem). The assumptions of the following propostion correspond to this list, but the compactness could be proven under weaker conditions.

Let $\mathcal{A}_{\infty}^{(0)}(r) \subset \mathcal{A}_{\infty}^{(0)}$ be the set of countable spingraphs where each vertex has degree not greater than r.

Proposition 1 *Assume S to be compact. Assume that for all N there exists a Gibbs family with potential Φ on $\mathcal{A}_N^{(0)}$. Assume that Φ has radius 1 and $\Phi(O(v)) = \infty$ if the degree of v is greater than some constant r. Then there exists a Gibbs family with potential Φ, with support on $\mathcal{A}_{\infty}(r)$.*

Proof. Consider the Gibbs family μ_N on $\mathcal{A}_N^{(0)}$. Let $\tilde{\mu}_N$ be an arbitrary probability measure on $\mathcal{A}_{\infty}^{(0)}$ such that its factor measure on $\mathcal{A}_N^{(0)}$ coincides with μ_N. Then the sequence $\tilde{\mu}_N$ of measures on the set $\mathcal{A}_{\infty}^{(0)}$ is compact, that can be proved by the standard diagonal process by enumerating all possible N-neighborhoods. Moreover, any limiting point of $\tilde{\mu}_N$ is a Gibbs family with potential Φ.

Generators To give examples of Gibbs families it is useful to introduce the following classes of graphs. Consider a set of "small" graphs $G_1, ..., G_k$ of the same radius r with respect to a specified vertex 0. Graph G is said to be generated by $G_1, ..., G_k$ if each vertex of G has a neighborhood (of radius r) isomorphic to one of $G_1, ..., G_k$. Let $\mathcal{G}(G_1, ..., G_k)$ be the set of all graphs generated by $G_1, ..., G_k$.

With each system $\mathcal{G}(G_1, ..., G_k)$ and fixed S we associate the following class $\mathcal{F}(G_1, ..., G_k)$ of potentials $\Phi = \Phi(G, s)$ if $G = G_i$ for some $i = 1, ..., k$, $\Phi(G, s) = \infty$ for all other graphs with the same radius r and $\Phi(G, s) = 0$ for all graphs with $R_0(G) \neq r$.

Nonexistence This class of potentials satisfies the previous compactness criteria if there exist Gibbs families on finite graphs. However this is not always the case.

For example, consider the case $k = 1, r = 1$. If G_1 is a finite complete graph then it is "selfgenerating" but there are no countable graphs generated by it. Introduce some graphs with radius 1: g_k with $k + 1$ vertices $0, 1, 2, 3, ..., k$ and k links $01, 02, ..., 0k$, and $g_{k,k}$ with $k + 1$ vertices $0, 1, 2, 3, ..., k$ and $2k$ links $01, 02, ..., 0k, 12, 23, ..., k1$. One can see after several iterations that the pentagon $G_1 = g_{5,5}$ cannot generate any countable graph.

Gibbs families and Gibbs fields ¿From the following simple result it becomes clear why the introduced distributions are called Gibbs families. Let μ be a Gibbs family with potential Φ. Assume that the conditions of the previous proposition hold. Consider the measurable partition of the space $\mathcal{A}_{\infty}^{(0)}$ of spin graphs: any element S_G of this partition is defined by a fixed graph G and consists of all configurations s_G on G.

Theorem 1 *For any given graph G the conditional measure on the set of configurations s_G is a.s. a Gibbs measure with the same potential Φ.*

Thus, any Gibbs family is a convex combination (of a very special nature) of Gibbs fields (measures) on fixed graphs, with the same potential, with respect to the measure $\nu = \nu(\mu)$ on the factor space $\mathcal{A}_{\infty}^{(0)} / \{S_G\}$ induced by μ.

The following lemma reduces this result to the corresponding result for Gibbs families on finite graphs where it is a straightforward calculation.

Lemma 1 *Let μ be a Gibbs family on $\mathcal{A}_{\infty}^{(0)}$ with potential Φ. Let $\nu_N(\gamma)$ be the measure on N-slices induced by μ. Then μ is a weak limit of Gibbs families on $\mathcal{A}_N^{(0)}$ with potential Φ and boundary conditions $\{\nu_N(\gamma)\}$.*

Topological phase transition We say that Gibbs family with potential Φ is a pure Gibbs family if it is not a convex combination of Gibbs families with the same potential. If there are more than one pure Gibbs family with potential Φ we say that Gibbs family with potential Φ is not unique. Intuitively, nonuniqueness of Gibbs families can be of two kinds: due to the structure of configurations (inherited from usual Gibbs fields) and due to topology of the graph. The following example (using Cayley graphs of abelian groups) gives an example of a "topological phase transition".

Let $U_{d,2}$ be the graph isomorphic to the 2-neighborhood of 0 on the lattice Z^d. Then the following graphs belong to $\mathcal{G}(U_{d,2})$. This is Z^d itself and any of its factorgroups $Z(k_1, ..., k_d)$ with respect to some subgroup of Z^d, generated by some vector $(k_1, ..., k_d), k_i \geq 1$.

Consider now Ising Gibbs fields on $Z(k_1, ..., k_d)$ with some potential from $\mathcal{F}(G_1)$. More exactly, take $S = \{-1, 1\}$ and introduce the following potential Φ. $\Phi(U_{d,2}, s_{U_{d,2}})$ is equal to the energy of the configuration $s_{U_{d,2}}$ corresponding to the nearest neighbor Ising model on $U_{d,2}$. For all other spingraphs α of radius 2 we put $\Phi(\alpha) = \infty$, we put $\Phi(\alpha) = 0$ if the radius of α is different from 2.

We know from statistical physics that for any such graph there is only one Gibbs field with potential Φ for β sufficiently small. The following theorem shows that the Gibbs family with potential Φ is not unique, however, each pure Gibbs family we found is a Gibbs field on the fixed graph.

Theorem 2 *For any β sufficiently small there are countable number of pure Gibbs families with potential Φ.*

Proof. An example (for $d = 2$) of a pure Gibbs family with potential Φ is the unique Gibbs field on $Z \times Z_n$ for any $n \geq 5$. Consider the (unique) Gibbs random field $\mu(n)$ with potential Φ on this infinite cylinder. Take the induced measure $\nu_N(\mu(n))$ on N-slices, that is on non-connected union of two circular strips of length n. Using Lemma 1 take Gibbs families with potential Φ and boundary conditions $\nu_N(\mu(n))$. Then the graph is uniquely defined (one can construct slice $N-1, N-2, ...$ by induction). Thus, this Gibbs family coincides the with the pure Gibbs field on the correponding graph. Another interesting possibility are Gibbs fileds on twisted cylinders, that is on the sets obtained from the strip $Z \times \{1, 2, ..., k\}$ if for each n we identify the points $(n, 1)$ and $(n + j, k)$.

3 Empirical correlation functions

We constructed probability measures on $\mathcal{A}_\infty^{(0)}$ in quite a standard way, using standard Kolmogorov approach with cylindrical subsets. However, one cannot use similar approach to define a probability measure on the set $\mathcal{A}_\infty$ of equivalence classes of countable connected spingraphs. The problem is that it is not at all clear how to introduce finite-dimensional distributions here, because the vertices are not enumerated (there is no coordinate system). Thus all Kolmogorov machinery fails. However one can propose an analog of finite-dimensional distributions. We call the resulting system an empirical distribution.

Assume S to be finite or countable. Let us consider systems of numbers

$$\pi = \left\{p(\Gamma), \Gamma \in \mathcal{A}^{(0)}\right\}, 0 \leq p(\Gamma) \leq 1, \tag{8}$$

i.e. Γ is an arbitrary finite spingraph with origin 0. We assume the following compatibility condition: for any $k = 0, 1, 2, ...$ and any fixed graph Γ_k of radius k we have

$$\sum_{\Gamma_{k+1}} p(\Gamma_{k+1}) = p(\Gamma_k), k = 0, 1, 2, ... \tag{9}$$

where the sum is over all Γ_{k+1} of radius $k+1$ such that $O_k(0, \Gamma_{k+1})$ is isomorphic to Γ_k. It is assumed that the summation is over equivalence classes of spingraphs.

We assume also the following normalization condition

$$\sum p(\Gamma_0) = 1 \tag{10}$$

where Γ_0 is the vertex 0 with any possible spin on it.

Definition 2 *Any such system π is called an empirical distribution.*

One can also rewrite compatibility conditions in terms of conditional probabilities

$$\sum_{\Gamma_{k+1}} p(\Gamma_{k+1} \mid \Gamma_k) = 1 \tag{11}$$

where the summation is as above and

$$p(\Gamma_{k+1} \mid \Gamma_k) = \frac{p(\Gamma_{k+1})}{p(\Gamma_k)} \tag{12}$$

Thus we consider $\mathcal{A}^{(0)}$ as a tree where vertices are spingraphs and a link between Γ_{k+1} and Γ_k exists iff Γ_k is isomorphic to the k-neighborhood of 0 in Γ_{k+1}.

Examples of empirical distributions can be obtained via the following limiting procedure. Let μ_N - probability measure on $\mathcal{A}_N$. For any N and any $\Gamma \in \mathcal{A}_k^{(0)}, \alpha \in \mathcal{A}_N$ put

$$p^N(\Gamma) = \left\langle \frac{n^N(\alpha, \Gamma)}{N} \right\rangle_{\mu_N} \tag{13}$$

where $n^N(\alpha, \Gamma)$ is the number of vertices in α having their k-neighborhoods isomorphic to Γ. Denote $\pi_N = \{p^N(\Gamma)\}$.

Lemma 2 *Assume that S is finite. Assume that for any r*

$$\mu_N(\min_{\alpha \in \mathcal{A}_N} R(\alpha) \leq r) \to 0 \tag{14}$$

as $N \to \infty$. Then any weak limiting point of π_N is an empirical distribution.

Proof. Note that for any n_0 there exists such $N_0 = N_0(n_0)$ such that for any $N > N_0$ numbers $p_N(\Gamma)$ satisfy compatibility conditions for all Γ with radius $n < n_0$. Finiteness of S implies compactness.

Interesting question is how one could characterize empirical distributions which can be obtained via this limiting procedure.

Empirical Gibbs families Note that for a given π the numbers $p(\Gamma)$ for all Γ with $R_0(\Gamma) = N$ define a probability distribution π_N on $A_N^{(0)}$. For any $n < N$ let $\pi_{N,n}(\gamma)$ be the probability that n-slice of Γ is equal to γ, and $\pi_{N,n}(\Gamma_n)$ be the probability that the neighborhood $O_n(\Gamma)$ of 0 is isomorphic to Γ_n.

For any γ and any Γ_n with n-slice γ introduce conditional probabilities

$$\pi_{N,n}(\Gamma_n|\gamma) = \frac{\pi_{N,n}(\Gamma_n)}{\pi_{N,n}(\gamma)} \tag{15}$$

An empirical distribution π is an empirical Gibbs family with potential Φ if for any $n < N, \gamma, \Gamma_n$ the distribution $\pi_{N,n}(\Gamma_n|\gamma)$ is the Gibbs family with potential Φ on the set spingraphs Γ_n with n-slice γ, that is

$$\pi_{N,n}(\Gamma_n|\gamma) = Z_n^{-1}(\gamma)\exp(-H(\Gamma_n)) \tag{16}$$

We see that in this case $\pi_{N,n}(\Gamma_n|\gamma)$ do not depend on N.

Two-dimensional quantum gravity (or string in zero dimension) In the following examples there is no spin, the free measue is trivial, and the partition function contains summation over the corresponding graphs.

We consider graphs corresponding to the set of all triangulations $\mathcal{T}_N$ of two-dimensional sphere with N triangles, where combinatorially isomorphic triangulations are identified. Another way to specify such graphs is to use the potential Φ such that $\Phi(O_1(v)) = \infty$ if $O_1(v) \neq g_{k,k}$ for some $, k \geq 2$.. Put also $\Phi(K_5) = \Phi(K_{3,3}) = \infty$, where K_5 is the complete graph with 5 vertices, $K_{3,3}$ is the graph with six vertices $1, 2, 3, 4, 5, 6$ and all links $(i, j), i = 1, 2, 3, j = 4, 5, 6$ (by Pontriagin-Kuratowski theorem this singles out planar graphs). In all other cases put $\Phi = 0$. This model is called in physics pure two-dimensional gravity or quantum string in zero dimensions.

Theorem 3 *There exists a unique limit* $p(\Gamma) = \lim_{N\to\infty} p_N(\Gamma)$. *Moreover,* $p(\Gamma)$ *is an empirical Gibbs family with potential* Φ.

Proof. We shall prove this by giving explicit expressions. Denote the number of triangulations with specified vertex 0 and fixed $O_d(0) = \Gamma_d$, let Γ_d have u triangles and k boundary links (with both vertices on the d-slice). The external part is the triangulation of the disk with $N - u$ triangles and also with k boundary links. The number $D(N - u, k)$ of such triangulations (this follows from results by Tutte, see [4], pp.21, 25) has the following asymptotics as $N \to \infty$

$$D(N-u,k) \sim k^{-1}\phi(k)N^{-\frac{5}{2}}c^{N-u} \tag{17}$$

Thus the probability of Γ (at vertex 0) is

$$P^N(\Gamma) \sim k^{-1}\phi(k)N^{-\frac{7}{2}}c^{N-u}C_0^{-1}(N) \tag{18}$$

where

$$C_0(N) \sim aN^{-\frac{5}{2}}c^N \tag{19}$$

is the number of triangulations with N triangles. The function ϕ and constants c, a are known explicitely but we do not need this. It can be proved (see similar statements in [4]) that probabilities for the case of specified origin and without it are asymptotically the same as $N \to \infty$.

Trees Trees cannot be characterized locally. That is they cannot be singled out from the set of all graphs via restrictions on subgraphs having radius not greater than some constant. It is interesting, however, that a local probabilistic characterization of trees can be given using Gibbs families. More exactly, one can define empirical Gibbs families with radius 1 potential and such that the distribution has its support on the set of trees. We give an example with p-regular trees.

Consider the set $\mathcal{A}_{N,p}$ of all p-regular graphs with N vertices, that is graphs with all vertices having degree p. Consider Gibbs family on $\mathcal{A}_{N,p}$ with potential $\Phi \equiv 0$. Otherwise speaking, we define Gibbs family on $\mathcal{A}_N$ with the following potential Φ : $\Phi = 0$ on all radius 1 graphs, where 0 has degree p, and $\Phi = \infty$ for all other radius 1 graphs.

Theorem 4 *$p^N(\Gamma_k)$ have limits $p(\Gamma_k) = 1$ for p-ary tree of height k, and 0 otherwise.*

Proof. The case $\Phi = 0$ can be reduced to combinatorics. To prove the theorem we need some techniques from graph enumeration.

Lemma 3 *The asymptotics of the number $C_p(N)$ of connected p-regular graphs with N vertices*

$$C_p(N) \sim \frac{(pN-1)(pN-3)...}{N!(p!)^N} \tag{20}$$

Proof. We use a combinatorial method to prove this lemma. Such method is known and was applied to other situations. It consists of several statements, see [1], section 9.4. We call graph labelled if both vertices and legs are labelled (enumerated). The scheme of the proof is the following:

1. Number of labelled p-regular graphs is $(pN-1)(pN-3)....$

2. Almost all labelled p-regular graphs are connected.

3. Almost all p-regular graphs are asymmetric.

4. It follows from 1-3 that almost all p-regular graphs are connected.

In the same way on can show that the number of p-regular graphs with no cycles of length q at vertex 0, is equal asymptotically to the number of all graphs. It follows that the probability that in a vertex there are no cycles tends to 1 as $N \to \infty$.

An interesting problem is to generalize this result for the case with $S = \{-1, 1\}$ and Ising type interaction.

References

[1] F. Harary, E. Palmer. Graphical Enumeration. Academic Press. 1973.

[2] R. Dobrushin. The description of random field by means of conditional probabilities and conditions of its regularity. Theory of Probability and Appl., 1968, v. 13, 197-224.

[3] O. Lanford, D. Ruelle. Observables at infinity and states with short range correlations in statistical mechanics. Comm. Math. Phys., 1969, v. 13, 174-215.

[4] V. Malyshev. Probability around Quantum Gravity. Russian Math. Surveys, 1999, v. 54, pp. 685-728.

[5] V. Malyshev. Random Grammars. Russian Math. Surveys, 1998, v. 53, pp. 345-370.

Trends in Mathematics, © 2000 Birkhäuser Verlag Basel/Switzerland

Generating functions with high-order poles are nearly polynomial

ROBIN PEMANTLE[1] *Department of Mathematics, Ohio State University, 231 W. 18th Avenue, Columbus, OH 43210*

Abstract. *Consider the problem of obtaining asymptotic information about a multi-dimensional array of numbers* $a_{\mathbf{r}}$, *given the generating function* $F(\mathbf{z}) = \sum_{\mathbf{r}} a_{\mathbf{r}}\mathbf{z}^{\mathbf{r}}$. *When* F *is meromorphic, it is known how to obtain various asymptotic series for* $a_{\mathbf{r}}$ *in decreasing powers of* $|\mathbf{r}|$. *The purpose of this note is to show that, when the pole set of* F *has singularities of a certain kind, then there can be only finitely many terms in such an asymptotic series. As a consequence, in the presence of a singularity of this kind, the whole asymptotic series for* $a_{\mathbf{r}}$ *is an effectively computable object.*

Key words. *Asymptotic, automatic coefficient extraction, meromorphic, multiple point, Cauchy integral formula, convolution.*

AMS subject classification. *Primary: 05A16l, 05A15 ; Secondary 14Q20, 32A20, 13P10.*

1 Introduction

Given a generating function $F(z) = \sum_{r=0}^{\infty} a_r z^r$, analytic in a neighborhood of the origin, it is usually possible to obtain a good explicit approximation for a_r. The transfer method of Flajolet and Odlyzko [FlaOdl90], for example, translates information about F near a singularity automatically into asymptotic information about a_r.

The corresponding problem in more dimensions, when r is replaced by a multi-index, $\mathbf{r}$, is much harder. Even rational functions, whose approximation theory in one dimension is trivial, are not well understood. The paper [CohElkPro96], for example, spends many pages deriving asymptotics by hand for an array $\{a_{rst}\}$ whose generating function is, up to minor changes,

$$F(x,y,z) = \sum_{r,s,t} a_{rst} x^r y^s z^t = \frac{1}{(1-yz)(1-\frac{x+x^{-1}+y+y^{-1}}{2}z+z^2)}.$$

The body of literature dealing with the problem of multivariate coefficient extraction in a systematic way is quite small. The purpose of this note is to shed some light on coefficient approximation for a class of meromorphic generating functions to be defined shortly.

[1] Research supported in part by National Science Foundation grant # DMS 9803249

The problem of finding an asymptotic expression for $a_{\mathbf{r}}$ falls naturally into two steps. The first is to find the correct exponential rate, namely a homogeneous function $\gamma(\mathbf{r})$ of degree 1 for which

$$\log a_{\mathbf{r}} = (1 + o(1))\gamma(\mathbf{r}) . \tag{1.1}$$

This step is geometric and amounts to finding an appropriate point on the variety $\mathcal{V}$ of poles of F. This step, which is not the main concern of this paper, will be discussed in Section 2.

If the first step can be carried out, the next step is to find an asymptotically valid expression (or better yet, a complete asymptotic series) for $a_{\mathbf{r}}$. This step is analytic. All known methods involve complex variable methods, namely contour integration or Fourier transforms. When this step can be carried out, one typically finds something like

$$a_{\mathbf{r}} \sim \exp(\gamma(\mathbf{r})) \sum_{j=0}^{\infty} b_j(\mathbf{r})$$

where γ is homogeneous of degree 1 and $\{b_j\}_{j\geq 0}$ is a sequence of homogeneous functions whose degrees decrease, typically as $(l-j)/2$ for some $l \in \mathbf{Z}$. An example of the leading term asymptotic (the $j = 0$ term) is given by $F(z, w) = 1/(1 - z - w - zw) = \sum a_{(r,s)} z^r w^s$ which generates the number of lattice paths from $(0, 0)$ to (r, s) that go north, east or northeast. Here the leading term asymptotic is given by

$$a_{rs} \sim \left(\frac{\sqrt{r^2+s^2} - s}{r}\right)^{-r} \left(\frac{\sqrt{r^2+s^2} - r}{s}\right)^{-s} \sqrt{\frac{1}{2\pi}} \sqrt{\frac{rs}{(r+s-\sqrt{r^2+s^2})^2 \sqrt{r^2+s^2}}} , \tag{1.2}$$

so $\gamma(r, s)$ is the logarithm of the first two terms and b_0 is the product of the last two terms, with $l = -1$.

Generally, if F is expressed as the quotient of analytic functions G/H, the function γ is determined by H, as is l in nondegenerate cases. As G varies, the space of possible asymptotic series will be quite large: even holding $\mathbf{r}$ fixed in projective space, any set of values for $b_0, \ldots, b_N$ will typically be possible, and the possible values of the sequence $\{b_j\}_{j\geq 0}$ will typically form an infinite-dimensional vector space. In some cases, however, the possible sequences $\{b_j\}_{j\geq 0}$ will form a finite vector space, and what is more, will consist of terminating sequences. Furthermore, each b_j will then be a polynomial function of $\mathbf{r}$, whence the whole asymptotic expansion up to terms of exponentially smaller order is a finite object. This is the topic of the present note.

The methods herein are more algebraic than geometric or analytic, and are not useful for computing the coefficients $\{b_j\}_{j\geq 0}$. The point is to find out *a priori* how many coefficients one has to compute for a complete asymptotic expansion, thus enabling computation algorithms such as those in [PemWil00b] to terminate.

This introductory section concludes with an imprecise statement of the main result of the paper. Section 2 gives some background on the determination of the correct exponential order. Section 3 sets forth the remaining notation and states the main theorem, Theorem 3.1, along with examples and corollaries. Section 4 shows how a local ring of analytic functions maybe be extended over a polydisk and characterizes when partial fraction expansions are available in the local and global rings. Section 5 finishes the proof of Theorem 3.1.

Some notation in use throughout the paper is as follows. Say that $g(\mathbf{r}) = o_{\exp} h(\mathbf{r})$ if $g(\mathbf{r})/h(\mathbf{r}) = O(e^{-\delta|\mathbf{r}|})$ for some $\delta > 0$. The function F is always taken to be analytic on a neighborhood of the origin in $\mathbf{C}^d$. The formal power series $\sum_{\mathbf{r}} a_{\mathbf{r}} \mathbf{z}^{\mathbf{r}}$ that represents F will then converge near the origin and its domain of convergence is denoted $\mathcal{D}$; here and throughout, $\mathbf{z}^{\mathbf{r}} := z_1^{r_1} \cdots z_d^{r_d}$. In order to accommodate functions such as the generating functions for self-avoiding walks and percolation probabilities, which are meromorphic near the origin but not necessarily near infinity [ChaCha86, CamChaCha91], the natural hypothesis that F be meromorphic will be weakened as follows. Let $\mathbf{z}$ be a point on the boundary of $\mathcal{D}$ where F is singular and let $D(\mathbf{z})$ be the closed polydisk $\{\mathbf{w} : |w_j| \le |z_j|, j = 1, \ldots, d\}$. Assume that F is meromorphic in a neighborhood Ω of $D(\mathbf{z})$, and express $F = G/H$ in this neighborhood (such a global choice is always possible). In particular, $\mathbf{z}$ is assumed to be a pole of F.

Let $\mathcal{V}$ denote the analytic variety where H vanishes. We say that $\mathbf{z}$ is a multiple point of $\mathcal{V}$ if near $\mathbf{z}$, $\mathcal{V}$ is locally the union of smooth manifolds. We say that the point $\mathbf{z}$ is a complete multiple point if in addition, the common intersection of these manifolds is locally the singleton $\{\mathbf{z}\}$. Define

$$\gamma(\mathbf{r}, \mathbf{z}) = -\sum_{j=1}^{d} r_j \log |z_j| \,. \tag{1.3}$$

As discussed later, when $\mathbf{z}$ is a multiple point of $\mathcal{V}$ on the boundary of the domain of convergence, $\mathcal{D}$, then

$$\frac{\log a_{\mathbf{r}}}{\gamma(\mathbf{r}, \mathbf{z})} \to 1$$

as $\mathbf{r}$ varies over a certain cone $\mathcal{C} \cap (\mathbf{Z}^+)^d$. The cone $\mathcal{C}$ depends on H but not on G.

The main result of this paper is as follows. Assume $\mathcal{C} = \mathcal{C}(H)$ has non-empty interior. Then there is a finite dimensional vector space $\mathbf{W}$ of polynomials in $\mathbf{r}$ such that for any G analytic in a neighborhood of $D(\mathbf{z})$ there is a $P \in \mathbf{W}$ with

$$a_{\mathbf{r}} = \exp(\gamma(\mathbf{r}, \mathbf{z})) \left(P(\mathbf{r}) + o_{\exp}(1) \right) \,. \tag{1.4}$$

Contrast this to (1.2), which is only the leading term, to see what complications are avoided when P is a polynomial. This result will be stated more precisely as Theorem 3.1 after the appropriate terminology has been introduced. This theorem does not address the possibility that P is always zero, but in fact this is ruled out by results of [PemWil00b, PemWil00c].

2 The exponential order of $a_{\mathbf{r}}$

The problem of determining the exponential order of $a_{\mathbf{r}}$ is completely solved only when $d \leq 2$ and the coefficients $a_{\mathbf{r}}$ are assumed to be nonnegative. This section summarizes most of what is known about determining the exponential order.

If one is interested only in those $a_{\mathbf{r}}$ with $\mathbf{r}$ on the diagonal, then relatively powerful results may be obtained. When $d = 2$, the generating function $\xi(z) = \sum_r a_{rr} z^r$ may be extracted analytically [HauKla71], reducing the problem to one dimension. A rational two-variable function has an algebraic diagonal [Fur67], so for nice two-variable functions, extraction of the diagonal is effective and asymptotics may then be obtained. In more than two dimensions, no analytic expression for the diagonal is available [Sta99], but the diagonal is still D-finite [Lip88] and a recursion for the diagonal may be effectively derived [ChySal96], which allows the derivation of asymptotics by solving difference equations with polynomial coefficients. This has in fact been implemented [LeyTsa00] and has no problem running on a standard laptop (circa 1999) when the inputs are reasonable. The methods used in these cases, though superficially analytic, are really algebraic and may be carried out over formal power series rings and modules over the Weyl algebra. The methods may, in theory, be applied to other rays such as $\{a_{rs} : s = 2r\}$, but unfortunately, they are inherently non-uniform in s/r, and may not therefore be applied when the direction of the ray is a changing parameter.

When the direction of $\mathbf{r}$ varies, all known results require analytic methods. To review what is known here, begin by defining a function $\mathbf{dir} = \mathbf{dir}_F$ on $\mathcal{V}$. The function $\mathbf{dir}$ takes values in $\mathbf{C}P^{d-1}$ and may be multi-valued.

If $\mathbf{z}$ is a simple pole and no z_j vanishes, then $\mathbf{dir}(\mathbf{z}) = \mathbf{dir}_F(\mathbf{z})$ is the single value $(z_1 \frac{\partial H}{\partial z_1}, \ldots, z_d \frac{\partial H}{\partial z_d})$, which is a nonzero element of $\mathbf{C}^d$ and thus defines an element of $\mathbf{C}P^{d-1}$. Under the additional assumption that $\mathbf{z}$ is on the boundary of $\mathcal{D}$, an equivalent definition of $\mathbf{dir}(\mathbf{z})$ is the normal to the support hyperplane at $(\log|z_1|, \ldots, \log|z_d|)$ of the (convex) logarithmic domain of convergence $\log\mathcal{D} := \{\mathbf{x} \in \mathbf{R}^d : (e^{x_1}, \ldots, e^{x_d}) \in \mathcal{D}\}$. If $\mathbf{z}$ is a manifold point of $\mathcal{V}$ but not a simple pole, then H is not square-free and $\mathbf{dir}(\mathbf{z})$ may be defined by replacing H with its radical. In the above cases, $\mathbf{dir}$ is single-valued. The remaining case is when $\mathbf{z}$ is not a manifold point of $\mathcal{V}$; in this case, define $\mathbf{dir}(\mathbf{z})$ as the set of limit points of $\mathbf{dir}(\mathbf{w})$ as $\mathbf{w} \to \mathbf{z}$ along manifold points of $\mathcal{V}$. When $\mathbf{z}$ is on the boundary of $\mathcal{D}$, this is again the set of normals to support hyperplanes $(\log|z_1|, \ldots, \log|z_d|)$ of the logarithmic domain of convergence, $\log\mathcal{D}$, but is now, in general, multi-valued.

An illustration will help to clarify the definition of $\mathbf{dir}$. Suppose $H = (1 - (2/3)z - (1/3)w)(1 - (1/3)z - (2/3)w)$ so that $\mathcal{V}$ is the union of two lines, as in figure 1. As $\mathbf{z}$ varies linearly from $(0,3)$ to $(1,1)$, not including $(1,1)$, the quantity $\mathbf{dir}(\mathbf{z})$ is single-valued and goes from slope ∞ to slope 2. The value of $\mathbf{dir}(1,1)$ is the cone of slopes between 2 and $1/2$. As $\mathbf{z}$ varies linearly from $(1,1)$ to $(3,0)$, the quantity $\mathbf{dir}(\mathbf{z})$ is once more single-valued and goes from slope $1/2$ to slope 0.

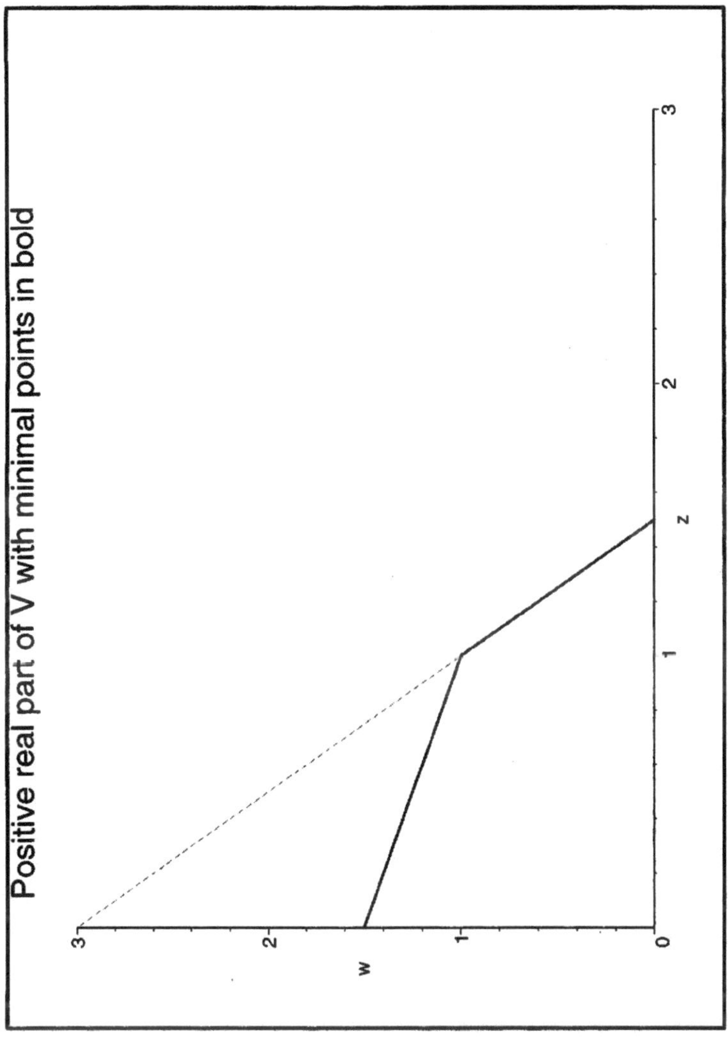

Figure 1:

The remaining points of $\mathcal{V}$ do not concern us, since they are not on the boundary of the domain of convergence of F.

For each $\mathbf{z} \in \mathcal{V}$, and each $\epsilon > 0$, the asymptotic inequality

$$a_{\mathbf{r}} = o\left((1+\epsilon)^{|\mathbf{r}|}\mathbf{z}^{-\mathbf{r}}\right) \tag{2.1}$$

is immediate from Cauchy's integral formula. In fact more is true:

Lemma 2.1 *If* $\mathbf{z} \in \mathcal{V}$ *is on the boundary of* $\mathcal{D}$ *and* $\mathbf{r}$ *varies over a compact subset of the complement of* $\mathbf{dir}(\mathbf{z})$*, then* $a_{\mathbf{r}} = o_{\exp}(\mathbf{z}^{-\mathbf{r}})$.

PROOF: Let $\mathbf{r}$ be a fixed direction not in $\mathbf{dir}(\mathbf{z})$. Since the hyperplane through $\mathbf{x} := (\log|z_1|, \ldots, \log|z_d|)$ normal to $\mathbf{r}$ is not a support hyperplane to $\log\mathcal{D}$, there is a $\mathbf{y}$ in the interior of $\log\mathcal{D}$ with $\mathbf{y} \cdot \mathbf{r} > \mathbf{x} \cdot \mathbf{r}$. From (2.1) we see that $a_{\mathbf{r}} = O((1+\epsilon)^{|\mathbf{r}|} e^{-\mathbf{y}\cdot\mathbf{r}})$ for any $\epsilon > 0$. Choosing ϵ small enough, the conclusion follows for fixed directions. The rate of decay (the δ in the definition of $o_{\exp}$) may be chosen continuously in $\mathbf{r}$, so the uniformity over compact sets follows. □

In particular, the true exponential rate for $a_{\mathbf{r}}$ is at most $\gamma(\mathbf{r}, \mathbf{z}) = -\sum_{j=1}^{d} r_j \log|z_j|$ for any $\mathbf{z}$. This is minimized precisely when $\mathbf{r} \in \mathbf{dir}(\mathbf{z})$. One might hope that the reverse inequality holds in this case, namely, that $\log a_{\mathbf{r}} = (-1+o(1)) \sum_{j=1}^{d} r_j \log|z_j|$ for $\mathbf{z} \in \mathcal{V} \cap \partial\mathcal{D}$ such that $\mathbf{r} \in \mathbf{dir}(\mathbf{z})$. Indeed this is conjectured always to hold when F has nonnegative coefficients, which is the case of greatest combinatorial interest. Results in this direction are as follows.

Pemantle and Wilson [PemWil00a, Theorem 6.3] show that if F has nonnegative coefficients, then for every $\mathbf{r}$ there is always a $\mathbf{z} \in \mathcal{V} \cap \partial\mathcal{D}$ with $\mathbf{r} \in \mathbf{dir}(\mathbf{z})$. The argument consists of generalizing the example in figure 1.

When $\mathbf{z} \in \partial\mathcal{D}$ is a smooth point of $\mathcal{V}$ and $\mathbf{r} \in \mathbf{dir}(\mathbf{z})$, there are several known proofs that $\gamma(\mathbf{r}, \mathbf{z})$ is the correct exponential rate for $a_{\mathbf{r}}$. Bender and Richmond [BenRic83] proved this in 1983, under some additional hypotheses, and also derived the leading term asymptotic. A different proof in a more general framework is given in [PemWil00a]; see also the book [FlaSed00]. When $\mathbf{z}$ is a singular point of $\mathcal{V}$, less is known. The preprint [PemWil00b] shows that $\gamma(\mathbf{r}, \mathbf{z})$ is the correct exponential order when $\mathbf{z} \in \partial\mathcal{D}$ is a multiple point and $\mathbf{r} \in \mathbf{dir}(\mathbf{z})$. As a consequence, if one assumes $a_{\mathbf{r}} \geq 0$, one has complete knowledge of the exponential order. Another consequence is that Theorem 3.1 is non-trivial (meaning that P is not identically zero). The case where $d \geq 3$ and $\mathbf{z}$ is some singularity other than a multiple point is addressed in a manuscript in preparation [CohPem00].

3 Statement of results

The definition of an isolated point may be made precise by the introduction of local rings. Let Ω be any open set containing a point $\mathbf{z}$ and let $\Re_{\mathbf{z}}$ be the complex algebra of germs of analytic functions in d complex variables, $w_1, \ldots, w_d$ at $\mathbf{z}$. This is naturally identified with the set of power series in $(\mathbf{w} - \mathbf{z})$ that converge in some neighborhood of $\mathbf{z}$ [GunRos65, Theorem 1 of Ch. II]. The ring $\Re_{\mathbf{z}}$ is a noetherian UFD and is a local ring, with the maximal ideal $\mathcal{M}$ consisting of functions vanishing at $\mathbf{z}$. Any $h \in \mathcal{M}$ may be factored uniquely into powers of irreducible factors $\prod_{j=1}^{k} h_j^{n_j}$ with each $h_j \in \mathcal{M}$. This corresponds to the decomposition of the variety

$V(h)$ locally as the union of V_j, where $\mathcal{V}_j$ is the zero set of h_j. Taking $h = H$, the denominator of F, the point $\mathbf{z} \in \mathcal{V} = V(H)$ is defined to be a multiple point if each h_j has non-vanishing linear part. It is a complete multiple point if, in addition, the common intersection of the sets $\mathcal{V}_j$ (locally a hyperplane arrangement) is the singleton $\{\mathbf{z}\}$.

Given complete multiple point, $\mathbf{z} \in V(H)$, let $\mathbf{dir}_j(\mathbf{z})$ be the limit of $\mathbf{dir}(\mathbf{w})$ as $\mathbf{w} \to \mathbf{z}$ in $\mathcal{V}_j$. This is the normal to $\mathcal{V}_j$ in logarithmic coordinates. The set $\mathbf{dir}(\mathbf{z})$ is simply the convex hull of $\{\mathbf{dir}_j(\mathbf{z}) : 1 \leq j \leq k\}$. Let $\mathcal{C}$ denote the cone $\mathbf{dir}(\mathbf{z})$ and for $S \subseteq \{1, \ldots, k\}$, let $\mathcal{C}(S)$ denote the convex hull of $\{\mathbf{dir}_j(\mathbf{z}) : j \in S\}$. Define $\mathcal{S}$ to be the family of sets $S \subseteq \{1, \ldots, k\}$ for which $\bigcap_{j \in S} \mathcal{V}_j \neq \{\mathbf{z}\}$, that is, the intersection is a variety $\mathcal{V}_S$ of dimension at least 1. By convention, $\emptyset \in \mathcal{S}$. It follows that if $S \in \mathcal{S}$, then for $j \in S$, each $\mathcal{V}_j$ contains $\mathcal{V}_S$, so each $\mathbf{dir}_j(\mathbf{z})$ is normal to $\mathcal{V}_S$ in logarithmic coordinates, implying that $\mathcal{C}(S)$ is a subspace of codimension at least 1. Let $U = \bigcup_{S \in \mathcal{S}} \mathcal{C}(S)$. The main result may now be stated as follows.

Theorem 3.1 *Let H be analytic on $D(\mathbf{z})$ and have a complete multiple point at $\mathbf{z}$ which is the only zero of H on the closed polydisk $D(\mathbf{z})$. Let $h_1^{n_1}, \ldots, h_k^{n_k}, \mathcal{V}_1, \ldots, \mathcal{V}_k$ be the local factorization of H at $\mathbf{z}$ and define $\mathbf{dir}_j, \mathcal{C}, \mathcal{C}(S)$ and $\mathcal{S}$ as above. Assume $\mathcal{C} = \mathbf{dir}(\mathbf{z})$ has non-empty interior. Then there exists a finite-dimensional complex vector space $\mathbf{W}$ of polynomials in $\mathbf{r}$ such that for every function G analytic in a neighborhood of $D(\mathbf{z})$ and any compact subcone $\mathcal{C}_1$ of $\mathcal{C} \setminus U$, the coefficients of $F := G/H$ are given by*

$$a_{\mathbf{r}} = \mathbf{z}^{-\mathbf{r}} \left(P(\mathbf{r}) + E(\mathbf{r})\right)$$

with $P \in \mathbf{W}$ and $E = o_{\exp}(1)$ uniformly on $\mathcal{C}_1$.

Remark: The set U contains the boundary of $\mathcal{C}$ but also possibly some hyperplanes in the interior of $\mathcal{C}$. Thus it is possible for $\mathcal{C} \setminus U$ to be disconnected. In this case, it can happen that $a_{\mathbf{r}}$ is approximated by two different polynomials on two different subcones of $\mathcal{C} \setminus U$.

The steps of the proof are as follows. (1) locally, F may be expanded by partial fractions when G is in a certain ideal, $\mathfrak{I}_{\mathbf{z}}$, the quotient by which is finite-dimensional (Lemma 4.1). (2) this is true globally in a neighborhood of $D(\mathbf{z})$ (Theorem 4.5). (3) the partial fraction summands are $o_{\exp}(\mathbf{z}^{-\mathbf{r}})$ on $\mathcal{C} \setminus U$ (Lemma 5.2). (4) the coset representatives for analytic functions modulo $\mathfrak{I}_{\mathbf{z}}$ have coefficients in a finite vector space of polynomials. Steps (1) and (2) are carried out in Section 4 and steps (3) and (4) are carried out in Section 5. Two examples serve to illustrate the use of the theorem.

Example 1 $(d = 2)$

When $d = 2$, any singular point $\mathbf{z} \in \partial\mathcal{D} \cap \mathcal{V}$ is a complete multiple point. This follows from the fact that the leading terms of the expansion of H near $\mathbf{z}$ are all of the same homogeneous degree, which is proved in [PemWil00a, Theorem 6.1]. Unless the branches of $\mathcal{V}$ near $\mathbf{z}$ all intersect tangentially, the cone $\mathbf{dir}(\mathbf{z})$ will have non-empty interior, and the hypotheses of Theorem 3.1 will be satisfied.

If k is the number of factors of H near $\mathbf{z}$, and $\mathbf{z}$ is a complete multiple point, then $k \geq d$. The simplest case, when $k = d$, is worth mentioning as a separate corollary. The proof will be given in Section 5 after the proof of Theorem 3.1.

Corollary 3.2 *Under the assumptions of Theorem 3.1, suppose $k = d$ and each $n_k = 1$. Then $\mathbf{W}$ is one-dimensional, consisting only of constants.*

Example 2 (crossing lines)

Consider the example in figure 1, where $H = (1 - (2/3)z - (1/3)w)(1 - (1/3)z - (2/3)w)$. The leading term asymptotic is computed in [PemWil00b] to be

$$a_{r,s} = 6 + O(|\mathbf{r}|)^{-1}$$

for $1/2 < r/s < 2$. Further terms are increasingly time-consuming to compute. From Corollary 3.2 we see that in fact there are no more terms of the same exponential order. In this case there is a more elementary method of obtaining a first-order approximation to a_{rs}. Because H factors globally, it is possible to represent $a_{r,s}$ as a two-dimensional convolution, resulting in a sum of products of binomial coefficients. A bivariate central limit approximation then recovers the leading term without too much trouble, but gives no indication that $a_{r,s}$ is in fact exponentially well approximated by the constant 6.

Example 3 (peanut)

Suppose $H = 19 - 20z - 20w + 5z^2 + 14zw + 5w^2 - 2z^2w - 2zw^2 + z^2w^2$. The real part of the zero set is shown in figure 2. The point $(1, 1)$ is on the boundary of $\mathcal{D}$, and $\mathbf{dir}(1, 1)$ is the set $1/2 \leq r/s \leq 2$. Thus by Corollary 3.2, we have again

$$a_{r,s} = C + o_{\exp}(1)$$

where the constant C is proportional to $G(1, 1)$. This example illustrates that it is the local nature of the singularity that allows us to apply Theorem 3.1 and its corollaries: H factors in the local ring at $(1, 1)$, but does not factor globally.

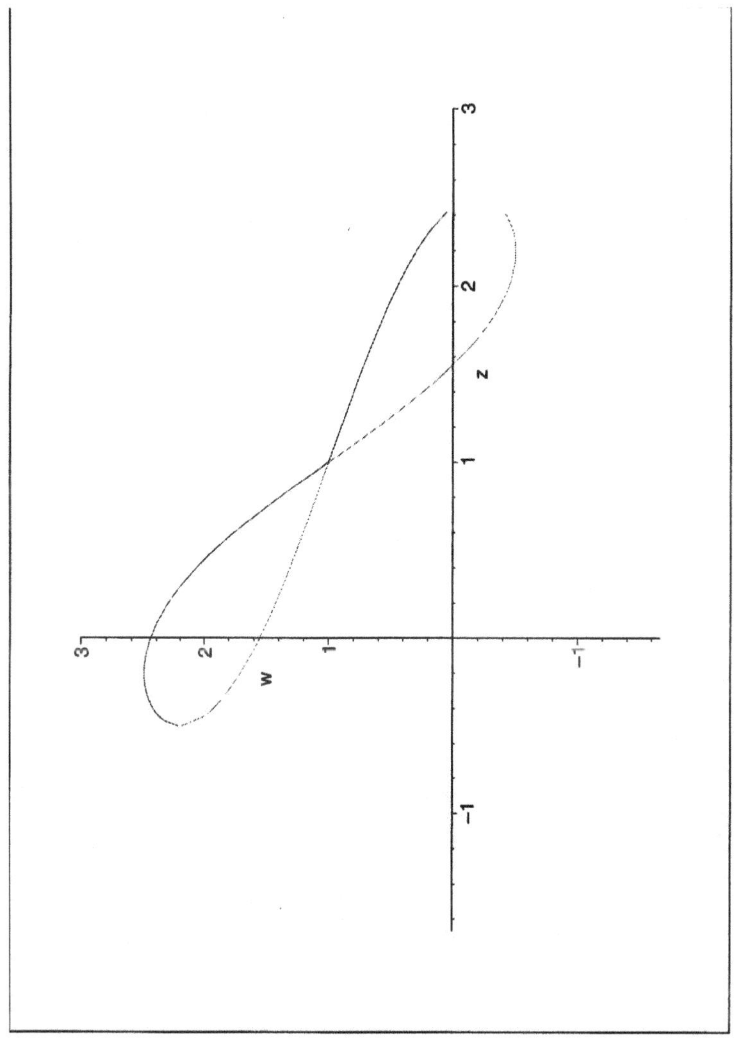

Figure 2:

4 Partial fraction expansions

If $F = G/H$ has a partial fraction representation as $\sum_{j=1}^{k}(g_j/h_j)$, then clearly G vanishes at $\mathbf{z}$. Amplifying on this, for $S \in \mathcal{S}$ we define $h_S = \prod_{j \in S^c} h_j^{n_j}$, so that

$$\frac{h_S}{H} = \frac{1}{\prod_{j \in S} h_j^{n_j}} .$$

Let $\Im_{\mathbf{z}}$ be the ideal in $\Re_{\mathbf{z}}$ generated by $\{h_S : S \in \mathcal{S}\}$. The following lemma is a local version of the main result of this section, namely the partial fraction representation, Theorem 4.5.

Lemma 4.1 *The quotient $\Re_{\mathbf{z}}/\Im_{\mathbf{z}}$ is a finite-dimensional complex vector space. Equivalently, there is a finite-dimensional vector space $\mathbf{V} \subseteq \Re_{\mathbf{z}}$ such that for all $g \in \Re_{\mathbf{z}}$,*

$$g = g_0 + \sum_{S \in \mathcal{S}} g_S h_S \tag{4.1}$$

with $g_0 \in \mathbf{V}$ and each $g_S \in \Re_{\mathbf{z}}$.

PROOF: First observe that $\mathbf{z}$ is an isolated element of $V(\Im_{\mathbf{z}})$. Indeed, if not, then some variety A of dimension at least 1 containing $\mathbf{z}$ is in $V(\Im_{\mathbf{z}})$. The (possibly empty) set S_A of j for which $A \subseteq \mathcal{V}_j$ is in $\mathcal{S}$, so $h_{S_A} \in \Im_{\mathbf{z}}$ and h_{S_A} does not vanish on A, contradicting $A \subseteq V(\Im_{\mathbf{z}})$.

The local ring $\Re_{\mathbf{z}}$ is noetherian [GunRos65, Theorem 9, Ch. IIB] and satisfies the Nullstellensatz (see the discussion after Corollary 16 of Ch. IIE on page 90 of [GunRos65]). From the Nullstellensatz, it follows that the radical of $\Im_{\mathbf{z}}$ is $\mathcal{M}$, the maximal ideal of $\Re_{\mathbf{z}}$. From the noetherian property, it follows that the radical of an ideal is finite-dimensional over the ideal, hence $\Re_{\mathbf{z}}/\Im_{\mathbf{z}}$ is finite-dimensional over $\Re_{\mathbf{z}}/\mathcal{M} \cong \mathbf{C}$. □

To transfer (4.1) to the global setting requires a formulation in terms of sheaves. Let ω be a neighborhood of $\mathbf{z}$ in which the factors h_j are analytic, and in which $\bigcap_{j=1}^k \mathcal{V}_j = \{\mathbf{z}\}$. Since $\mathcal{V}$ does not intersect the interior of $D(\mathbf{z})$, the intersection of $\mathcal{V}_j$ with $\partial\omega$ is disjoint from $D(\mathbf{z})$ and it follows that we may choose a neighborhood Ω of $D(\mathbf{z})$ containing no such intersection point.

Lemma 4.2 *Fix any $\mathbf{x} \in D(\mathbf{z}) \setminus \{\mathbf{z}\}$. There are functions $h_j^{\mathbf{x}}$ analytic on Ω for which the following hold:*

(1) each $h_j^{\mathbf{x}}$ is analytic on Ω;

(2) $h_j^{\mathbf{x}} = u \cdot h_j$ with u a unit in $\Re_{\mathbf{z}}$;

(3) $h_j^{\mathbf{x}}(\mathbf{x}) \neq 0$.

PROOF: Fix j for the entire proof. Let $\mathcal{F}_j$ be the sheaf over Ω of ideals $< h_j >$. That is, when $\mathbf{w} \in \omega$ and $h_j(\mathbf{w}) = 0$, then $(\mathcal{F}_j)_{\mathbf{w}}$ is the germs of functions divisible by h_j at $\mathbf{w}$, while when $\mathbf{w} \notin \omega$ or $\mathbf{w} \in \omega$ with $h_j(\mathbf{w}) \neq 0$, then $(\mathcal{F}_j)_{\mathbf{w}}$ is all analytic germs at $\mathbf{w}$. The definition of $(\mathcal{F}_j)^{\mathbf{w}}$ is potentially ambiguous when $\mathbf{w} \in \partial\omega$ is in the interior of Ω, but since h_j is nonzero here, there is no problem.

The sheaf $\mathcal{F}_j$ is a subsheaf of the structure sheaf $\mathcal{O}$, hence coherent, so by Cartan's Theorem A (see [GraRem79, page 96-97]) there is a map ψ_j from some $\mathcal{O}^l$ onto $\mathcal{F}_j$, where $\mathcal{O}$ is the sheaf of germs of analytic functions (the structure sheaf) on Ω and $l \geq 1$. Denote the l generators of $\mathcal{O}^l$ by $\mathbf{1}_i, i \leq l$.

Surjectivity of ψ_j is a local property, but since each $\mathbf{1}_i$ is a global section of $\mathcal{O}^l$, each $f_{ij} := \psi_j(\mathbf{1}_i)$ is an analytic function defined globally on Ω. Surjectivity at $\mathbf{z}$ implies that h_j is in the image of ψ_j, which is the ideal generated by the functions f_{ij} at $\mathbf{z}$ as i varies. Thus for some functions u_i in a neighborhood of $\mathbf{z}$,

$$\psi_j(\sum_i u_i \mathbf{1}_i) = \sum_i u_i f_{ij} = h_j\,. \tag{4.2}$$

Surjectivity at any other point implies that f_{ij} do not simultaneously vanish anywhere that h_j does not. By definition of ψ_j, each f_{ij} is in the ideal generated by h_j and hence may be written as $u_{ij}h_j$ in $\mathfrak{R}_{\mathbf{z}}$. If each $u_{ij} \in \mathcal{M}$, then each $f_{ij} \in \mathcal{M}\cdot < h_j >$ contradicting the fact that h_j is in the ideal generated by the f_{ij}. Thus for some i, $f_{ij} \notin \mathcal{M}\cdot < h_j >$.

Given $\mathbf{x}$, if there is an i with $f_{ij} \notin \mathcal{M}\cdot < h_j >$ and $f_{ij}(\mathbf{x}) \neq 0$, then the lemma is proved with $h_j^{\mathbf{x}} := f_{ij}$ and $u = u_{ij}$. If not, then choose i and i' so that $f_{ij} \notin \mathcal{M}\cdot < h_j >$ and $f_{i'j}(\mathbf{x}) \neq 0$. Since it was not possible to choose $i = i'$, we know that $f_{ij}(\mathbf{x}) = 0$ and $u_{i'j} \in \mathcal{M}$. It follows that $u_{ij} + u_{i'j} \notin \mathcal{M}$ and the lemma is proved with $h_j^{\mathbf{x}} := f_{ij} + f_{i'j}$. □

Corollary 4.3 *There is a finite collection $\{h_\alpha : \alpha \in A\}$ analytic on a neighborhood of $D(\mathbf{z})$ such that for each $S \in \mathcal{S}$ and each $\mathbf{w} \in D(\mathbf{z}) \setminus \{\mathbf{z}\}$ there is an $\alpha \in A$ with*

$$h_\alpha(\mathbf{w}) \neq 0 \text{ and } h_\alpha = h_S u \tag{4.3}$$

with u a unit of $\mathfrak{R}_{\mathbf{z}}$.

PROOF: Fix $S \in \mathcal{S}$. The function $h_S^{\mathbf{x}} := \prod_{j \in S^c} (h_j^{\mathbf{x}})^{n_j}$ satisfies (4.3) for all $\mathbf{w}$ in some neighborhood $\mathcal{N}_{\mathbf{x}}$ of $\mathbf{x}$. It also satisfies (4.3) for all $\mathbf{w}$ in some neighborhood $\mathcal{N}$ of $\mathbf{z}$. By compactness of $D(\mathbf{z})$, we may choose finitely many $\mathbf{x}$ for which the collection of sets $\mathcal{N}_{\mathbf{x}}$ covers $D(\mathbf{z}) \setminus \mathcal{N}$. Taking the union of such collections over $S \in \mathcal{S}$ proves the corollary. □

Lemma 4.4 *Let Ω be a polydisk containing z and let $\{h_\alpha : \alpha \in A\}$ be a finite collection of functions analytic in Ω. Suppose an analytic function g on Ω is represented as $\sum_\alpha g_\alpha^{\mathbf{x}} h_\alpha$ in a neighborhood of each $\mathbf{x}$ where each $g_\alpha^{\mathbf{x}}$ is analytic. Then*

$$g = \sum_\alpha g_\alpha^* h_\alpha$$

with g_α^* analytic in Ω.

PROOF: This is a straightforward application of Cartan's Theorem B. A sketch of the argument is as follows. Define sheaves over Ω by $\mathcal{F} = \mathcal{O}^{|A|}$ and $\mathcal{G} =< h_\alpha : \alpha \in A >$. The map $\eta : \mathcal{F} \to \mathcal{G}$ defined by $\eta(f_\alpha : \alpha \in A) = \sum_\alpha f_\alpha h_\alpha$ is a surjection of sheaves. The space of global sections of a sheaf is the zeroth cohomology group, and $H^0(\Omega, \mathcal{F})$ maps onto $H^0(\Omega, \mathcal{G})$ only if the coboundary map from $H^0(\Omega, \mathcal{G})$ to $H^1(\Omega, \mathcal{E})$ is trivial, where $\mathcal{E}$ is the kernel of η. By Cartan's Theorem B ([GunRos65, Theorem 14, Ch. VIIIA]), since $\mathcal{E}$ is a subsheaf of $\mathcal{O}^{|A|}$ and Ω is a Stein space, the cohomology groups $H^q(\Omega, \mathcal{E})$ vanish when $q \geq 1$. Hence the coboundary map is trivial, and there is a global section $(g_\alpha : \alpha \in A)$ of $\mathcal{F}$ mapping by η to g. □

Let $\Re$ now denote the ring of functions analytic beyond $D(\mathbf{z})$, that is, functions f for which there exists a neighborhood of $D(\mathbf{z})$ on which f is analytic. Putting together the lemmas of this section yields the following result.

Theorem 4.5 *Let H be analytic on $D(\mathbf{z})$ and have a complete multiple point at $\mathbf{z}$ which is the only zero of H on the closed disk $D(\mathbf{z})$. Let $h_1^{n_1}, \ldots, h_k^{n_k}, \mathcal{V}_1, \ldots, \mathcal{V}_k$ be the local factorization of H in a neighborhood of $\mathbf{z}$, and let $\mathcal{S}$ be the family of subsets S of $\{1, \ldots, k\}$ such that $\mathbf{z}$ is not isolated in $\bigcap_{S \in S} \mathcal{V}_j$. Then there is a finite subset $\{h_\alpha : \alpha \in A\}$ of $\Re$, each localizing to h_S times a unit in $\Re_\mathbf{z}$ for some $S \in \mathcal{S}$, and having the following property.*

There is a finite-dimensional vector subspace $\mathbf{V}^*$ of $\Re$ such that each $G \in \Re$ may be written as

$$g_0^* + \sum g_\alpha^* h_\alpha \tag{4.4}$$

with $g_0^* \in \mathbf{V}^*$ and the dimension of $\mathbf{V}^*$ equal to the dimension of $\Re_\mathbf{z} / \Im_\mathbf{z}$.

PROOF: The $\{h_\alpha : \alpha \in A\}$ is constructed in Corollary 4.3. Choose coset representatives for a basis of $\Re_\mathbf{z} / \Im_\mathbf{z}$ and let $\mathbf{V}^*$ be their span. We need only to verify the representation property (4.4). By construction, if $G \in \Re$, then G may be written as $g_0^* + g$ with $g_0^* \in \mathbf{V}^*$ and the germ $(g)_\mathbf{z}$ in $\Im_\mathbf{z}$. Evidently, the dimension of $\mathbf{V}^*$ is equal to the dimension of $\Re_\mathbf{z} / \Im_\mathbf{z}$, which is finite by Lemma 4.1.

We now verify the hypotheses of Lemma 4.4. In a neighborhood of $\mathbf{z}$, we know from (4.3) that the functions $\{h_\alpha : \alpha \in A\}$ generate $\Im_\mathbf{z}$. Hence there is a representation $g = \sum g_\alpha^\mathbf{z} h_\alpha$. In a neighborhood of any other $\mathbf{x} \in D(\mathbf{z})$ some h_α is nonzero, so there is trivially a representation $g = \sum g_\alpha^\mathbf{x} h_\alpha$. Applying Lemma 4.4, it follows that $g \in \Im$. □

5 Finite-dimensional shift-invariant spaces of arrays must be polynomial

For this section we fix a compact subcone $\mathcal{C}_1 \subseteq \mathcal{C} \setminus U$ with non-empty interior.

Let $\mathbf{W}(d)$ denote the set of complex valued functions on $(\mathbf{Z}^+)^d$. For each $G \in \mathfrak{R}$, let $b_{\mathbf{r};G}$ be the coefficients of the expansion

$$\frac{G(\mathbf{w})}{H(\mathbf{w})} = \sum_{\mathbf{r}} b_{\mathbf{r};G} \mathbf{z}^{\mathbf{r}} .$$

Denote by q_G the element of $\mathbf{W}(d)$ mapping $\mathbf{r}$ to $b_{\mathbf{r};G}$. Thus q is a correspondence between certain meromorphic functions and coefficient arrays. If S is a subset of $\mathfrak{R}$, let q_S denote $\{q_f : f \in S\}$. Let $X \subseteq \mathbf{W}(d)$ denote the vector space $q_{\mathfrak{R}}$ and let $E \subseteq X$ denote the subspace of functions from $(\mathbf{Z}^+)^d$ to $\mathbf{C}$ that are $o_{\exp}(1)$ uniformly on $\mathcal{C}_1$. For $1 \leq j \leq d$, define a linear map $\sigma_j : \mathbf{W}(d) \to \mathbf{W}(d)$ by $\sigma_j b(\mathbf{r}) = b(\mathbf{r} - e_j)$ where e_j is the vector whose i^{th} component is δ_{ij} and $f(\mathbf{r})$ is defined to be zero if $\mathbf{r}$ has a negative component.

Recall that $\mathfrak{I}$ denotes the ideal in $\mathfrak{R}$ generated by $\{h_\alpha : \alpha \in A\}$. The next two lemmas state the properties that will be used of the correspondence q. The proof of the first one is trivial and is omitted.

Lemma 5.1 *The map $g \mapsto q_g$ is linear over* $\mathbf{C}$ *and*

$$\sigma_j(q_g) = q_{w_j g} .$$

□

Lemma 5.2

$$q_{\mathfrak{I}} \subseteq \mathbf{z}^{-\mathbf{r}} E .$$

PROOF: Since E is a vector space, it suffices to show that $q_{gh_\alpha} \in \mathbf{z}^{-\mathbf{r}} E$ for each $\alpha \in A$. This is equivalent to $b_{\mathbf{r};gh_\alpha/H} = o_{\exp}(\mathbf{z}^{-\mathbf{r}})$ on $\mathcal{C}_1$. Each h_α is chosen as $h_S^{\mathbf{x}}$ for some $S \in \mathcal{S}$ and $\mathbf{x} \in D(\mathbf{z})$. For such an S, the pole set of the meromorphic function gh_α/H is a subset of the set $\mathcal{V}$. Thus $\mathbf{z}$ is on the boundary of the domain of convergence of gh_α/H. In a neighborhood of $\mathbf{z}$, the pole set of gh_α/H is simply the union of the sheets $\{\mathcal{V}_j : j \in S\}$. Thus $\mathbf{dir}_{\frac{gh_\alpha}{H}}(\mathbf{z}) = \mathcal{C}(S)$. We have chosen the cone $\mathcal{C}_1$ to avoid $\mathcal{C}(S)$, so the conclusion of Lemma 2.1 yields the exponential decay of $\mathbf{z}^{\mathbf{r}} q_{gh_\alpha}$ on $\mathcal{C}_1$, and establishes the lemma. □

The ideal $\mathcal{M}$ contains all functions of the form $1 - w_j/z_j$. Write $(1 - \mathbf{w}/\mathbf{z})^{\mathbf{r}}$ to denote $\prod_{j=1}^d (1 - w_j/z_j)^{r_j}$ and similarly write $(1 - \sigma/\mathbf{z})^{\mathbf{r}}$ to denote the product of the operators $(1 - \sigma_j/z_j)^{r_j}$. Since $\mathcal{M}$ is the radical of $\Im_{\mathbf{z}}$, some power of each $(1 - w_j/z_j)$ annihilates $\Re_{\mathbf{z}}/\Im_{\mathbf{z}}$ and hence the set $F := \{\mathbf{r} : (1 - \mathbf{w}/\mathbf{z})^{\mathbf{r}} \notin \Im_{\mathbf{z}}\}$ is finite. From Lemmas 5.1 and 5.2, we see that for any $\mathbf{r} \notin F$ and any $G \in \Re$,

$$(1 - \sigma/\mathbf{z})^{\mathbf{r}} q_G \in E. \tag{5.1}$$

The final lemma is as follows.

Lemma 5.3 *Let $Y \subseteq \mathbf{W}(d)$ be a finite-dimensional subspace such that there is a finite set F for which $\mathbf{r} \notin F$ implies $(1 - \sigma/\mathbf{z})^{\mathbf{r}} Y \subseteq \mathbf{z}^{-\mathbf{r}} E$. Then for each $f \in Y$ there is a polynomial g whose terms have multidegrees in F, and for which $f - g\mathbf{z}^{-\mathbf{r}} \in \mathbf{z}^{-\mathbf{r}} E$.*

Assuming this for the moment, the proof of Theorem 3.1 can be finished as follows.

PROOF OF THEOREM 3.1: Let Y be the space $q_{\mathbf{V}^*}$, where $\mathbf{V}^*$ is as in the conclusion of Theorem 4.5. By the conclusion of that theorem, for any $G \in \Re$, we may write $Gu = g_0^* + \sum_{S \in \mathcal{S}} g_S^* h_S^*$. By linearity of q, we have written q_G as the coefficients of $g_0^* / \prod_{j=1}^k (h_j^*)^{n_j}$ plus terms of the form $q_{g_S^* h_S^*}$. By Lemma 5.2 these latter terms are in $\mathbf{z}^{-\mathbf{r}} E$. According to (5.1), the hypotheses of Lemma 5.3 are satisfied, and the conclusion of this lemma then proves Theorem 3.1. □

PROOF OF COROLLARY 3.2: The dimension of $\mathbf{V}^*$ is constructed in the proof of Theorem 4.5 to equal the dimension of $\Re_{\mathbf{z}}/\Im_{\mathbf{z}}$. This is at most the cardinality of F, though it may be less. In the case where each $n_j = 1$ and the surfaces $\mathcal{V}_j$ intersect transversely at $\mathbf{z}$, the ideal $\Im_{\mathbf{z}}$ contains d independent linear polynomials, so is equal to $\mathcal{M}$. Hence $|F|$ is the singleton $\{\mathbf{0}\}$ and $\mathbf{W}$ contains only constants. □

It remains to prove Lemma 5.3.

PROOF OF LEMMA 5.3: Replacing each function f in $\mathbf{W}(d)$ by $\mathbf{z}^{-\mathbf{r}} f$, it suffices to prove the lemma for the case $\mathbf{z} = \mathbf{1}$.

Proceed by induction on $|F|$. If $|F| = 1$ then $F = \{\mathbf{0}\}$. In this case, for each $f \in Y$ and $i \leq k$, the function $\mathcal{E}_i := (1 - \sigma_i) f$ is in E. The cone C_1 has nonempty interior, which implies that $C_1 \cap \mathbf{Z}^d$ has a co-finite subset C' which is a connected subgraph of the integer lattice. For any $\mathbf{r} \leq \mathbf{s} \in C_1$, there is an oriented path $\gamma_0, \gamma_1, \ldots, \gamma_l$ connecting $\mathbf{r}$ to $\mathbf{s}$ in C', where $l = \sum_{i=1}^k (s_i - r_i)$. (An oriented path takes steps only in the increasing coordinate directions.) Then

$$f(\mathbf{s}) - f(\mathbf{r}) = \sum_{j=1}^{l} f(\gamma_j) - f(\gamma_{j-1}) = \sum_{j=1}^{l} \mathcal{E}_{m(j)}(\gamma_{j-1})$$

where $m(j) = i$ if γ_{j-1} and γ_j differ by e_i. Sending $\mathbf{s}$ to infinity, we see that $\lim_{\mathbf{s}\to\infty} f(\mathbf{s})$ exists and

$$f(\mathbf{r}) = \lim_{s\to\infty} f(\mathbf{s}) + \sum_{j=1}^{\infty} f(\gamma_j) - f(\gamma_{j-1}) = -\sum_{j=1}^{l} \mathcal{E}_{m(j)}(\gamma_{j-1})$$

where γ connects $\mathbf{r}$ to infinity. Thus on C', f is a constant plus a tail sum of functions in E, and the conclusion is true with $g = \lim_{s\to\infty} f(\mathbf{s})$, the constant polynomial.

The induction step is similar. Let $F_i = \{\mathbf{r} : \mathbf{r} + e_i \in F\}$. Fix $f \in X$. The space $(1 - \sigma_i)X$ satisfies the hypotheses of the lemma with F_i in place of F. Since $|F_i| < |F|$, we may apply the induction hypothesis to conclude that $(1 - \sigma_i)f = g_i + \mathcal{E}_i$ where g_i is a polynomial with multi-degrees in F_i and $\mathcal{E}_i \in E$. For any $\mathbf{r} \leq \mathbf{s} \in (\mathbf{Z}^+)^d$, and any oriented path γ from $\mathbf{r}$ to $\mathbf{s}$, we have

$$\begin{aligned} f(\mathbf{s}) - f(\mathbf{r}) &= \sum_{j=1}^{l} f(\gamma_j) - f(\gamma_{j-1}) \\ &= -\sum_{j=1}^{l} g_{m(j)}(\gamma_{j-1}) - \sum_{j=1}^{l} \mathcal{E}_{m(j)}(\gamma_{j-1}) . \end{aligned}$$

If $\mathbf{r}, \mathbf{s} \in C'$ then we have already seen that, as a function of $\mathbf{s}$, the last contribution $\sum_{j=1}^{l} \mathcal{E}_{m(j)}(\gamma_{j-1})$ is equal to a function $C(\mathbf{r})$ plus a term decaying exponentially in $\mathbf{s}$. Fixing $\mathbf{r} \in C'$ so that the set of $\mathbf{s} \in C_1$ not greater than or equal to $\mathbf{r}$ is finite, it remains to show that

$$p(\mathbf{s}) := f(\mathbf{r}) + \sum_{j=1}^{l} g_{m(j)}(\gamma_{j-1})$$

defines a polynomial in $\mathbf{s}$ whose terms have multidegrees in F.

We see from the equation

$$(1 - \sigma_i)(1 - \sigma_j)f = (1 - \sigma_j)(1 - \sigma_i)f \tag{5.2}$$

that

$$(1 - \sigma_i)g_j = (1 - \sigma_j)g_i + \mathcal{E}$$

where $\mathcal{E} = (1 - \sigma_j)\mathcal{E}_i) - (1 - \sigma_i)\mathcal{E}_j$. By (5.2), $\mathcal{E}$ is a polynomial, and since it is exponentially small it must vanish entirely. Thus for $\mathbf{x} \in \mathbf{Z}^d$, we see that $g_i(\mathbf{x}) + g_j(\mathbf{x}+e_i) = g_j(\mathbf{x}) + g_i(\mathbf{x}+e_j)$. It follows that the sum defining p is invariant under switching the order of two steps in the path γ, and hence is independent of the choice of γ. Choosing γ to take first $s_1 - r_1$ steps in direction e_1, then $s_2 - r_2$ steps in direction e_2 and so on, we may write

$$p(\mathbf{s}) = f(\mathbf{r}) + \sum_{j=1}^{k} \sum_{t=1}^{s_i - r_i} -g_j(s_1, \ldots, s_{j-1}, r_i + t - 1, r_{j+1}, \ldots, r_k) .$$

Each of the inner sums is the sum to $s_i - r_i$ of a polynomial with multi-degrees in F_i, which is well known to be a polynomial with multi-degrees in F. Hence $p(\mathbf{s})$ is a polynomial with multi-degrees in F and the proof is done. □

Acknowledgements: Many thanks to Andrei Gabrielov for patiently explaining to me everything in Section 4. Thanks to Henry Cohn for several conversations and to Bernd Sturmfels for further help with the algebra.

References

[BenRic83] Bender, E. and Richmond, L. B. (1983). Central and local limit theorems applied to asymptotic enumeration II: multivariate generating functions. *JCT A* **34** 255 - 265.

[CamChaCha91] Campanino, M., Chayes, J.T. and Chayes, L. (1991). Gaussian fluctuations in the subcritical regime of percolation. *Prob. Th. Rel. Fields* **88** 269 - 341.

[ChaCha86] Chayes, J. T. and Chayes, L. (1986). Ornstein-Zernike behavior for self-avoiding walks at all non-critical temperatures. *Comm. Math. Phys.* **105** 221 - 238.

[ChySal96] Chyzak, F. and Salvy, B. (1998). Non-commutative elimination in Ore Algebras proves multivariate identities. *J. Symbol. Comput.* **26** 187 - 227.

[CohElkPro96] Cohn, H., Elkies, N. and Propp, J. (1996). Local statistics for random domino tilings of the Aztec diamond. *Duke Math. J.* **1985** 117 - 166.

[CohPem00] Cohn, H. and Pemantle, R. (2000) A determination of the boundary of the region of fixation for some domino tiling ensembles. *In Preparation.*

[FlaOdl90] Flajolet, P. and Odlyzko, A. (1990). Singularity analysis of generating functions. *SIAM J. Disc. Math.* **3** 216 - 240.

[FlaSed00] Flajolet, P. and Sedgewick, R. (2000). The average case analysis of algorithms. *Book Preprint.*

[Fur67] Furstenberg, H. (1967). Algebraic functions over finite fields. *J. Algebra* **7** 271 - 277.

[GraRem79] Grauert, H. and Remmert, R. (1979). Theory of Stein spaces. Translated from the German by Alan Huckleberry. Grundlehren der Mathematischen Wissenschaften 236. Springer-Verlag: Berlin.

[GunRos65] Gunning, R. and Rossi, H. (1965). Analytic functions of several complex variables. Prentice-Hall: Englewood Cliffs, NJ.

[HauKla71] Hautus, M. and Klarner, D. (1971). The diagonal of a double power series. *Duke Math. J.* **23** 613 - 628.

[LeyTsa00] Leykin, A. and Tsai, H. (2000). *D-modules library for Macaulay 2.* Available at http://www.math.umn.edu/ leykin.

[Lip88] Lipshitz, L. (1988). The diagonal of a D-finite power series is D-finite. *J. Algebra* **113** 373 - 378.

[PemWil00a] Pemantle, R. and Wilson, M. (2000a). Asymptotics for multivariate sequences, part I: smooth points of the singular variety. *In preparation.*

[PemWil00b] Pemantle, R. and Wilson, M. (2000b). Asymptotics for multivariate sequences, part II: multiple points. *In preparation.*

[PemWil00c] Pemantle, R. and Wilson, M. (2000c). Asymptotics for multivariate sequences, part III. *In preparation.*

[Sta99] Stanley, R. (1999). Enumerative combinatorics, vol. 2. Cambridge Studies in Advanced Mathematics no. 62. Cambridge University Press: Cambridge.

Trends in Mathematics, © 2000 Birkhäuser Verlag Basel/Switzerland

Ultrahigh Moments for a Brownian Excursion

JOEL H. SPENCER *Courant Institute, New York University, 251 Mercer Street, New York, NY 10012.* spencer@cs.nyu.edu

Note: This is a somewhat speculative report, describing approaches to a problem which have not been put on a rigorous foundation.

1 The Exact Problem

Let $X_1, \ldots, X_n$ be i.i.d., each with distribution minus one plus a Poisson with mean one. Set $S_0 = 1$, $S_i = S_{i-1} + X_i$, the result of a walk with steps X_i beginning at 1. We condition on $S_n = 0$ and $S_i > 0$ for $i < n$, that the excursion first hits 0 at time n. Set $M = \sum_{i=1}^{n}(S_i - 1)$. We seek an asymptotic formula for $E[\binom{M}{k}]$ where $k = k(n)$.

The application is to graph theory. Let $c(n,k)$ denote the number of connected labeled graphs with n vertices and $n-1+k$ edges. Then [2] $E[\binom{M}{k}] = n^{n-2}c(n,k)$. An asymptotic formula for $c(n,k)$ was found by Bender, Canfield and McKay [1] in 1990. Hopefully one can get an alternate (simpler?) proof of this formula from the straight probability problem and also finding where the "weight" of $E[\binom{M}{k}]$ comes from gives insight into the nature of the random connected graph.

If $k > (\frac{1}{2} + \epsilon)n \ln n$ then a classic result of Erdös and Rényi gives that almost all graphs on n vertices, k edges are connected so that $E[\binom{M}{k}] \sim n^{2-n}\binom{N}{k}$ with $N = \binom{n}{2}$. Hence we restrict ourselves to $k < (\frac{1}{2} + \epsilon)n \ln n$.

We shall consider the asymptotics of the k-th moment, $E[M^k]$. This is asymptotic to $E[(M)_k]$ for $k = o(n)$ and differs from $E[(M)_k]$ by a calculatable constant when $k = \Theta(n)$. For $E[M^k]$ there is no longer a natural upper bound for k.

There are four basic regions: k constant, $k \to \infty$ but $k = o(n)$, $k = \Theta(n)$, $n \ll k$.

2 The region k constant

This was done in [2]. We scale time by n and distance by $n^{1/2}$ getting a brownian excursion with $f(t) = S_{nt}n^{-1/2}$. Then $M \sim Ln^{3/2}$ where $L = \int_0^1 f(t)dt$ is the mean distance from the origin in a brownian excursion. Then $E[M^k] \sim n^{3k/2}E[L^k]$ where the $E[L^k]$ are the moments of L which have been calculated by G. Louchard in 1984. This matches a known 1977 paper of E.M. Wright in which the asymptotic number of connected graphs with n vertices, $n-1+k$ edges was found.

3 The region $k \to \infty$ but $k = o(n)$

Now scale time by n and distance by $n^{1/2}k^{1/2}$ so that an excursion is associated with the function $f(t) = S_{nt}n^{-1/2}k^{-1/2}$. Now moving from $f(t)$ to $f(t+dt) = f(t) + f'(t)dt$ corresponds to the original walk moving $f'(t)(nk)^{1/2}dt$ in $n \cdot dt$ steps.

With $k = o(n)$ this is not a very large deviation and the step size can be considered just to have mean zero and standard deviation one so that the probability is $\exp[-(f'(t)^2 k/2) \cdot dt]$ for the walk to go this distance. Letting $\Pr[f]$ denote (non-rigorously!) the probability that the excursion follows path f we have $\Pr[f] \sim \exp[-k \int_0^1 f'(t)^2/2 \cdot dt]$. Such paths have value $n^{3k/2} k^{k/2} \exp[k \ln[\int_0^1 f(t)dt]]$. Ignoring the scaling terms, taking logs, and dividing by k we set

$$\Psi(f) = \ln[\int_0^1 f(t)dt] - \int_0^1 f'(t)^2/2 \cdot dt$$

so that the bigger $\Psi(f)$ is the larger the contribution to $E[M^k]$ of excursions of shape f.

This leads to a calculus of variations problem. Fixing $\int_0^1 f(t)dt$ we want to minimize $\int_0^1 f'(t)^2/2 \cdot dt$. The solution is a parabola $f(t) = at(1-t)$. Such f have $\Psi(f) = \ln(a/6) - \frac{a^2}{6}$ which is maximized at $a = \sqrt{3}$. Plugging back in this f gives that $E[M^k]$ is roughly $n^{3k/2} k^{k/2} (1/12e)^{k/2}$.

Suppose now $k \to \infty$ slowly. We outline an argument to give an asymptotic formula for $E[M^k]$ and thus an asymptotic formula for $c(n,k)$. However, making this argument rigorous presents a daunting technical challenge and we should note that we were guided by the already calculated value of $c(n,k)$.

We shall calculate probabilities for the unrestricted random walk with step size Poisson of mean one minus one and introduce the conditioning at the end. Split the excursion of time n into s equal parts. Define σ by $n/s = \sigma^2$ for convenience. Set $a_0 = 1, a_s = 0$ and, for $0 < i < s$, $a_i = \sqrt{3}\sqrt{k}\sqrt{n}\frac{i}{s}(1 - \frac{i}{s})$ and consider those walks that at time iN/s are at position a_i. (That is, the walk follows the parabola given by the Calculus of Variations solution. We ignore integrality here and throughout this outline.) Now consider in general a walk of length M with each step of distribution Poisson of mean one, minus one. For a wide range of m the probability that the total distance is m is asymptotic to $(2\pi M)^{-1/2} e^{-m^2/2M}$. This is natural from the approximation by Brownian motion but also can be computed directly as the total distance is Poisson of mean M minus M. Then the probability of the walk passing through these points is asymptotically $(2\pi\sigma^2)^{-s} \exp[\sum_{0<i\leq s} -(a_i - a_{i-1})^2/2\sigma^2]$.

Of course, the paths don't have to go precisely through the $a_i's$. Set $SUM = \sum_{i=0}^s a_i$. We consider those paths such that the sum of their values at the iN/s is *precisely* SUM. We parametrize them by considering the walks that at time iN/s are at position $a_i + z_i$ and requiring $z_0 = z_s = 0$ (so the path begins and ends at the right place) and $\sum_{i=0}^s z_i = 0$ (so that SUM remains the same. These probabilities are as above except that $\exp[-(a_i - a_{i-1})/2\sigma^2]$ is replaced by $\exp[-(a_i - a_{i-1} + z_i - z_{i-1})^2/2\sigma^2$. Now in the cross terms z_i will have a coefficient of $-a_{i+1} - 2a_i + a_{i-1}$. As the a_i give a parabola this coefficient is constant (i.e., independent of i) and so with $\sum_i z_i = 0$ the total contribution of the cross terms is a factor of one! [This is not serendipitous but rather reflects the parabola being the solution of the Calculus of Variations problem.] One is left with an additional factor of $\exp[\sum_i -(z_i - z_{i-1})^2/2\sigma^2]$. To calculate this set $b_i := z_i - z_{i-1}$ for $1 \leq i \leq s$ so that the factor is $\exp[\sum_i -b_i^2/2\sigma^2]$. As an unrestricted sum over all possible integers

$b_1, \ldots, b_s$ this would split into s identical products, each of which is asymptotically $\sigma\sqrt{2\pi}$ to give $(2\pi\sigma^2)^s$, which conveniently cancels the factor when all $z_i = 0$. But the sum is now restricted to $\sum_i b_i = 0$ (so that $z_s = 0$) and $\sum_i ib_i = 0$ (so that $\sum_i z_i = 0$). We may think of the b_i as weighted with a normal distribution with variance σ^2. Then $\sum_i b_i$ has variance $s\sigma^2 = n$ and is precisely zero with weight $(2\pi n)^{-1/2}$. The variable $\sum_i (i - \frac{s+1}{2})b_i$ is then orthogonal and has variance $\sigma^2 \sum_i (i - \frac{s+1}{2})^2 \sim \sigma^2 s^3/12 = s^2 n/12$ so that it is precisely zero with weight $(2\pi n s^2/12)^{-1/2}$. Together, the total probability of all paths running through these points is $(2\pi n)^{-1} s^{-1}\sqrt{12}$ times $\exp[\sum_i -(a_i - a_{i-1})^2/2\sigma^2]$.

A path going through such points is likely to have M close to $SUM\frac{N}{s}$, effectively approximating the integral (as M is the sum over all values) by the trapezoidal rule. We'll make the assumption (which we do not justify rigorously) that we can asymptotically replace M by $SUM \cdot \frac{N}{s}$. Now we are in the Brownian calculation and the contribution is $n^{3k/2}(k/12e)^{k/2}$ (the main term) times the $(2\pi n)^{-1} s^{-1}\sqrt{12}$ factor.

The sum of the values at the iN/s need not, of course, be precisely SUM and this gives another factor. Suppose SUM is replaced by $SUM(\sqrt{3}+\epsilon)/\sqrt{3}$. This changes the parabola by replacing $\sqrt{3}$ by $\alpha := \sqrt{3}+\epsilon$. The main factor has a term $\alpha^k e^{-k\alpha^2/6}$ (the remaining terms independent of α) which is maximized at $\alpha = \sqrt{3}$. The logarithm divided by k is then $\ln\alpha - \alpha^2/6 \sim c_0 - \frac{1}{3}\epsilon^2$ by Taylor Series, with c_0 the value at the maximum. When SUM is multiplied by $(\sqrt{3}+\epsilon)/\sqrt{3}$ the contribution is then multiplied by $\exp[-k\epsilon^2/3]$. Setting $\gamma := \epsilon(2k/3)^{1/2}$ we have that when SUM has $\gamma n^{1/2} s(3/2)^{1/2}/6$ added to it the contribution is multiplied by $\exp[-\gamma^2/2]$. This would give an extra factor of $(2\pi)^{1/2}$ but with the scaling factor the extra factor is $(2\pi)^{1/2} n^{1/2} s(3/2)^{1/2}/6 = s(\pi n/12)^{1/2}$. Note this cancels the previous s^{-1} factor and now the total contribution is $n^{3k/2}(k/12e)^{k/2}$ (the main term) times $(n\pi)^{-1/2}/2$.

While we have required our paths to begin at one and end at zero we have not yet introduced the requirement that they not otherwise touch the X-axis. This factor comes in at the beginning and at the end. At the beginning we have conditioned essentially on slope $\epsilon := \sqrt{3}\sqrt{k}n^{-1/2}$ so it is as if each step was Poisson with mean $(1+\epsilon)$ minus one. The probability that such a walk never hits the origin is [not an easy problem!] asymptotically 2ϵ so this gives an additional factor of $2\sqrt{3k/n}$. At the end, looking backwards, we start at zero and each step is one minus Poisson with mean $1-\epsilon$. Here the probability that such a walk never returns to the origin (i.e., goes positive and stays positive forever) is asymptotic to ϵ giving an additional factor of $\sqrt{3k/n}$ - so the total additional factors are $6k/n$ giving now a total contribution of $n^{3k/2}(k/12e)^{k/2}$ times $3k\pi^{-1/2}n^{-3/2}$.

Finally, the actual expectation is in the space conditional on the walk being an excursion so we must divide by the probability that the unrestricted walk really is an excursion. Remarkably, this has an exact value in relatively simple form. First, the total distance is Poisson with mean n minus n and so the probability this is precisely -1 is the probability that Poisson of mean n has value $n-1$ which is $e^{-n}n^{n-1}/(n-1)!$. Now we claim that given the walk ends at the origin the probability that it is an excursion (i.e., hadn't hit the origin before) is precisely $1/n$. We may think of balls labelled $1, \ldots, n-1$ each being independently and uniformly places on one of the positions $1, \ldots, n$ and letting X_i be the number of balls in position i. We set $W_0 = 1$ and $W_i = W_{i-1} + X_i - 1$. There are precisely n^{n-2} cases when this is an excursion as they are in bijective correspondence with labelled trees T on $0, 1, \ldots, n-1$ as follows: apply breadth-first search to T starting at 0, adding new vertices in numerical order. When vertex j is discovered at "time" i place ball j into position j. As there are n^{n-1} possible placements of the balls the probability is $1/n$ as claimed. Thus the exact probability that the unrestricted walk is an excursion is $e^{-n}n^{n-1}/n!$. This is asymptotic to $(2\pi)^{-1/2}n^{-3/2}$ by Stirling's formula.

Dividing by this final term, $E[M^k] \sim n^{3k/2}(k/12e)^{k/2} \cdot 3\sqrt{2}k$.

This yields Corollary 2 of [1] (noting their $w_k \to 1$ and that their $c(n, n+k)$ is our $c(n, k+1)$) as

$$c(n,k) = n^{n-2}E[\binom{M}{k}] \sim n^{n-2}n^{3k/2}(e/12k)^{k/2} \cdot 3k^{1/2}\pi^{-1/2}$$

4 A Parking Approach

Here we examine a somewhat different approach which in some rough sense attempts to move from a Brownian Bridge to a Brownian Excursion. Place balls $1, \ldots, n-1$ independently and uniformly into boxes $1, \ldots, n$. Let Y_j, $1 \le j \le n-1$, be the position of the j-th ball. Let X_i, $1 \le i \le n$, the number of balls in the i-th box minus one, the number of j with $Y_j = i$ minus one. Set $S_0 = 1$ and $S_i = 1 + \sum_{l=1}^{i} X_l$. Let $PARK$ be the event that $S_i > 0$ for all $1 \le i < n$ - that for each such i there are at least i balls in the first i boxes. Note that the joint distribution of the X_i conditional on $PARK$ is identical to the distribution defined at the top of this paper. [Generally, if $W_1, \ldots, W_a$ are independent Poissons of mean one and we condition on $W_1 + \ldots + W_a = b$ it is equivalent to throwing b balls into a boxes and letting W_i be the number of balls in the i-th box.] We set $M = \sum_{i=1}^{n}(S_i - 1)$ as before. We observe that

$$M = \sum_{j=1}^{n} \left(\frac{n}{2} - Y_j\right)$$

(Both formulae give $M = 0$ when there is one ball in each of the first $n-1$ boxes. Both formulae go down (up) by one when a single ball is moved one space to the left (right). Hence both formulae are always equal.)

Our original problem is then reformulated as that of estimating $E[\binom{M}{k}|PARK]$. For $PARK$ to hold there can be no ball in the final, n-th, box. Hence we may think of the $n-1$ balls being placed independently and uniformly in the first $n-1$ boxes, so that the Y_j are independent and uniform on $1, \ldots, n-1$. Observe that M now is the sum of independent identically distributed distributions which have a simple form and are of zero mean. Thus the calculation of the factorial moments of M is attackable by standard techniques. The S_i form a bridge, with $S_0 = S_{n-1} = 1$. The event $PARK$, that all intermediate $S_j \ge 1$, turns this into an excursion. What is the affect on these moments of the conditioning by $PARK$?

`Conjecture:` Let $k = k(n)$ satisfy $k(n) \to \infty$ and $k(n) = o(n)$. Then

$$E\left[\binom{M}{k}\chi(PARK)\right] \sim E\left[\binom{M}{k}\right](2\epsilon)(e\epsilon)$$

where $\epsilon = (3k)^{1/2}n^{-1}$. (Here $\binom{M}{k}$ is understood to be zero if M is negative.)

Here is the motivation. The calculus of variation discussed previously gives that the contribution to the k-th moment of M is concentrated around the excursion $S_j = (3k)^{1/2}j(n-j)n^{-2}$. In the region around zero this curve has slope ϵ. The number $1 + X_l$ of balls in the l-th box is averaging $1 + \epsilon$. It is "as if" the $1 + X_l$ were independent Poisson Poisson distributions with mean $1 + \epsilon$. As such the escape probability (in an infinite process with step size X_l (which is Poisson mean $1 + \epsilon$ minus one) where you start at one and die if you hit zero) would be $\sim 2\epsilon$. Around $n - 1$ (the right hand end) the curve has slope $-\epsilon$ and it is "as if" the $1 + X_l$ had independent Poisson distributions with mean $1 - \epsilon$. As such the escape probability (in an infinite process with step size $-X_l$ (which is one minus Poisson mean $1 - \epsilon$) where you start at one and die if you go hit or cross zero – looking at the process in reverse time) would be $\sim e\epsilon$. This gives the two extra factors.

5 The region $k = \Theta(n)$

We set $k = cn$ and consider c a positive constant, $n \to \infty$.

We scale time by n and distance by n so that an excursion is associated with the function $f(t) = S_{nt}/n$. (Note $f(t) \leq 1$ tautologically as $X_i \geq -1$ and $S_n = 0$.) Now moving from $f(t)$ to $f(t+dt) = f(t) + f'(t)dt$ corresponds to the original walk moving $f'(t)n \cdot dt$ in $n \cdot dt$ steps. This is in the realm of large deviations and we can no longer think of the steps X_i as simply having mean zero and standard deviation one. Letting $U_m = Y_1 + \ldots + Y_m$ with the Y_i i.i.d. with distribution minus one plus Poisson of mean one, standard methods give $\Pr[U_m \sim \alpha m] = \exp[(h(\alpha) + o(1))m]$ where $h(\alpha) = \alpha - (\alpha+1)\ln(\alpha+1)$ and the domain of definition is $\alpha \geq -1$. In our case this is $\exp[nh(f'(t))dt]$ so that $\Pr[f] \sim \exp[n\int_0^1 h(f'(t))dt]$. Such paths have value $n^{2cn}\exp[cn\ln[\int_0^1 f(t)dt]]$. Ignoring the scaling terms, taking logs, and dividing by n we set

$$\Psi(f) = c\ln[\int_0^1 f(t)dt] - \int_0^1 f'(t) - (f'(t)+1)\ln(f'(t)+1)dt$$

This leads to a calculus of variations problem. Fixing $\int_0^1 f(t)dt$ we want to minimize $\int_0^1 f'(t) - (f'(t)+1)\ln(f'(t)+1)dt$. The solution is a function of the form

$$f(t) = \frac{1 - e^{-bt}}{1 - e^{-b}} - t$$

where b is positive. (One can check that as $b \to 0$ this curve approaches a parabola symmetric about $t = \frac{1}{2}$ so that this solution meshes with the $k = o(n)$ solution.)

This gives

$$\Psi(f) = c\ln\left[\frac{1}{1-e^{-b}} - \frac{1}{2} - \frac{1}{b}\right] - \ln\left[\frac{b}{1-e^{-b}}\right] + 1 - \frac{be^{-b}}{1-e^{-b}}$$

We select $b = b(c)$ to maximize the right hand side, it does not appear to have a closed form, and let $z = z(c)$ denote the maximal value. Then $E[M^{cn}]$ is roughly $n^{2cn}e^{nz(c)}$.

6 The region $n \ll k$

Here $E[M^k] = [\frac{n^2}{2}(1+o(1))]^k$. We note $M \leq \frac{n^2}{2}$ tautologically, since the best M can do is jump to $n-1$ on the first step and slide back down to zero one step at a time. Conversely, for any fixed s the probability of having $\sim n/(s+1)$ steps s followed by $ns/(s+1)$ steps -1 is β^{-n} for some calculatable $\beta = \beta(s)$ but then its k-th root is negligible and $M = \frac{n^2}{2}\frac{s}{s+1}$ which is within an arbitrary factor of $\frac{n^2}{2}$.

This region meshes with the $k = cn$ region as when $c \to \infty$ $b = b(c) \sim 2c \to \infty$ and f approaches the spike function $f(t) = 1 - t$, the original walk jumping from $S_0 = 1$ to $S_i \sim n$ with $i = o(n)$.

7 Two Questions

1. Can the above be made rigorous?
2. Can the estimates be improved to give an asymptotic formula?

References

[1] Ed Bender, E. Rodney Canfield and B.D. McKay, The Asymptotic Number of Labeled Connected Graphs with a Given Number of Vertices and Edges, Random Structures & Algorithms 2 (1990), 127-169

[2] J. Spencer, Enumerating Graphs and Brownian Motion, Communications on Pure and Appl. Math. 50 (1997), 293-296

Trends in Mathematics, © 2000 Birkhäuser Verlag Basel/Switzerland

A zero-one law for random sentences in description logics

BERNARD YCART *UFR Math-Info, 45 rue des Saints-Pères, 75270 Paris Cedex 06, France.* ycart@math-info.univ-paris5.fr
AND MARIE-CHRISTINE ROUSSET *LRI, UMR 8623 CNRS, University of Paris-Sud, 91405 Orsay Cedex, France.* mcr@lri.fr

Abstract. *We consider the set of sentences in a decidable fragment of first order logic, restricted to unary and binary predicates, expressed on a finite set of objects. Probability distributions on that set of sentences are defined and studied. For large sets of objects, it is shown that under these distributions, random sentences typically have a very particular form, and that all monotone symmetric properties of these random sentences have a probability close to* 0 *or* 1.

1 Introduction

Description logics (DL) are decidable fragments of first order logic with equality, which have recently received significant attention because of their relevance for modeling a wide variety of applications. A description logic is restricted to unary predicates (referred to as *concepts*), representing sets of objects, and binary predicates (called *roles*), representing relationships between objects. Various logical constructors can be used for defining concepts by imposing some restrictions on the number of fillers of a role, or on the concepts to which fillers of a certain role have to belong. Inference problems within DLs have been extensively studied and a whole family of DL systems have been implemented and used for building a variety of applications (see e.g. [16] for a survey on DLs, and http://dl.kr.org/ for the official description logics home page). In this paper, we consider sets of DL ground sentences that are referred to as *Aboxes* in the DL community. Given a set of concept definitions (called a *Tbox*), an Abox asserts facts about objects in the form of *concept facts* and *role facts.* A *concept fact* $C(o)$ states that the object specified by the constant o, is an instance of the concept C, while a *role fact* $R(o, o')$ states that the two objects o, o' are related by the role R.

For example, the following definition of a concept named C models the set of persons such that all of their children are male, and have at most 3 children, in the form of the sentence:

$$C := person \sqcap (\forall child.male) \sqcap (\leq 3\, child).$$

This definition illustrates some standard DL constructors: conjunction, number restrictions and type restrictions. Relatively to that concept definition, we may have an Abox containing the facts $child(Bob, Jacky)$ and $C(Bob)$: $child(Bob, Jacky)$ is a role fact saying that Jacky is Bob's child, and $C(Bob)$ is a concept fact saying that Bob is an instance of the concept C. From those facts, other facts can be logically inferred: for instance, we can infer the new concept fact $male(Jacky)$ as a logical consequence of the definition of concept C.

This study was motivated by the need for generating *random benchmarks* for DL systems. The advantage of generating random benchmarks is twofold. Firstly, it is a way to provide a great number of artificial knowledge bases whose structure and distribution can be tuned in order to possibly account for some known real-world distributions. Secondly, it makes it possible to perform average-case complexity analysis for DL algorithms, also called *probabilistic analysis* [11].

In this paper, we define and study probability distributions on the set of all satisfiable Aboxes associated to a given Tbox and a given set of objects. These distributions are obtained by conditioning products of Bernoulli measures on facts (definition 3.1). In practical applications the number of objects is usually larger by several order of magnitudes than the number of concepts and roles. It is therefore natural to study the asymptotics of our distributions as the number of objects tends to infinity. The concept definitions appearing in a Tbox entail exclusion rules between facts of any satisfiable Abox associated with it. It turns out that exclusion rules have a dramatic effects on the structure of 'typical' random Aboxes, for large object sets. Some concept facts totally disappear while the Abox tends to be almost deterministic, in the sense that all reasonable properties are either true or false with a probability tending to one. This *zero-one law* is the main result of the article. Similar results were known to hold for random graphs or other structures involving a high degree of independence (see for instance [17, 10]).

The paper is organized as follows. In section 2, we define Aboxes and their main logical property, satisfiability. The probability distributions on satisfiable Aboxes are the object of section 3. In section 4, we introduce a particular case that turns out to provide an asymptotic pattern for the general case. This particular case concerns Aboxes for which exclusion rules involve only concept facts, and are therefore called *role-free*. A zero-one law for probability distributions on role-free sets of Aboxes is stated in theorem 4.1, which is our main result. Finally the asymptotic properties of general random Aboxes is studied in section 5. There it will be shown that a typical random Abox with a large number of objects resembles a role-free random Abox, and therefore has almost deterministic properties.

2 Satisfiable Aboxes

We now formally define the syntax and semantics of the Aboxes that we consider in this paper. They are related to a fixed Tbox $\mathcal{T}$ which is a set of concept definitions in the $\mathcal{ALCN}$ description logic which is one of the most expressive decidable implemented description logic.

Concept expressions in $\mathcal{ALCN}$ are defined using the following syntax (A denotes a primitive concept name, C and D represent concept expressions and R denotes a role name):

$C, D \rightarrow$	$A \mid$	(primitive concept)
	$C \sqcap D \mid C \sqcup D \mid$	(conjunction, disjunction)
	$\neg C \mid$	(complement)
	$\forall R.C \mid$	(type restriction)
	$\exists R.C \mid$	(existential quantification)
	$(\geq n\, R) \mid (\leq n\, R)$	(number restrictions)

A Tbox is a set of *concept definitions*, which are statements of the form $CN := D$, where CN is a concept name and D is a $\mathcal{ALCN}$ concept expression. We assume that a concept name appears on the left-hand side of at most one definition. A concept name A is said to *depend* on a concept name B if B appears in the concept definition of A. We consider concept definitions that do not have cycles in the dependency relation.

The semantics of the constructors is defined via *interpretations* in the standard way of first order logic. An interpretation I contains a non-empty domain $\mathcal{D}^I$. It assigns a unary relation over $\mathcal{D}^I$ to every concept name, and a binary relation over $\mathcal{D}^I \times \mathcal{D}^I$ to every role name. The interpretations of concept expressions are uniquely defined by the following equations: ($\sharp S$ denotes the cardinality of a set S):

$$(C \sqcap D)^I = C^I \cap D^I,$$
$$(C \sqcup D)^I = C^I \cup D^I,$$
$$(\neg C)^I = \mathcal{D}^I \setminus C^I,$$
$$(\forall R.C)^I = \{d \in \mathcal{D}^I \mid \forall e[(d,e) \in R^I \rightarrow e \in C^I]\}$$
$$(\exists R.C)^I = \{d \in \mathcal{D}^I \mid \exists e[(d,e) \in R^I \wedge e \in C^I]\}$$
$$(\geq n R)^I = \{d \in \mathcal{D}^I \mid \sharp\{e \mid (d,e) \in R^I\} \geq n\}$$
$$(\leq n R)^I = \{d \in \mathcal{D}^I \mid \sharp\{e \mid (d,e) \in R^I\} \leq n\}$$

An interpretation I is a model of a Tbox $\mathcal{T}$ if $C^I = D^I$ for every concept definition $C := D$ in $\mathcal{T}$. We say that C *is subsumed by* D w.r.t. $\mathcal{T}$ if $C^I \subseteq D^I$ in every model I of $\mathcal{T}$.

Given a fixed finite set of objects (constants) $\mathcal{O}$, and given a Tbox $\mathcal{T}$, an Abox is a set of facts of the form $C(o)$ (concept facts) or $R(o, o')$ (role facts), such that $o, o' \in \mathcal{A}$. Given an Abox $\mathcal{A}$, we extend an interpretation with a mapping ψ^I from the objects appearing in $\mathcal{A}$ to $\mathcal{D}^I$. An interpretation I is said to be a *model* of $\mathcal{A}$ if whenever $C(a) \in \mathcal{A}$ then $\psi^I(a) \in C^I$, and whenever $R(a,b) \in \mathcal{A}$ then $(\psi^I(a), \psi^I(b)) \in R^I$.

An Abox is *satisfiable* if it has a model, i.e, there exists an interpretation in which the conjunction of its facts is true. In the rest of the paper, we focus on satisfiable Aboxes since only satisfiable Aboxes are meaningful to capture realistic properties of data.

It is important to note that the concept definitions appearing in a Tbox entail exclusion rules between facts of any satisfiable Abox associated with $\mathcal{T}$. For instance, the presence of the concept definition:

$$C := person \sqcap (\forall child.male) \sqcap (\leq 3child)$$

in a Tbox imposes exclusion rules between the fact $C(o)$, the role facts $child(o, o')$, and possibly the concept facts involving the object o'. For instance, in a satisfiable Abox, we cannot have simultaneously the facts $C(o)$, $child(o, o_1)$, $child(o, o_2)$, $child(o, o_3)$, $child(o, o_4)$. Similarly, the facts $C(o)$, $child(o, o_1)$, $\neg male(o_1)$ cannot appear together in a satisfiable Abox.

Finally, note that since the semantics of DLs is defined in terms of interpretations in the standard way of first order logic, they obey the so-called *open world assumption*. The open world assumption leaves the possibility open that other unknown facts might be true besides those explicitly mentioned in the Abox.

3 Probability distribution on Aboxes

The set of objects $\mathcal{O} = \{o_1, \ldots, o_n\}$ is fixed for the rest of this section.

Let c and r be the numbers of concepts and roles in $\mathcal{T}$. ¿From now on, $\widetilde{\mathcal{A}}$ will denote the set of all 2^{cn+rn^2} Aboxes, and $\mathcal{A} \subset \widetilde{\mathcal{A}}$ will denote the set of all *satisfiable* Aboxes, relative to the Tbox $\mathcal{T}$ and the set of objects $\mathcal{O}$. We want to define a family of product-type probability distributions on $\mathcal{A}$. To this end, we associate to each concept C a probability p_C and to each role R a probability p_R, all being strictly positive and lesser than 1. The parameters p_C and p_R may depend on the number of objects n. The set $\widetilde{\mathcal{A}}$ is first endowed with a product of Bernoulli distributions with parameters p_C and p_R, then the distribution on $\mathcal{A}$ is defined by conditioning.

Definition 3.1 *Let us consider on $\widetilde{\mathcal{A}}$ the product of Bernoulli distributions, denoted by $\widetilde{\mathbb{P}}$, and defined for all $A \in \widetilde{\mathcal{A}}$ by:*

$$\begin{aligned} \widetilde{\mathbb{P}}[A] &= \prod_{C,o} p_C^{\mathbb{1}_A(C(o))} (1-p_C)^{1-\mathbb{1}_A(C(o))} \\ & \prod_{R,o,o'} p_R^{\mathbb{1}_A(R(o,o'))} (1-p_R)^{1-\mathbb{1}_A(R(o,o'))} . \end{aligned} \tag{3.1}$$

We call conditional product distribution (CPD) with parameters p_C and p_R, $C, R \in \mathcal{T}$ the distribution $\mathbb{P}$ on $\mathcal{A}$ defined for all $A \in \mathcal{A}$ by:

$$\mathbb{P}[A] = \frac{\widetilde{\mathbb{P}}[A]}{\widetilde{\mathbb{P}}[\mathcal{A}]} .$$

In formula (3.1), the first product extends over all concepts $C \in \mathcal{T}$ and all objects $o \in \mathcal{O}$. The second product extends over all roles $R \in \mathcal{T}$ and all couples of objects $(o, o') \in \mathcal{O} \times \mathcal{O}$. The notation $\mathbb{1}_A(.)$ is standard for the indicator function of the set A (1 if the element belongs to A, 0 else). Consider the event "$C(o) \in A$". Under the distribution $\widetilde{\mathbb{P}}$, it has probability p_C, whatever $o \in \mathcal{O}$. Similarly the event "$R(o, o') \in A$" has probability p_R, whatever (o, o'). Moreover all these events are mutually independent. When exclusion rules are present, $\mathcal{A}$ is only a subset of $\widetilde{\mathcal{A}}$, and independence between facts no longer holds. Notice that due to combinatorial explosion, $\widetilde{\mathbb{P}}[\mathcal{A}]$ cannot be explicitly computed in general. Nevertheless we shall see that explicitly computing $\widetilde{\mathbb{P}}[\mathcal{A}]$ can be avoided in an asymptotic description (section 5).

In the particular case where all the probabilities p_R and p_C are equal to $1/2$, then all satisfiable Aboxes have the same probability under $\mathbb{P}$: $\mathbb{P}[A] = 1/\sharp\mathcal{A}$.

4 Role-free sets

This section deals with very particular Aboxes, namely those for which exclusion rules only concern concept facts.

Definition 4.1 *Let $\mathcal{T}$ be a Tbox, and $\mathcal{A}$ be the set of Aboxes associated to $\mathcal{T}$ and a given set of objects $\mathcal{O}$. The set $\mathcal{A}$ is said to be* role-free *if any role fact $R(o, o')$ can be added to a satisfiable Abox, while preserving satisfiability.*

In particular, if $\mathcal{A}$ is role free, then $\mathcal{T}$ cannot contain any concept defined by an upper-bound cardinality restriction ($C =\leq kR$), nor by a type restriction ($C' = \forall RC$). If $\mathcal{A}$ is role-free, then the distribution $I\!P$ of Definition 3.1 has a particularly simple product-type structure, which is described in Proposition 4.2. The main result, Theorem 4.1 proves that any symmetric monotone property of a random role-free Abox is either true or false with a probability close to 1 as the number of objects tends to infinity. Obviously, role-free sets of Aboxes are not the most interesting from the point of view of applications. However, it will be proved in section 5 that as the number of objects tends to infinity, most CPD's on general sets of Abox tend to resemble those described here in the role-free case.

We first need to define the *decomposition* of an Abox.

Definition 4.2 *Let A be an Abox.*

1. *Let o be an object. We call* object-component of A associated to o *the Abox, denoted by $A^{(o)}$, which is the subset of A containing only those facts of type $C(o)$, for all C.*

2. *Let R be a role. We call* role-component of A associated to R *the Abox, denoted by A_R, which is the subset of A containing only those facts of type $R(o, o')$, for all o, o'.*

Note that in Definition 4.2, some of the components may be empty. Obviously

$$\{ A^{(o)} , o \in \mathcal{O} ; A_R , R \text{ role } \in \mathcal{T} \} ,$$

is a partition of A, and A is uniquely defined by the $(r+n)$-tuple of its components. So the mapping that associates to a given Abox the $(r+n)$-tuple of its components is one-to-one, and it will be referred to as *decomposition.* For each object o, denote by $\mathcal{A}^{(o)}$ the set of all satisfiable Aboxes containing only concepts facts of type $C(o)$. For each role R, denote by $\mathcal{A}_R$ the set of all 2^{n^2} Aboxes containing only role facts of type $R(o, o')$. Role-free sets of Aboxes are actually Cartesian products of $\mathcal{A}^{(o)}$'s and $\mathcal{A}_R$'s.

Proposition 4.1 *The set $\mathcal{A}$ is role-free if and only the decomposition maps it onto the Cartesian product of all $\mathcal{A}^{(o)}$'s and $\mathcal{A}_R$'s. Moreover, Let $\phi^{(o,o')}$ be the mapping defined on $\mathcal{A}^{(o)}$, with values in $\mathcal{A}^{(o')}$, defined by:*

$$C(o) \in A \Longleftrightarrow C(o') \in \phi^{(o,o')}(A) .$$

If $\mathcal{A}$ is role-free, then all mappings $\phi^{(o,o')}$ are one-to-one.

Regarding CPD's, Proposition 4.1 yields probabilistic independence of components. Proposition 4.2 below describes the structure of CPD's on role-free sets of Aboxes.

Proposition 4.2 *For each object o, denote by $\mathbb{P}^{(o)}$ the conditional probability distribution of $\mathbb{P}$ over $\mathcal{A}^{(o)}$. For each role R, denote by $\mathbb{P}_R$ the conditional probability distribution of $\mathbb{P}$ over $\mathcal{A}_R$. If $\mathcal{A}$ is role-free then:*

1. *If $A \in \mathcal{A}$ is any satisfiable Abox with decomposition $\{A^{(o)}\,,\, o \in \mathcal{O}\,,\, A_R\,,\, R\,role \in \mathcal{T}\}$, then:*

$$\mathbb{P}[A] = \prod_{o \in \mathcal{O}} \mathbb{P}^{(o)}[A^{(o)}] \prod_{R \in \mathcal{T}} \mathbb{P}_R[A_R]\,.$$

2. *Let o and o' be two objects. Then for all $A \in \mathcal{A}^{(o)}$:*

$$\mathbb{P}^{(o)}[A] = \mathbb{P}^{(o')}[\phi^{(o,o')}(A)]\,.$$

3. *Let R be a role. Let $A \in \mathcal{A}_R$ be any Abox containing m facts. Then:*

$$\mathbb{P}_R[A] = p_R^m (1 - p_R)^{n^2 - m}\,.$$

Thus in the decomposition of a random Abox under $\mathbb{P}$, all components are independent random Aboxes (item *1*). All object-components are identically distributed, up to identification of objects (item *2*). In each role-component, any role fact $R(o, o')$ occurs with probability p_R and all these facts are independent (item *3*).

In order to further understand the structure of CPD's on role-free sets of Aboxes, some properties of the marginal distributions $\mathbb{P}^{(o)}$ and $\mathbb{P}_R$ are worth pointing out. Let us begin with $\mathbb{P}^{(o)}$.

Let $\mathcal{C}$ be the set of concepts in $\mathcal{T}$. To an Abox $A \in \mathcal{A}^{(o)}$, one can uniquely associate a boolean vector, indexed by $\mathcal{C}$, that will be denoted by $\mathbb{1}_A$. For all $C \in \mathcal{C}$,

$$\mathbb{1}_A(C) = \begin{cases} 1 & \text{if} \quad C(o) \in A\,, \\ 0 & \text{if} \quad C(o) \notin A\,. \end{cases}$$

Since the distributions $\mathbb{P}^{(o)}$ do not depend on o (item *2* of Proposition 4.2), they all induce the same distribution, denoted by $\overline{\mathbb{P}}$, on boolean vectors.

$$\overline{\mathbb{P}}[\mathbb{1}_A] = \frac{1}{Z} \prod_{C \in \mathcal{C}} p_C^{\mathbb{1}_A(C)} (1 - p_C)^{1 - \mathbb{1}_A(C)}\,,$$

where Z denotes the normalizing constant, i.e.:

$$Z = \sum_{A \in \mathcal{A}^{(o)}} \prod_{C \in \mathcal{C}} p_C^{\mathbb{1}_A(C)} (1 - p_C)^{1 - \mathbb{1}_A(C)}\,.$$

In the particular case where no exclusion rule involves concepts, all Aboxes are satisfiable, and the distribution $\overline{\mathbb{P}}$ is a product of Bernoulli distributions on $\{0, 1\}^{\mathcal{C}}$. In the general case, $\overline{\mathbb{P}}$ is only a truncation of that product of Bernoulli distributions. It has already appeared in a different context as the reversible measure of some resource sharing models (see [7]). Indeed, one can interpret concepts as components in a system, and a boolean vector as a pattern of activity: $\mathbb{1}_A(C) = 1$ if component C is active, $\mathbb{1}_A(C) = 0$ if it is inactive. The parameter p_C should be seen as the proportion of time that component C spends in the active state, in the absence of constraints. However, components need their share of common resources to be active, so that in certain subsets, all components cannot be active at the same time. These constraints are the exclusion rules of the Tbox.

The particular case where all exclusion rules are binary, and all parameters p_C are equal, has been studied in more depth ([18, 8, 9]). It is connected to an interesting combinatorial problem, that of finding stable sets of an undirected graph. By "binary exclusion rules", we mean that two concepts may be mutually exclusive but it is impossible to find $k \geq 2$ concepts, any two of them are compatible but all k are not. Define an undirected graph structure, the vertices of which are the concepts in $\mathcal{T}$. The edges join two concepts if and only if they are mutually exclusive. This graph will be referred to as the exclusion graph. Now if A is a satisfiable Abox, then a concept can be represented in A only if its neighbors on the exclusion graph are not. In other terms, the boolean vector $\mathbb{1}_A$ is a "stable vector" of the graph ([9]). Determining maximal stable vectors (satisfiable Aboxes of maximal cardinality) was proved to be an NP-hard problem by Karp [13]. When all p_C's are equal, the probability distribution $\overline{\mathbb{P}}$ is the reversible measure of the so called Philosophers' Process (cf. [18]) derived from a classical resource sharing model of Dijkstra ([5]). It has been completely characterized in the case of ladder graphs in [8]. In the case where all p_C's are equal to $1/2$, the normalizing constant Z is proportional to the total number of stable sets of the exclusion graph. Recursive combinatorial methods allow to compute it for special classes of graphs (see [9]). Apart from these classes of graphs, very little can be said in general of the distribution $\overline{\mathbb{P}}$.

It is understood in our models that the number of concepts is relatively small compared to the number of objects. It is reasonable to expect that the cardinality of the $\mathcal{A}^{(o)}$'s is not too high, and that it is possible to explicitly compute the distribution $\overline{\mathbb{P}}$, at least numerically. Once this has been done, then due to the independence of components, the probability of most events concerning concept facts can be explicitly computed. As an example, fix a concept C and consider the event "$C(o) \in A$". Let π_C denote its probability under $\mathbb{P}^{(o)}$. Consider now in a random Abox with distribution $\mathbb{P}$, the total number of facts of type $C(o)$, for all $o \in \mathcal{O}$. That number is a random variable. Under the role-free assumption, it is easy to derive from proposition 4.2 that its distribution is binomial, with parameters n and π_C. If n is large, it can be approximated by a Gaussian distribution with mean $n\pi_C$, and variance $n\pi_C(1 - \pi_C)$.

Let us now turn to role components. Under the role-free hypothesis, the situation is quite simple. The set $\mathcal{A}_R$ contains all possible 2^{n^2} Aboxes formed with role facts involving R only. It will be interesting to identify such an Abox to the edge set of a directed graph (or *digraph*) with vertex set $\mathcal{O}$. If $A \in \mathcal{A}_R$, and $o, o' \in \mathcal{O}$, define an edge pointing from o to o' if and only if $R(o, o') \in A$. Thus, to each element of $\mathcal{A}_R$ can be associated in a one-to-one manner a directed graph with vertex set $\mathcal{O}$. (Notice that in our case, contrarily to other models, loops are not excluded). Through that correspondence, the distribution $\mathbb{P}_R$ can be read as that of a random digraph, with independent edges and edge probability p_R. Since the founding paper by Erdös and Rényi [6], random undirected graphs have been extensively studied and all sorts of asymptotic results are now available (see Bollobás [1] or Spencer [17] as general references). Random digraphs, in spite of their numerous applications, have aroused lesser interest. Part of the reason is the belief, supported by some general results, that most asymptotic properties known for random graphs can be adapted to random digraphs. One of the most famous examples is the emergence of the so called giant component around $p_R = 1/n$ (see [12] for graphs and [14, 15] for digraphs).

Random graphs and digraphs are well known examples of the so called "$0-1$ laws", according to which most propositions on a large random system with a high degree of independence are either true or false with probability close to 1. At first, properties were restricted to first order logics. The celebrated Glebskii-Fagin theorem states that in a random (undirected) graph with fixed edge probability, any property of first order logics has a probability tending either to 1 or to 0 as n tends to infinity. That theorem was later extended by Shelah and Spencer to random graphs with edge probability $p = kn^{-\alpha}$, if α is any irrational (positive) number.

We shall rather focus here on *monotone* properties for which some recent results by Friedgut and Kalai [10] and Bourgain [3] equally apply to random graphs and digraphs, as well as other problems such as the phase transition in the k-satisfiability of random clauses. Let N be an integer and consider the set $E = \{0, 1\}^N$. A subset F of E is said to be monotone if:

$$\eta \in F \Longrightarrow \zeta \in F\,, \quad \forall \zeta \geq \eta\,,$$

where the ordering is taken componentwise. A property on E is monotone if the set F of those elements of E for which it is true is itself monotone. Translated in terms of Aboxes, a property is monotone if, being true for A, it remains true for any other Abox containing A. Under the open world hypothesis, non-satisfiability is a monotone property.

Another requirement imposed to our properties is *symmetry* with respect to permutations of objects. If σ is such a permutation, and A is an Abox, we define the Abox $\sigma.A$ by:

$$C(o) \in A \Longleftrightarrow C(\sigma(o)) \in \sigma.A$$

and

$$R(o, o') \in A \Longleftrightarrow R(\sigma(o), \sigma(o')) \in \sigma.A\,.$$

A set of Aboxes is said to be *symmetric* if it contains all $\sigma.A$'s as long as it contains A. A property is symmetric if the set of Aboxes for which it is true is symmetric.

Although the recent results of Bourgain [3] seem to indicate that symmetry could be superfluous in some cases, we believe that it is natural enough to impose it for properties of Aboxes. Notice that we do not impose symmetry through permutations of concepts or roles. Here are a few examples of monotone symmetric properties.

1. P_1: All objects are related by R to at least k other objects.

2. P_2: For any two objects o and o', there exists a path of relations of length k linking o to o'.

3. P_3: There exists k objets all related to each other by role R and for which concepts C_1 and C_2 are true.

The zero-one law that will be stated below concerns mainly the case where the parameters p_R and p_C are constant. However in that case random Aboxes contain mostly role facts, and are very unrealistic. It would be of greater interest to let the parameters depend on n. This will be the object of future work. Theorem 4.1 is essentially an extension of Theorem 1.1 of Friedgut and Kalai [10].

Theorem 4.1 *Assume that $\mathcal{A}$ is role free. Let P be any symmetric monotone proposition. For $\epsilon > 0$, denote by V_ϵ the subset of those elements p_C, p_R of $[0,1]^{c+r}$ such that, under the CPD $\mathbb{P}$, the probability that a random Abox satisfies P is between ϵ and $1-\epsilon$. There exists a constant γ, depending only on the Tbox, such that the diameter of the set V_ϵ is lesser than $\gamma \log(1/2\epsilon)/\log n$.*

Since the diameter of the set V_ϵ tends to 0 as n tends to infinity, it turns out that for most values of the parameters, the probability that a random Abox satisfies P will be either close to 0 (lesser than ϵ), or close to 1 (larger than $1-\epsilon$).

5 Random Aboxes with large object sets

The CPD of definition 3.1 depends on parameters associated to the elements of the Tbox, namely the values of the probabilities p_C and p_R. In the Tbox $\mathcal{T}$, some concepts may be defined by cardinality or type restrictions. The cardinality restrictions do not depend on the number of objects. In this section, we shall study the asymptotic properties of a random Abox under distribution $\mathbb{P}$, as the number n of objects tends to infinity. Recall from Definition 3.1 that $\mathbb{P}$ is the conditional measure on $\mathcal{A}$ of a product of Bernoulli distributions $\widetilde{\mathbb{P}}$. For the measure $\widetilde{\mathbb{P}}$, the average number of concept facts is $n \sum p_C$, whereas the average number of role facts is $n^2 \sum p_R$. This is an indication that role facts should be predominant in a random Abox. The predominance of role facts has dramatic consequences on the structure of random Aboxes. Some concepts, in particular those defined by cardinality restrictions ($\leq kR$), disappear whereas the others tend to be independent of roles. Actually, the distribution $\mathbb{P}$ is close to another CDP, on a particular role-free set of Aboxes. The precise approximation statement is given in Theorem 5.1, which is the main result of this section. Before that, we need to define an intermediary role-free set of Aboxes, between $\mathcal{A}$ and $\widetilde{\mathcal{A}}$ (the notations are those of section 3).

Definition 5.1 *Let $\mathcal{A}$ be a set of Aboxes, $\mathcal{A}^{(o)}$ and $\mathcal{A}_R$ its images through the coordinates of the decomposition operator of Definition 4.2. We call* role-free completion *of $\mathcal{A}$ the image through inverse decomposition of the Cartesian product:*

$$\mathcal{A}' = \times_{o \in \mathcal{O}} \mathcal{A}^{(o)} \times_{R \in \mathcal{T}} \mathcal{A}_R \,.$$

The idea of role-free completion is to separate exclusion rules between concepts, from constraints between concepts and roles. Denote by $\mathcal{A}_1$ the subset mapped by decomposition onto the Cartesian product $\times_o \mathcal{A}^{(o)}$: it could have been defined as the set of those elements of $\mathcal{A}$ that contain no role fact. Exclusion rules between concepts reduce $\mathcal{A}_1$ to a strict subset of $\{0,1\}^{cn}$. On the other hand, no restriction appears between roles in the $\mathcal{A}_R$'s: each of them has all possible 2^{n^2} elements.

We first need to identify those concepts that tend to disappear.

Definition 5.2 *A concept C is called* evanescent *if the probability that a random Abox under distribution $\mathbb{P}$ contains at least one concept fact $C(o)$ tends to 0 as n tends to infinity.*

The following result describes evanescent concepts.

Proposition 5.1

- *Any cardinality-restriction concept $\leq kR$ is evanescent.*
- *Any type-restriction concept $\forall RC$ is evanescent.*
- *If C is evanescent, then $\exists RC$ is evanescent.*
- *If C_1 and C_2 are evanescent, then $C_1 \sqcup C_2$ is evanescent.*
- *If C_1 or C_2 are evanescent, then $C_1 \sqcap C_2$ is evanescent.*

Evanescent concepts tend to disappear from random Aboxes, whereas non evanescent ones tend to be independent of roles. More precisely, the distribution $\mathbb{P}$ becomes close to another CPD $\mathbb{P}'$ on the role free completion of $\mathcal{A}$, in which evanescent concepts are absent.

Theorem 5.1 *Let $\mathcal{T}$ be a Tbox, and p_C, p_R be the parameters of a CPD $\mathbb{P}$ on $\mathcal{A}$. Define new parameters p'_C by:*

$$p'_C = \begin{cases} 0 & \textit{if } C \textit{ is evanescent,} \\ p_C & \textit{else.} \end{cases}$$

Denote by $\mathbb{P}'$ the CPD on the role-free completion $\mathcal{A}'$ of $\mathcal{A}$, with parameters p'_C and p_R. Then as n tends to infinity, the distribution $\mathbb{P}$ tends to $\mathbb{P}'$ in the sense that:

$$\lim_{n\to\infty} \sup_{\mathcal{B}\in\mathcal{A}} |\mathbb{P}[\mathcal{B}] - \mathbb{P}'[\mathcal{B}]| = 0\,.$$

6 Conclusion

A natural probability distribution on sets of Aboxes associated to a given Tbox has been defined. Its parameters are fixed probabilities attached to the elements of the Tbox. As the number of objects tends to infinity, some concepts tend to disappear from random Aboxes, the others become independent of roles. The result is quasi-deterministic, in the sense that any monotone symmetric assertion on the random Abox is either true or false with a probability tending to one.

Under the fixed parameters hypothesis studied here, random Aboxes contain mostly role facts and they are not very realistic from the point of view of applications. A more interesting situation could be obtained by letting the role probabilities p_R tend to 0 as the inverse of the number of objects. This will be the theme of future researches.

References

[1] B. Bollobás. *Random Graphs.* Academic Press, London, 1985.

[2] B. Bollobás and A. Thomason Threshold functions. *Combinatorica 7, 35–38,* 1986.

[3] J. Bourgain. On sharp thresholds of monotone properties. *To appear*, 1999.

[4] G. Chartrand and L. Lesniak. *Graphs and Digraphs. Second ed.* Wadsworth & Brooks/Cole, Mathematics Series, 1986.

[5] E.W. Dijkstra. The structure of the multiprogramming system. *Comm. ACM 11 341–346*, 1968.

[6] P. Erdös and A. Rényi. On the evolution of random graphs. *Mat. Kutató Int. Közl., 5 17–61*, 1960.

[7] F. Forbes, O. François, and B. Ycart. Stochastic comparison for resource sharing models. *Markov Proc. Rel. Fields 2(4) 581–605*, 1996.

[8] F. Forbes and B. Ycart. The Philosophers' process on ladder graphs. *Comm. Stat. Stoch. Models 12(4) 559–583*, 1996.

[9] F. Forbes and B. Ycart. Counting stable sets on Cartesian products of graphs. *Discrete Mathematics*, 186 105–116, 1998.

[10] E. Friedgut and G. Kalai. Every monotone graph property has a sharp threshold. *Proc. Amer. Math. Soc. 124 2993–3002*, 1996.

[11] M. Hofri. *Probabilistic analysis of algorithms.* Springer-Verlag, New York, 1987.

[12] S. Janson, D.E. Knuth, T. Łuczak, and B. Pittel. The birth of the giant component. *Random Structures and Algorithms, 4(3) 233–258*, 1993.

[13] R.M. Karp. Reducibility among combinatorial problems. In *Complexity of computer computations, R.E. Miller and J.W. Thatcher ed., 85–103,* Plenum Press, New-York, 1972.

[14] R.M. Karp. The transitive closure of a random digraph. *Rand. Structures and Algo. 1(1) 73–94*, 1990.

[15] T. Łuczak. The phase transition in the evolution of random digraphs. *J. Graph Theory 14(2) 217–223*, 1990.

[16] D. Nardi, F.M. Donini, M. Lenzerini, and A. Schaerf. Reasoning in description logics. In *Principles of Artificial Intelligence.* G. Brewska (ed.), Springer-Verlag, New York, 1995.

[17] J.H. Spencer. Nine lectures on random graphs. In P.L. Hennequin, editor, *Ecole d'été de probabilités de Saint-Flour XXI-1991*, L.N. in Math. 1541, 293–347. Springer-Verlag, New York, 1993.

[18] B. Ycart. The Philosophers' Process: An ergodic reversible nearest particle system. *Ann. Appl. Probab. 3(2), 356–363*, 1993.

List of Authors

GPSR Compliance
The European Union's (EU) General Product Safety Regulation (GPSR) is a set of rules that requires consumer products to be safe and our obligations to ensure this.

If you have any concerns about our products, you can contact us on

ProductSafety@springernature.com

In case Publisher is established outside the EU, the EU authorized representative is:

Springer Nature Customer Service Center GmbH
Europaplatz 3
69115 Heidelberg, Germany

www.ingramcontent.com/pod-product-compliance
Ingram Content Group UK Ltd.
Pitfield, Milton Keynes, MK11 3LW, UK
UKHW021531300726
14060UKWH00011B/264

* 9 7 8 3 7 6 4 3 6 4 3 0 4 *